ichim

ichim01

International Cultural Heritage Informatics Meeting

CULTURAL HERITAGE and TECHNOLOGIES in the THIRD MILLENNIUM

[Volume 2]
Short Papers/Posters
Demos

http://www.ichim01.polimi.it
http://www.archimuse.com/ichim01

Edited by David Bearman and Franca Garzotto

ichim01
International Cultural Heritage Informatics Meeting
CULTURAL HERITAGE and TECHNOLOGIES in the THIRD MILLENNIUM

Politecnico di Milano, Italy
September 3-7, 2001

http://www.ichim01.polimi.it *or* http://www.archimuse.com/ichim01
e-mail: ichim01@polimi.it *or* ichim01@archimuse.com

ISBN 1-885626-24-X
Printed in Italy

[Volume 2]

page 11. Short Papers

.Building a New Culture

.Data Models

.Restoration and Preservation

.Case-studies

Short Papers

Building a New Culture

New Economy and Digital Culture – the REGNET project

Walter Koch (*) *and Gerda Koch* (*)

(*) AIT Angewandte Informationstechnik Forschungsgesellschaft mbH
E-mail: kochw@ait.co.at
E-mail: kochg@ait.co.at

ABSTRACT

The concept of a European Digital Library cannot only be based on a technical framework enabling access to digital goods – there is also a need to introduce new ways of cooperation between different stakeholders (Cultural Organisations, Industries, Administrations, etc.) as well as to "re-engineer" traditional business processes in the light of globalisation and world wide markets. The main activities within a support environment for I(nternet) Markets comprise: Content Engineering, Platform Engineering, and Enterprise Engineering.

The REGNET - Cultural Heritage in REGional NETworks project - targets to all of the three areas. The project was introduced under the Action Line 'Access to digital collections of cultural and scientific content' of the European Union IST-Information Society Technologies Programme and 23 partners, cultural organisations (museums, libraries and archives) and IT-industry representatives, from 10 European Union states as well as Bulgaria and Russia are participating.

ENVISAGED ACHIEVEMENTS

REGNET will set up a functional network of cultural service centres through Europe which will provide IT-services dedicated to cultural heritage organisations. A technical and legal framework, the REGNET system, for such a service infrastructure will be developed. This will offer services like data entry, search and retrieval, and e-Business. It will be based mainly on integration work using state-of-the-art components. The network will integrate multi-media industries, content providers and service centre operators. Existing cultural infrastructures will be exploited and new infrastructures should be developed where necessary.

The technical infrastructure should allow to set up even low cost service centres. The membership concept of the service centres will generate a critical mass of digital or physical goods contained in Content Provider's organisations.

THE PLAYERS WITHIN THE NETWORK

The four players within the network are the content providers, the service centre operators, the system developers and end users. The content providers (museums, libraries, archives etc.) will provide access (via wired and wireless communication) to their digital contents, services and products and offer them to their clients (B2C). In return they can use the REGNET facilities for multimedia productions and data base management, or cooperate with other REGNET partners during the creation of data bases, generation of multimedia products or creation of a virtual exhibition (B2B). The service centre operators will generate income by providing the technical infrastructure (software/hardware) to content providers and other partners within the REGNET network. They offer additional IT-services and consultancies. And the system developers are selling the REGNET system to other cultural service centres and content providers. They implement additional components for the REGNET software system (additional 'nodes' like an 'exhibition creator', etc), and will generate income via licence fees for the REGNET system. For the end user the system will offer easy and wide access to cultural heritage data information and the purchase of CH related goods and services at one point, with stress on the production of personalized goods (e.g. CDROM) and services.

STANDARDS AND METADATA

REGNET will also provide a unique facility to meta data definitions for both CH-related data and e-business related data via the search entry in the REGNET – Portal and the REGNET - Ontology subsystem. Here not only meta data related to objects is addressed, but also definitions of work flows, business processes, etc. are included which lead to the concept of Ontology within the Cultural Heritage domain. Within the project a data entry facility using WEB-browsers and adaptable to different needs within different domains (Museums, Library, Archives) will be established. This facility will support existing standards (ICOM/CIDOC, UNIMARC, ISAD(G), etc) and is configurable by the REGNET System user. At the e-Business level REGNET provides the generation of a customised shopping cart systems within the business to consumer (B2C) framework.

MOBILE ENTRY

It will be examined how next generation mobile networks can be exploited to widen the potentiality of WEB services in the field of Cultural Heritage. Appropriate gateway functions and interworking units will be envisaged to interface the REGNET architecture with the UMTS (Universal Mobile Telecommunication Systems) structure and ensure a seamless provisioning of REGNET services to mobile customers.

ELECTRONIC CATALOGUES

Already existing electronic catalogues (OPACS: Online Public Access Catalogues) referring to cultural & scientific objects contained in libraries, museums, archives, and galleries, as well as to goods and services will be used. Where there are no catalogues yet new catalogues will be developed in order to establish a 'virtual union' catalogue of all OPACS and product/service catalogues held locally.

BUSINESS ENGINEERING

In addition the work includes the outlining of necessary 'supply chains' and the connected business processes and functions to deliver digital and physical goods. All the business transactions on the B2B and B2C level (containing payment features, copyright systems, authentication control, etc) will be guided by a legal framework that will be worked out accordingly. The reorganisation of existing processes and introduction of computerised functions which make transactions for low money goods worth while (e.g. small copyright fees for digital images) is an essential part of the project.

THE REGNET SYSTEM ARCHITECTURE

The building blocks which constitute the REGNET-System are supporting access to cultural & scientific information as well as to product & service information offered by different organizations. The building blocks are outlined in the figure below. The REGNET-Portal enables entry of remote data, distributed search and e-business functions. The Cultural Heritage Data Management allows the search in distributed metadata repositories connected to data of cultural heritage content. The e-Business Data Management gives access to distributed goods and services catalogues via an e-business system. And the Ontology Checker contains specifications of metadata, terminologies etc. which are used in the e-business and cultural heritage field.

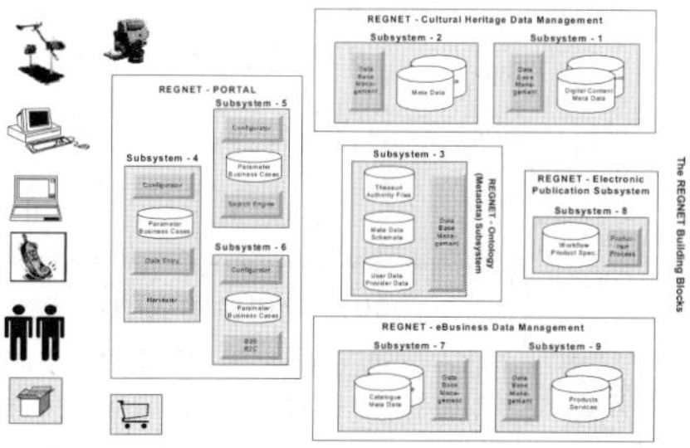

Finally the Electronic Publisher allows the production of personalized digital products based on standardized metadata and work flows.

REGNET extends the traditional access facilities to distributed heterogeneous catalogues by integration of product and service catalogues managed by e-business systems. Because of the large REGNET consortium real world tests covering 6 European regions (a dozen states) are possible.

The project will observe and take into account the ongoing and recent work of CEN/ISSS (Electronic Commerce Workshop) but will also look into recent activities like RosettaNet, eCo-Framework, OBI, etc.

REGNET will evaluate a networked organisation model (comparable to the Art Museums Image Consortium/AMICO in the United States) and the integration possibilities of tools and systems for co-operative working.

CONTENT CREATION, PLATFORM MANAGEMENT, ENTERPRISE ENGINEERING

Within the REGNET system there are three building blocks which can be considered as vertical functions:

1) Content creation and management is based on actual standards efforts in the field of the different organisations involved. The inclusion of so called 'Dublin Core' meta data is as well an actual topic as the Z39.50/XML-related standardisation work is, which is targeted to a harmonised search and retrieval facility across different domains. REGNET may influence the development of a so called Z39.50/XML-application profile: the Bath-Profile. This technique, to make Online Public Access Catalogues (OPACs) interoperable, is essential to REGNET, since by this way the catalogue of the REGNET shopping system is generated dynamically according to the request of a user.

2) The platform management is based on up-to-date internet technology and is the basis for the middle ware being the agent between content and service supplier and the requester (user).

It is expected that during the implementation of REGNET the first large trials of systems following the OBI reference architecture will deliver results (RosettaNet).

3) The enterprise engineering will focus on some selected business processes and functional units: access to distributed catalogues, a shopping cart system, creation of a personalised catalogue based on retrieved data from the 'virtual catalogue' (in printed and electronic form), an internet auction system (e.g. offering duplicates of posters), and a delivery system for physical goods (e.g. goods from museum stores).

On a 'horizontal' basis the XML/XSL-technology is used to structure data semantically and physically. This affects the creation of meta data, describing real (primary) objects (artefacts, naturefacts,...), media objects (photos, videos, ...) or bibliographic type objects (literature in the broadest sense). On the other hand all information within business transactions is wrapped within XML structures: Order, Invoice, Despatch, Report, ...). The recent developments in the field of XML/EDI standardisation will be used (ebXML). Another topic will be the definition of information products by appropriate document type definitions and style sheets. This should enable the 'non-media-professional' end user to easily generate catalogues or even CD-ROMs on demand. This might be the first step into the direction to create virtual exhibitions on demand by users themselves.

USER SCENARIOS

The scenarios outlined below can be considered as a first reference to e-Business processes only.

1. Business to Consumer (B2C):

A tourist wants to buy articles related to the cultural heritage of a region; he/she is interested in physical goods from one or more museum shops as well as in specific surrogate (images) of cultural objects located in museums, archives, or libraries in the region.

He/she can use a terminal at a tourist office where he/she searches the REGNET virtual catalogue for relevant images and information, browses for articles located in museum stores and places orders.

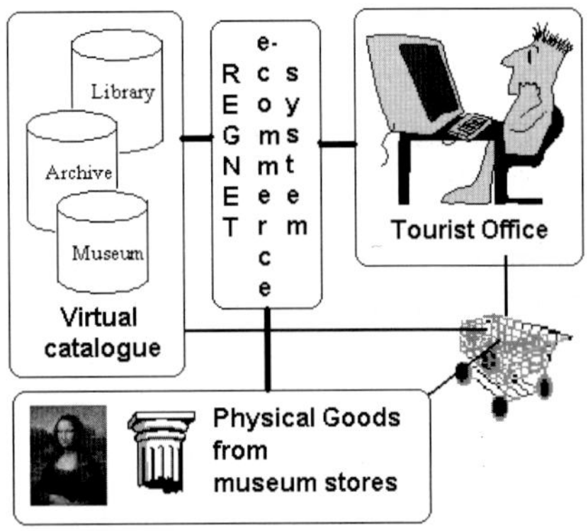

2. Business to Business (B2B):

A museum wants to produce a CDROM containing information about specific objects that he/she is interested in and which are related to a region.

The curator at the museum's side selects relevant information (text, images, films); he/she contacts a media producer, selects the basic layout/storyboard for the CDROM and works out together with the industrial partner - eventually with the help of other expert(s) - the final storyboard and the work flow necessary for the production process. He/she signs a contract for the production process and receives the master copy of the CDROM.

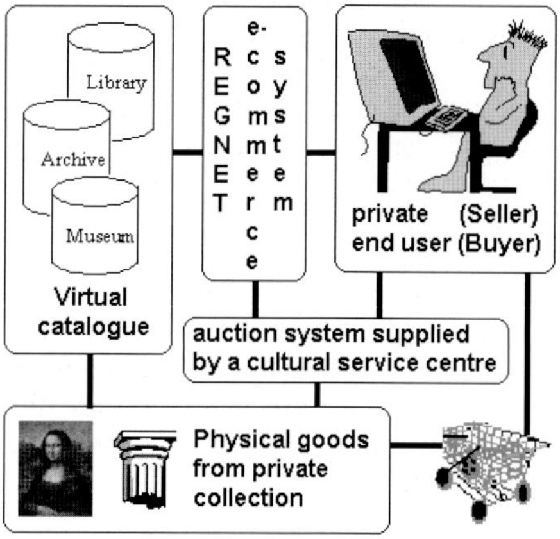

3. **Consumer to Consumer (C2C):** The owner of a private collection wants to sell some of his/her assets on a maximum prize. He/She thinks about putting the offers on the internet.

He/She digitises with the help of a REGNET-Partner (e.g. Cultural Service Centre/CSC) surrogates of the goods which will be sold. Additional information and expertises are gained after consultation of the Virtual Catalogue. An dedicated Internet Auction System is set up at an Cultural Service Centres (CSC) site and offers

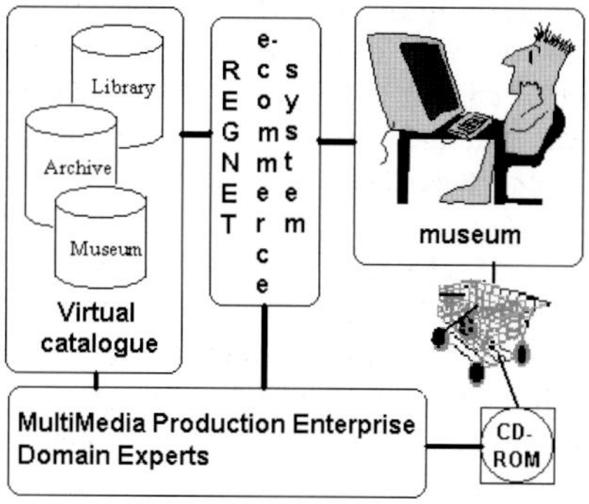

bidding to interested user communities. The marketing of this auction is supported by the CSC. This business case can be considered as combination of B2C and B2B functions and demonstrates that a stakeholder can obtain different roles in different business-cases.

MARKET SITUATION

Electronic commerce, notably business to business e-commerce, is booming and world-wide e-commerce sales are expected to grow 40 times between 1998 and 2003 by which time they will become over 15% of all sales. The Internet and e-commerce are also leading to a surge in new company creation. The market for REGNET systems and services can to some extent be reckoned by the number of museums, libraries and archives in Europe. A rough estimation would deliver a potential number of 3.000 (15% of 20.000 museums) museum shops throughout Europe being candidates to be partner of the REGNET network!

The project is a research and development project with a demonstration phase. It started April 1st 2001 and will last for two years. The whole project budget is above 5 Mio.

EUR and to 48,5% funded by the European Union. Project Coordinator is AIT Angewandte Informationstechnik Forschungsgesellschaft mbH. http://www.regnet.org/

Partners within REGNET
For efficient testing of the developed REGNET System a wide area distribution of partners with regional importance is crucial. Therefore 23 partners and 2 subcontractors from twelve different European countries are included in the REGNET project.

The following map shows the geographical distribution of the partners.

ABOUT THE AUTHORS

Dr.Walter Koch received a PhD in Mathematics and Physics from University of Graz in 1970. He is shareholder and director for AIT Research Ltd. (Angewandte Informationstechnik Forschungsgesellschaft mbH.), and is an Associate Professor at the University of Graz and a Guest Professor at the University of Krems.
Prof. Koch's relevant experience includes:

- From 1976 to 1998: Head of different research institutes at JOANNEUM RESEARCH Ltd. In Graz:
- Consultant to different national and international organizations (eg UNESCO, DFG, EC, Austrian ministries, Graz municipality)
- Member of several international and international scientific associations (eg. ICOM, ÖGDI, ONORM, VÖB, UNESCO)
- Project experience: bibliographic information, information systems, IT-management, EU Projects (TAP, Raphael, TenTelecom; IST), ESA, CIMI
- Overall Project Co-ordination and management of the EU-project REGNET, Cultural Heritage in Regional Networks
- Personal consultant to the EU-Project DIGICULT
  Prof. Koch has published more than 70 papers and technical reports, and presented papers in four continents at least at 100 conferences, seminars and workshops.

Address: AIT Angewandte Informationstechnik Forschungsgesellschaft mbH, Hans-Sachsgasse 14/3, A-8010 Graz, Austria.
Email: kochw@ait.co.at

Mag. Gerda Koch received a Master of Arts in English with a combination of studies in Economics and Law at the University of Graz, and also received an academic degree in a course of studies on Multimedia at the University of Graz. Since 1991 Mag. Koch is employed at AIT, Applied Information Technique Ltd. There she worked for diverse informatics projects as eg. a production automation system for a window-producing company, or the database publishing production of the yearly published Austrian Statistics on Non-University Education. She also collaborated in the European Union TenTelecom project MOSAIC (Museums Over States And vIrtual Culture) for the Austrian partner of the project. Since 1998 Mag. Koch is managing director of AIT Ltd. She now focuses on cultural digital content provision as she worked for the national project MODOK (Modular Documentation System) for the Austrian Ministry of Science and Traffic and the EU-project COVAX (Contemporary Culture Virtual Archive in XML). Within the EU-project REGNET, Cultural Heritage in Regional Networks Gerda Koch is also responsible for co-ordination and project management.

Address: AIT Angewandte Informationstechnik Forschungsgesellschaft mbH, Hans-Sachsgasse 14/3, A-8010 Graz, Austria.
E-mail: *kochg@ait.co.at*

Strategic combination of electronic and traditional services

Ellen Fleurbaay (*)

(*) Amsterdam Municipal Archives, the Netherlands
www.gemeentearchief.amsterdam.nl
E-mail: efleurbaay@gaaweb.nl

ABSTRACT
Two years ago the Amsterdam Municipal Archives launched a new website. The aim of the site was to increase the Archives' clientele. This objective has been pursued by stimulating traditional services via the website, as well as electronic services. To date we are extremely satisfied with the results: the number of hits has exceeded our expectations, while 2000 saw a reversal of the decline in the number of visitors to the exhibitions and reading room of previous years. We believe that it was our analysis of our clients' questions and the products we provide prior to the construction of the website that laid the groundwork for this excellent result. In this brief article I shall attempt to describe how the Amsterdam Archives came to choose a combination of electronic and traditional services in order to achieve the objective of increasing the clientele. First, I shall examine the Amsterdam Municipal Archive a little more closely. Then I shall outline the results of the research into our products and clients. Finally, I shall conclude by showing how we put theory into practice with two selections from the website.

KEYWORDS: product – client - internet strategy

THE AMSTERDAM ARCHIVES

The Amsterdam Municipal Archives are a historical documentation centre in the widest sense. It houses the archives of the local authority departments, but also of countless private institutions, companies and families. In addition, it has long kept a substantial library and many collections of images, including photographs, drawings, maps, posters, as well as recent acquisitions of audio products and moving images, including gramophone records, cds, film and video. The Archives, which has existed more than 150 years, has been called Amsterdam's memory. In terms more appropriate to today's digital age, the Amsterdam Archives is the content manager of Amsterdam's history. Around 45,000 people visit the Archives each year.

The above is the classic description of our institution, emphasising our **cultural-historical value** to our **visitors**. Today, however, government institutions are viewed increasingly as businesses, that is from an economic perspective. To achieve the objective outlined in my introduction - that is, to increase the clientele - it is necessary to approach a cultural institution as a business. What follows is therefore a description of the Amsterdam Archives as a company delivering **products** to **clients**.

PRODUCTS

The products delivered by the Amsterdam Archives fit into two categories: demand-oriented and supply-oriented. **Demand-oriented products** are (a) historical information and (b) reproductions of documents in archives and collections. Clients are able to collect historical information themselves by consulting the archives and collections in the reading room free of charge. For an hourly fee staff are also available to do historical research on request. Reproductions of documents can be obtained in the reading room for a basic fee. But reproductions can also be delivered to order, although the cost is considerably higher.

The Archives' **supply-oriented products** include all kinds of events, exhibitions, educational activities and various types of publication. The aim is to disseminate knowledge about the history of Amsterdam among as wide a range of people as possible. Today the accent is on reaching new target groups, especially among Amsterdam's more recent arrivals from abroad.

CLIENTS

The **clients** who currently consume archival products are not particularly diverse in age, social status or geographical origin. Most of our clients are middle-aged (between 40 and 70), college or university trained and born in or near to Amsterdam. An expansion of the clientele could either focus on attracting more of the same or on finding new clients from other sections of society: younger, less highly-educated or born elsewhere in Europe or further afield.

To make the right choices it is essential to discover the **reasons** why clients want archival products. To this end it is necessary to establish which **values** clients attach to our products, preferably related to turnover per product. Although each client has his or her own personal reasons, analysis brought to light that there is one value all clients in common attach to an archival visit: they visit the Archives for pleasure.

VALUES

What pleasure do clients derive from these products? Firstly, of course, there is the actual **possession of the product**. Supplying demand-oriented products that consist of information clearly satisfies a client's desire for historical knowledge if the question relates to a family matter, the home or any subject about which information is required. When the product we supply is a reproduction, f.i. a photograph, the client feels satisfied when it is framed and displayed on the wall at home.

But apart from possession of a product, other aspects also play a role. For clients in the reading room **the search itself** can be part of the enjoyment: certainly if the search is difficult and unexpectedly brings more or new information to light than originally anticipated. In the reading room clients can communicate with fellow-clients and more important: there is trained staff that offers assistance free of charge.

Another kind of pleasure is enjoyed by clients being able to examine **authentic historical documents** either in exhibitions or personally in the reading room. Imagine the emotion you feel when you open an old book and find a letter, even a brief note in the handwriting of an admired or a beloved deceased person. Or the emotion evoked when viewing a fine exhibition. Indeed a visit to an Archive exhibition is often part of a day out, or a weekend in Amsterdam. This too can be part of the pleasure of recreation.

INTERNET-STRATEGY

The principal conclusion drawn by the Amsterdam Archives from this analysis regarding the Archives' Internet strategy was that the Internet, as a new medium, would never be able to replace a visit to the Archives' reading room or exhibition gallery. Indeed, the desire to hunt for treasure and the enjoyment of a genuine, authentic product cannot be communicated through the Internet. So we decided, on the one hand, to use the Internet to facilitate a quick and efficient delivery of products to clients, and on the other hand, to use to Internet to stimulate the use of traditional services.

A quick and efficient delivery of products to clients contributes considerably to the client's pleasure in possessing the product. That is why the Amsterdam Archives has decided, wherever possible, to couple an **order function** linked to the products offered at every section of the site. Accordingly, the site has a webshop selling books and videos, and all the photographs (those not subject to copyright) are linked to an order button, as is the search system in the databases. At this moment we are building an image database that shall contain 20,000 photographs next year, and it will be possible to order all photographs in different formats.

To **facilitate the traditional services**, of course the Internet can contribute to the provision of correct information and communication with clients. The most obvious information is of course opening times, parking facilities, tariffs, an events diary and a news service. Information about our archives and collections is another key aspect The Survey of Archives and Collections provides primary information about the archives and collections we store, what clients can find in them and whether they are accessible and open to the public. This allows clients to prepare their visit to the reading room. The next step, for which preparations are now in full swing, is the provision of secondary access to the archives and collections, for example linking inventories and catalogues to the option to reserve items via the Internet so that they are already in the reading room when the client arrives.

Finally, the Internet can be used to **promote** the use of **traditional services**. For example, to encourage people to visit the reading room, exhibitions or other events at the Archive complex on the Amsteldijk in Amsterdam. The following example demonstrates how we have designed this both for demand-oriented and supply-oriented products.

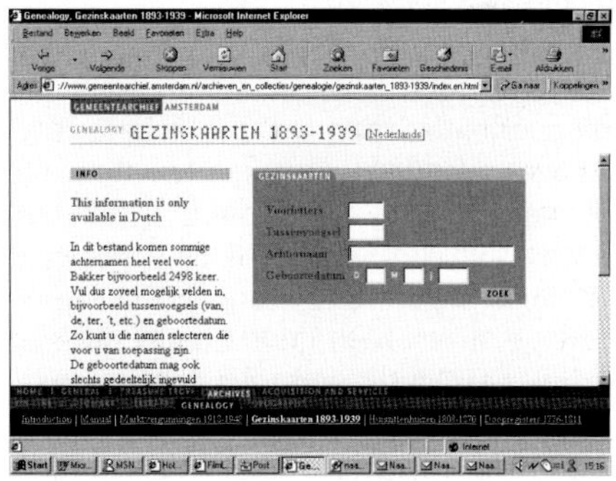

THE FAMILY CARD INDEX: A DEMAND-ORIENTED PRODUCT

Genealogical research is without doubt the most commonly practised form of historical research in our reading room. For many people, finding out about their background is vitally important. And, to judge from the genealogists in our reading room, family-tree research is addictive. So we have made a major archive source available on the Internet to this, the largest group of clients - a group with a potential for enormous growth.

This is the register of all the families living in Amsterdam between 1893 and 1939 (in all some 600,000 families). A small database of names on the website allows clients to find out whether we have information about their family, although this data is not itself available through the Internet. Clients are advised in the accompanying explanation to consult the data in our reading room. Here we can help them decipher the difficult handwriting and show them how to pursue their research. Clients can make their own copies of their family's registration in the reading room at ∈ 0,34 a copy. Access to the reading room and the assistance of our staff is all free of charge. In fact clients can also order copies of archive items relating to their family via the website. This costs ∈ 7,94 (almost twenty times as much).

As we expected, this section became the most popular part of the whole site. Around 80,000 visitors to the site viewed a total of around 1,200,000 pages, and an average of 28.9 per cent of these page views were in the family card section. Over 5,000 clients ordered prints of family cards via the site to a total value of around ∈ 18,500. We were also surprised that these orders were not limited to clients living outside Amsterdam. Many residents of the city used the Internet service too. In fact, one client lived only four doors away from the Archive! Many clients subsequently consulted the family cards in the reading room after having ordered them via or read about them on the Internet. Exact figures are not available because the databases consulted by visitors to the reading room are not recorded. I do know, however, that the total number of visitors to the reading room rose last year and that after the introduction of the family card database we were forced to increase the number of computers in the reading room. The sale of prints in the reading room grew last year by fifteen per cent.

We therefore conclude that the clients we have been able to supply via the Internet have largely been those who wished to acquire a print quickly, while we have also managed to reach new clients via the website who have discovered the pleasures of treasure hunting in the archives, the joy of doing their own historical research.

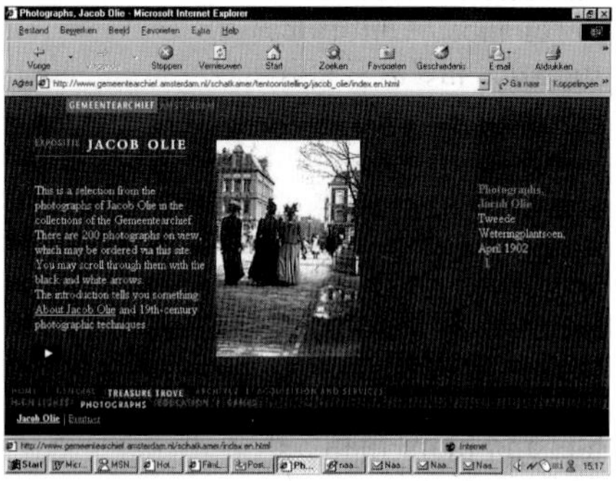

THE JACOB OLIE-EXHIBITION: A SUPPLY-ORIENTED PRODUCT

A similar tale attaches to the organisation of a simultaneous photograph presentation on the Internet and an exhibition of original prints at the Archives' exhibition gallery. The exhibits were unique photographs of Amsterdam taken in the latter half of the nineteenth century. Of course it is always delightful to be able to see these old views of the city with a single click on the screen, and even more fun, a few clicks further, to be able to order a photograph of the street where your parents or grandparents once lived. However, for the connoisseur of nineteenth-century photography, nothing can replace the original prints. In this case, these were small pictures, printed in a light-brown tint with a crisp sharpness showing all the details.

Again, the electronic and traditional services pertaining to this supply-oriented product turned out each to attract their own clients, yet both types of service exercised a positive influence on the other. Clients surprised by a preview of the exhibition on the website subsequently visited the show itself, while others ordered a print of a photograph via the Internet after having seen the exhibition, since searching and comparing on screen makes selecting far easier. These are examples I heard of by chance. Unfortunately, no accurate statistics are available since no research has been done on the subject. Perhaps next time.

CONCLUSIONS

The aim of the new website of the Amsterdam Archives was to increase the Archives' clientele. The website counted in 2000 more than 1,200,000 hits and more than 5,000 clients ordered reproductions via the site to a total value of around ∈ 18,500. The same year saw a reversal of the decline of visitors to the reading room of previous years, and an increase of 13 % reproductions sold in the reading room's self-service.

The website has proved to be an invaluable medium to get new clients interested in archival products. 'Virtual' clients are prepared to pay a relatively high price for efficient delivery of products via electronic services. At the same time more 'traditional' clients are attracted to the exhibition gallery and reading room, where the authentic documents can be seen, where products are sold for a basic fee and where trained staff offers free advice.

After a careful analysis of our own products and of our clients' wishes the Amsterdam Archives planned its internet-strategy of strategic combination of electronic and traditional services. We think the success of the website has proved this to be the right thing to do.

ABOUT THE AUTHOR

Ellen Fleurbaay is Head of the Division of Public Services and Deputy-director of the Amsterdam Municipal Archives.
E-mail: *efleurbaay@gaaweb.nl*

Cultural Heritage in the Age of the Web: How New Media Can Affect Art Fruition

Licia Calvi(*)

(*) Department of Computing Science, Eindhoven, The Netherlands
E-mail: l.calvi@tue.nl

ABSTRACT
Cultural heritage information and adaptive hypermedia have so far been pursuing completely different goals. Only recently, a research trend has emerged which has pointed out the potential advantages of exploiting adaptive methodologies to the delivery of cultural information. In this paper, we present an ongoing project that shows how these two research areas can come to terms.

KEYWORDS: adaptive hypermedia, cultural heritage, personalised content delivery

INTRODUCTION

The advent of new technologies has deeply influenced the way in which cultural heritage information is approached, accessed and fruited.

At a closer look, the contamination of art and technology is historically older than what the last years of Web explosion may seem to suggest: at the end of the nineteenth century, for instance, Rimbaud's fascination for photography was influencing his poetical style in determining which words to select, how to construct sentences, and how to juxtapose them in a visually effective way[1].

An analysis of how technology, and in particular the new media, can influence cultural heritage access and fruition, which ultimately entails an understanding of the philosophical underpinnings present in the relationship between art and technology, has therefore to be coupled with an attempt to develop *ad hoc* tools for the conservation and public use of such information (via, e.g., adaptive systems), not to mention its possible enhancement: "electronic argumentation" [16], the advent of new literary *genres*, "art mutations" [8], all this can indeed be seen as a way to enhance the effective use of the Web by exploiting its pragmatics and rhetoric.

According to the Musea Project [12], *Cultural Heritage* is an umbrella term under which different entities may be included (Fig. 1).

This paper deals with *documentary*[2] cultural heritage. It presents an ongoing project that aims at the digitalization and encoding of a corpus of (Italian) literary texts into a hypermedia system with the inherent goal of fostering their preservation, enhancement, and public access. Its basic assumption is that it

---

[1] See, for instance, in [8].

[2] *Documentary* objects are normally understood in a wide definition of the term, among which not just textual documents but also images and sound documents are included [12].

should be possible to integrate the knowledge about the users with respect to the information presented (see further), the knowledge about the application area and the processes playing in it, and the knowledge about the medium itself to deliver to users personalised information.

But no new artifact lives in the vacuum: it is always somehow related to the ones which have preceded it and which surrounds it. This is why it could be useful to review the history of this evolution to better understand how new media can impact on cultural heritage fruition.

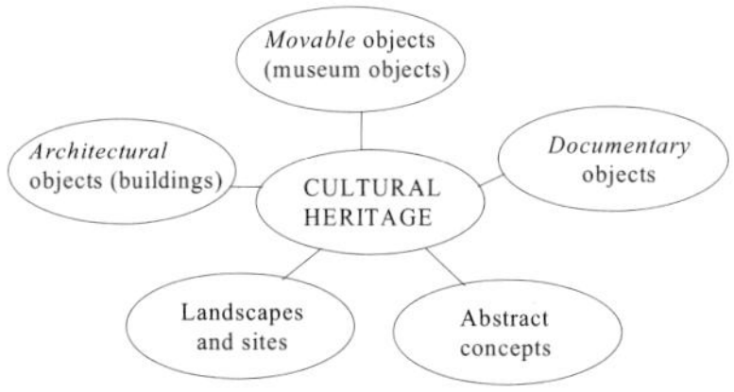

Figure 1:

Cultural heritage entities from [12]

ART AND NEW MEDIA: IS IT REALLY THE DEATH OF ART?

The transformations imposed on art by the advent of technology have ultimately annulled its distance from science to the extent that art has acquired a new epistemological status: "art has modified its own essence and has become the place for the *preventive domesticity of the possible existentials*" ([8]: 28).

Costa [8] indeed identifies four logical movements at the basis of such a modification. Such movements are:

1. *Fascination/imitation*: this is the first step in the transformation of art. Such a scenario comprises an artist who has to face and come to terms with his/her fascination for a technological domain. This was, for instance, the case of Rimbaud who was fascinated by photography, or by Calvino who was fascinated by combinatorial analysis (so, the "writing machine"). Because of such a fascination, the artist is urged to start imitating, to take over, some of the modes of functioning of the technology s/he is approaching;

2. *Change*: as a second step, the acquisition of some of the procedures of the technology has a first impact on the artistic product, but only at the more superficial level, i.e., the level of its content, whereas its essence, the medium itself, remains untouched. Calvino, for instance, during his OULIPO years, had transferred his fascination for the combinatorial calculus into his novels. But the novel remains a novel, and its recipient, the book, a book;

3. *Crisis*: it is only in this third phase that the modifications, which so far had remained at a mainly shallow level, start to impact the essence of the artistic product, to question its identity. Questions over the status of the book, in the case of Calvino, over a possibly new poetic agenda, in the case of Rimbaud, become pressingly unavoidable;

4. *Mutation*: ultimately, answering these questions does affect art in its own form: a book can no longer be a book to carry the newly emerged content; poetry has to find other forms to express the same artistic power. A mutation has taken place.

This is a sort of "death of the art", where "technical simulation resolves the problem of truth and reality with an uncontrolled production of images where not much to be seen remains" ([9]: 181). Or, to cite Baudrillard, "we live in a world where the supreme function of the sign is to make reality disappear and, at the same time, to disguise this disappearance" ([1]: 9, as quoted in [9]: 181).

This paper places itself at the level of the third logical movement in art transformation, i.e., the level of crisis described above. The overcoming of such a crisis will require the adoption of a new artistic agenda, i.e., one that exploits the rhetoric and pragmatics of the new media.

PERSONALISED CONTENT DELIVERY OF DOCUMENTARY OBJECTS

In [4], we introduced our long-term project to investigate how documentary objects could be made publicly accessible through adaptive hypertext. This project entails two ultimate goals: in the first place, to encode a corpus of literary works into an online archive; in the second place, to develop an adaptive tool to write hyperfiction. Encoding text ultimately requires a redefinition of what its status is [5] and developing a tool to deliver personalised literary content presupposes an understanding of the dynamics operating in electronic reading and writing, as well as in knowledge representation and content extraction.

This project started with the development of an online adaptive course where the convergence between hypertext theory and practice was investigated. This course was intended to highlight the processes in hyper-reading and writing and was considered as a test bed for using adaptive methods to deliver cultural material. As additional feature (and as a preliminary attempt at creating such a wide documentary repository), a pointer to an online archive of contemporary literary works where such a convergence is expressed would be added. The course, called Postmodern Web[3], is split into two modules, i.e., a propaedeutic module and the core module [4]. In the propaedeutic module, the basic principles of hypertext as a tool are introduced, whereas the core module shows how such principles can be derived from the theory of literature on the basis of real examples drawn from existing paper-based and electronic novels. The separation into these two modules is not only functional, but also design-related: the propaedeutic module is structured hierarchically, i.e., it has a starting and an ending point and a guiding principle to determine navigation that exploits one of the well-known adaptation techniques, i.e., link removal, disabling or annotation [2]. The core module, instead, is intended to be structured horizontally, in the sense that the absence of starting or ending points as well as of guiding principles and of reading priorities would make it strictly flexible. Flexibility would imply here the necessity of adapting the content to the user's reading history: the content material would be accessed as a result of the actions performed, i.e., on the nodes read, by the individual reader and their content changed accordingly.

Michael [13], as B.Sc. final year project, contributed to the further development of the Postmodern Web, mainly by augmenting the propaedeutic module

[3] The system is currently available at the following URL: *http://wwwis.win.tue.nl/~calvi/pmw_project.*

with specific functionalities (see further). Initially, however, he conceived two different navigation scenarios: whereas the original Postmodern Web site was supposed to present only adaptive link visibility, Michael [13] introduced, along with an adaptive pathway based on the basic principles described above, a strictly linear scenario for those readers who are not too at ease with the use of hypertextual tools. This linear option is offered on every adaptive page, so that the user who finds him/herself in difficulty can always resort to this solution. However, the linear navigation still results in an updating of the user model by the addition of knowledge values (see below), although it does not really modify the presentation of content information. But the simple provision of these two alternatives, which ultimately refines the readers' profiles entailed in the adaptive version of the system, mirrors a more fundamental distinction among readers based on their learning/reading style, and results in a separation between more experimental users, who would choose for the adaptive version of the system and who would undergo a progressive learning development from novice to expert users [7], and the less experimental ones, who would follow the linear alternative and would not profit from such a personalised content delivery.

Once the user has registered, s/he can access the information. The first operation s/he is required to do is actually to read an instruction page, where guidelines are given on how to use the system and explanations on how it has been conceived. This is particularly relevant for the readers who decide to follow the adaptive version. After this operation has been accomplished, indeed, two options opened up for the readers: to access the information in a linear fashion or to opt for the personalised content delivery (see above). This last one presupposes a fundamental learning assumption: if readers access information in the order foreseen by the designer, then their knowledge can increment and they themselves evolve with respect to it, from the level of having no knowledge on that topic to the level of mastering it [7]. This modelling makes sense because the system is in the first place a didactical course. We will see below how it can also make up for a more general framework to deliver personalised information.

Michael's contribution to the long-term Postmodern Web project [13] consists in adjusting the navigation procedures of the AHA architecture [10] according to the guidelines sketched above by introducing thresholds. Thresholds are approximations on a continuous interval of the value assigned to the knowledge. The knowledge is represented in the domain model of the system and can be defined at either the concept level or the document level. A document is really the smallest operative knowledge entity and corresponds to a node. But documents can ultimately be further decomposed into fragments, which are the smallest theoretical knowledge unit, the atom of knowledge. More documents normally contribute to the definition of a concept, which is therefore a higher-level entity. So, for instance, the documents *art*, *comics*, and *hypertext* in the Postmodern Web project generate the concept *culture*. Concepts in the system often correspond to a structural section. The system is conceived in such a way that reading a node, i.e., a document, augments the knowledge attained by the user and stored in his/her user model. Reading is taken in this context as a very broad notion: no estimate of the actual time the user has "spent" on the page is computed; rather, the simple fact that a user has selected a certain link and accessed the corresponding page is considered enough to update the knowledge in his/her user model. The thresholds assigned at document level determine the hierarchical access to information in this guided fashion (see above). The combination of more documents determines the threshold that is necessary to access the subsequent concept, much like in [15]. Whereas not all the documents composing the precedent concept need to be "read" before accessing the following concept, a minimum threshold is set to navigate the concept hierarchy bottom-up. This value consists of the highest document threshold augmented by a unit, normally 10, to ensure that at least more that one document is accessed before progressing further. Propagating knowledge between documents occurs similarly, in a straightforward fashion. Such thresholds are specified in the document list (i.e., the *requirement list*) of the AHA architecture [10], where the dependencies between documents and concepts are specified. Moreover, they also appear in the user model, together with the thresholds assigned to the documents. The advantage of this solution compared to the existing implementation of AHA presented in [10] consists in freeing document access from a straightforward navigation, where the link removal technique shows the documents to access in a "linear" fashion, depending on the one preceding and the one following each of them. The introduction of thresholds, instead, makes document access really dependent on what the user has read, rather than on what is structurally (and conceptually) just before each document from a design point of view.

All pages still start with two HTML comments (see also in [10]), which indicate whether concepts are *required* or *forbidden* for accessing that page. All pages other than the index page are normally specifying a condition for accessibility. This can take the form of

<!-- requires 20 -->

[13]. According to the requirement list in the Postmodern Web, this statement indicates that this particular threshold (20) must be true for the corresponding page (i.e., the document *introduction*) to be accessed [13]. Such a value must ultimately appear in the user model as well.

The second HTML comment at the beginning of the page indicates which concepts are *generated* after visiting that page. So, for instance, the statement

<!-- generates 30, intro -->

indicates the threshold that becomes known and will contribute to the propagation of knowledge in the user model, so as to make other documents within the same concept potentially available [13]. The page being generated is again the introduction page. Generated concepts are however added to the student's knowledge only after their corresponding pages have been read.

The importance of this technique will become more evident once the core module of the Postmodern Web will be finalized in the flexible way described above, where thresholds and knowledge propagation will make document access strictly dependent on the user's reading history.

USERS' PROTOTYPING AND ACTIONS

In an attempt at augmenting the existing AHA architecture [10], the Postmodern Web project described so far aims at developing an adaptive system to deliver personalised cultural information that presents the following features:

1. It supports users' diversity;

2. It promotes users' interactivity.

The distinction discussed above between the two alternative navigation scenarios and which we explained as determined by the possible learning/reading styles performed by the users can also be seen as accounting for different users' profiling based on their tasks. We cannot talk properly of a task model, as in [14], for instance, because in our case the choice of one of the two options is not justified by an analysis of the tasks each target group may wish to undertake with the system. In our case, however, we could still assume a distinction between students and occasional readers: the former are given an information presentation which is more structured and satisfies certain pedagogical assumptions [7]; the latter, instead, are not concerned with issues such as knowledge acquisition and updating. The interface differs in these two cases only in the way information is presented to readers, in how far the knowledge level attained by the users is taken into account, which results in a varying link functionality and in providing local guidance[4] [2], whereas the occasional reader receives no local supports at all. Global navigation support is however offered to both kinds of readers by a conceptual map of the site that distinguishes between pages already accessed and pages still to access.

As for what concerns the second aspect, there is essentially one way for users to interact with the present version of the system, i.e., by writing an essay in hypertextual form. Each essay is stored into a central repository and made available for consultation to other users once they reach the core module. Essays can cover a set of predefined topics, four in total, each of which is associated to a unique page to compose and post essays and to view them. Such an opportunity is also adaptively determined by the system. Each document points to one only topic for essay composition and consultation, whereas only the beginning, index page points to all four topics but only as the user has reached the core module [13]. Such essays would achieve the twofold goal of verifying the extent to which the course topic has been understood in a learning-by-doing way and would at the same time allow students to access each other's essays, which would, in its turn, further extend one's proficiency in the subject [4].

Another form of interactivity is the contact with the tutor who supervises the entire learning process and who can intervene at any moment during its progression. According to the form of help or request asked by one reader, the tutor can decide to keep the other

[4] This was achieved by Michael by introducing the JavaScript ONMOUSEOVER function [13]. This facility informs the reader in the window status bar about the name of the node that will be accessed by selecting that link, about the section of the node s/he will be pointed to, and about its status just by positioning the cursor on the link itself.

readers posted and organise a sort of forum or discussion group on the matter.

DOES THIS MODEL ACCOUNT FOR THE PERSONALISED DELIVERY OF CULTURAL INFORMATION?

The prototype described in the paper exploits the notion of *knowledge* to provide readers with adequate support in navigation and in information extraction. The term knowledge is used in this context to include readers' learning styles, their level of mastery of the information and experience in the domain content. Although it directly addresses the well-defined field of education and ultimately accounts for its specific requirements, the enhancements in knowledge delivery achieved by means of adaptive techniques can be generalized and extended to other domains, like, for instance, information publication and diffusion in the cultural and public sphere.

Indeed, from the experience described in the paper, despite its, to a certain extent, restricted character, there are at least two lessons that can be drawn:

1. Knowledge delivery should rely on user's profiling in order to be effective [3, 11];

2. Adaptivity in knowledge delivery can improve task performance (see also in [6] for more experimental results).

Our system is indeed relatively user-friendly for readers: once they have understood how the presentation is changed adaptively, they mostly follow the guidelines suggested by the system and seem to rely on its interface design [6]. It is nevertheless not as user-friendly for authors, especially if they do not possess a thorough understanding of programming and of computer in general: this is why it would be necessary to develop an authoring environment to facilitate all possible authors in writing their own application using this framework.

Clearly, a generalization and extension of this model cannot be performed without a prior analysis of the specific mechanisms and processes present in other fields. This is how the study on hypertext rhetoric and practice carried on in the online course has contributed to a better understanding of the dynamics enacted in hyperfiction by readers and writers. An analysis, among other things, of users' reactions at using the system with its present functionalities and adaptivity features will become necessary. This knowledge will constitute the backbone for the creation of an online archive for documentary literary objects.

CONCLUSION

In [10], we have announced the possibility of extending the then existing AHA architecture to include numeric ratings (thresholds) assigned to concepts and to allow for knowledge propagation. In this paper, we have presented such a possibility within the framework of an ongoing project aiming at the effective delivery of personalised cultural information. The project has now undergone a preliminary phase where the dynamics present in electronic reading and writing, as well as in knowledge representation and content extraction were investigated. This was achieved by developing an adaptive course that was seen as a test bed for using adaptive methods to deliver cultural material. Next, it will evolve towards the creation of an online hypermedia archive of cultural information and in a system to author cultural, mainly literary, documents.

ACKNOWLEDGMENTS
The author would like to thank Ramy Michael for his implementation efforts during his B.Sc. thesis.

REFERENCES

1. Baudrillard, J. Le crime parfait. Éditions Galilée, Paris, 1995
2. Brusilovsky, P. Methods and Techniques of Adaptive Hypermedia. In User Modeling and User-Adapted Interaction, 6, 1996, 87-129
3. Brusilovsky, P. and Eklund, J. A Study of User Model Based Link Annotation in Educational Hypermedia. In Journal of Universal Computer Science, 4(4), Springer Science Online, 1998, 429-448, *http://www.iicm.edu/jucs_4_4/a_study_of_user/paper.html*
4. Calvi, L. The Post-Modern Web: An Experimental Setting. In McCarty, W. and Fiormonte, D. Eds. Computer, Literature and Philology, Humanities Computing Unit, Oxford. In press
5. Calvi, L. Text and Hypertext: Always a Binary Relationship?. In Proceedings of the Eleventh ACM Conference on Hypertext '00, ACM Press, 2000a, 218-219
6. Calvi, L. On the Formative Evaluation of CALLware. In Brusilovsky, P., Stock, O. and Strapparava, C., Eds. Adaptive Hypermedia and Adaptive Web-Based Systems, Proceedings of the AH'00 Conference, Lecture Notes in Computer Science, Spriger-Verlag, 2000b, 276-279
7. Calvi, L. and De Bra, P. Using Dynamic Hypertext to Create Multi-Purpose Textbooks. In Müldner, T. and Reeves, T.C., Eds. ED-MEDIA'97, Proceedings of the World Conference on Educational Multimedia and Hypermedia, AACE Press, 1997, 130-135
8. Costa, M. L'estetica dei media. Avanguardie e tecnologia. Castelvecchi, 1999
9. De Angelis, V. Arte e linguaggio nell'era elettronica. Bruno Mondadori, 2000
10. De Bra, P. and Calvi, L. AHA! An open Adaptive Hypermedia Architecture. In The New Review of Hypermedia and Multimedia, 4, 1998, 115-139
11. Eklund, J. and Brusilovsky, P. The Value of Adaptivity in Hypermedia learning Environments: A Short Review of Empirical Evidence. In Brusilovsky, P. and De Bra, P., Eds. Proceedings of the Second Workshop on Adaptive Hypertext and Hypermedia, Computing Science Reports, Eindhoven University of Technology, 1998
12. Makemson, C., Clark, R., Jørgensen, P.H., Hogenboom, J. Standards for Cultural Heritage Information On-line. Final Report, 1997, *http://www.portia.dk/musea/index.htm*
13. Michael, R. The Postmodern Web. An Adaptive Courseware. B.Sc. Final Year Project Report, Computer Science, Trinity College Dublin, 2000

14. Paternò, F. and Mancini, C. Effective Levels of Adaptation to Different Types of Users in Interactive Museums Systems. In Journal of the American Society for Information Science, 51(1), 2000, 5-13

15. Pilar Da Silva, D., Van Durm, R., Duval, E., Olivié, E. Adaptive Navigational Facilities in Educational Hypermedia. In Proceedings of the Ninth ACM Conference on Hypertext '98, ACM Press, 1998, 291-292

16. Williams, S.D. Structured Dissonance and the Art of Building Arguments for the World-Wide Web, *http://www.nyu.edu/classics/keefer/waoe/williams.html*

ABOUT THE AUTHORS

Licia Calvi is Assistant Professor at the Department of Computing Science at the Eindhoven University of Technology. Her research interests include multimedia for cultural heritage and education, adaptive hypermedia systems for distant learning, philosophy of culture and digital media and art, AI and philosophy of science.
Contact address: Dept. of Computing Science, Eindhoven University of Technology, P.O. Box 513, 5600 MB Eindhoven, The Netherlands
E-mail: *l.calvi@tue.nl*

Hypertext and word of mouth interact in theatre for the transmission of cultural heritage

Anna Gentile (*) *and G. Ilenia Paladini* (#)
(*) Department of Innovation Engineering, University of Lecce, Italy
E-mail: anna.gentile@unile.it
(#) Innovative Materials and Technologies, ISUFI of Lecce, Italy
E-mail: ilenia.paladini@unile.it

ABSTRACT
Hypertext are proposed to be adopted in theatre as the "cultural" alternative to the "physical" means of interaction such as multivision or tridimensional effects. A particular class of hypertexts, oriented to multimedial shows on the cultural heritage of some peoples, has been designed and experimented in several performances. Hypertexts enrich and are enriched from the performance of the actors and musicians, themselves constituting, in this ways, a means of preservation of the cultural heritage of peoples.

KEYWORDS: Hypertext, Multimedia, Multimedia Shows, Cultural Heritage

INTRODUCTION

The culture of many peoples was for centuries handed down from generation to generation by word of mouth. In time, this oral tradition gave birth to language forms that were extremely rich and expressive, as well as closely linked to the tone of voice and the gestures used, to pictures and symbols that integrated the linguistic medium. Whatever trick or device that might fix words and ideas in the memory - including rhythm and rhyme, and the intonation of the voice - were spontaneously brought into play. Now, if we compare the oral traditions of early societies with the medium of information technologies of the present day, it is immediately clear that the primary concern of both is communication: with all their differences, the cultures of both past and present societies grow up around and thanks to moments, places and environment in which communication is of the essence and to be jealously cultivated. [1]

What is more, just as the oral tradition relied on non-linguistic contributions for maximum effect, so the multi-medial potential of computers helps develop and transmit multicultural and multi-discipline themes which succeed in appealing to several senses and many different types of intelligences, thus activating processes which fix new knowledge in ever-growing cognitive networks.

At the same time, however, the kind of technology we have at our disposal nowadays is not only stimulating in its complexity and many-sided potential, it is also as easily adapted to the creation of spectacular, almost hypnotic effects, as it is to the promotion of social and cultural communication (see, for example, Imax [2], Multivision [3] and some aspects of Virtual Reality [4]). Real communication between users requires that there exist a shared vision and acceptance of the underlying cultural message upon which interaction can develop: without such common ground, attempts at communication might easily remain isolated, unconnected to wider contexts, while open and harmonious sharing allows contributions to be made by a whole range of actors, each integrating the others.

This is the spirit with which we have been working: bringing together the oral traditions of peoples and the communicative power of the new media with a mind to spreading to a wider public the experiences and the cultural heritage of entire societies. We have tried to do this through the creation of multimedial performances where poetry, music, myth and legend are brought together in the suggestive framework and texture of images created by a hypertext and shown on screen by a projector connected to a computer.

At first sight there may seem to be substantial differences, indeed contradictions, between the flexibility of a hypertext, the essential characteristic of which lies in its adaptability to the interests of the user, and the apparently fixed structure of a theatrical performance. Yet, it is the very adaptability of a well-made hypertext that makes for variously organised performances – what is shown can be modified, in terms of content or of juxtaposition of images and sounds, depending on the context and the cultural message to be transmitted.

The context of the performance, the overall visual and theatrical impact, is created by the hypertext. There is an actor who lends his voice to chosen texts, almost reviving the oral tradition; and there are musicians who respond to verbal texts, giving a body of sound to the words and the pictures around them. Context, then, but also catalyst, shared space and communicative medium – the hypertext is all of these, extending, of course, out to the spectators, who are drawn into a world which communicates to all their senses. Not a world that is repeated as a fixed series of moments, but which changes in response to external situation and demands, and internal dynamics. Multimedial hypertexts produce, then, communicative contexts and means, and, from a different point of view, offer a valid form of technology as entertainment and interaction with spectators and users.

MULTIMEDIA SHOWS

The multimedia shows we are speaking of are the fruit of long-term research and experimentation dealing with, on the one hand, the oral and written traditions of given peoples, and, on the other, how to go about

designing a hypertext that might satisfy the kind of requirements mentioned in the introduction.

Oral and written traditions

We decided to develop our research into the historical and cultural roots of peoples by exploring two different dimensions: first of all, what might be called the "vertical" dimension in that we were looking down history to the origins of the Afro-American people, and secondly a "crosswise" dimension that helped connect the different cultures of peoples distant from each other in time and space but who shared ideals of freedom in addition to a sense of harmony in relation to the world of creation.

These traits and qualities are what give the hypertext and the consequent theatrical performance their essential substance. Performance is interpretation, never the same and always in a process of being created anew; the hypertext enables the culture of a people to be passed on in a more genuine form because that culture is represented multi-dimensionally and not, for example, just on paper.

"Poetry and music of Black America: spirituals blues jazz" [5-6-7], in collecting and collating the poetical works of Afro-American writers, attempts to tell the cultural history of a people whose music, originating in the rhythms of Africa, inspires the music of all the world. The narration of their history and the sights encountered in the hypertext (Figure 1, Figure 2) recreates the lives of the American black people who, even in the midst of a hostile environment, managed to retain such a creative energy and a firm hold on the preciousness of life that these features of their culture stand out even in the spirituals, blues and jazz of today.

Figure 1: Poet's photo and poem

Figure 2: Musician's full screen photo

This cultural journey starts off from the melancholy notes of the blues, which tell of the hardships of the slaves working in the cotton plantations, arrives at the emergence of be-bop, when young musicians began to feel they were true artists and not simply performers of another's music; the journey passes through the period of cool jazz, which marks the arrival of jazz in California, and then meets the beat generation, before arriving at the free jazz of the sixties. The poetical works we chose have been translated in such a way as to present sound patterns similar to those of blues and jazz, and also to respect the spirit with which the originals were composed. These works are accompanied in the hypertext by photographs, documents of the period, and newspaper articles.

"Myths Poetry Legends and Music of the World" [8] tries to reveal the secret kinship of myths and legends belonging to peoples from all over the world (Figure 3, Figure 4).

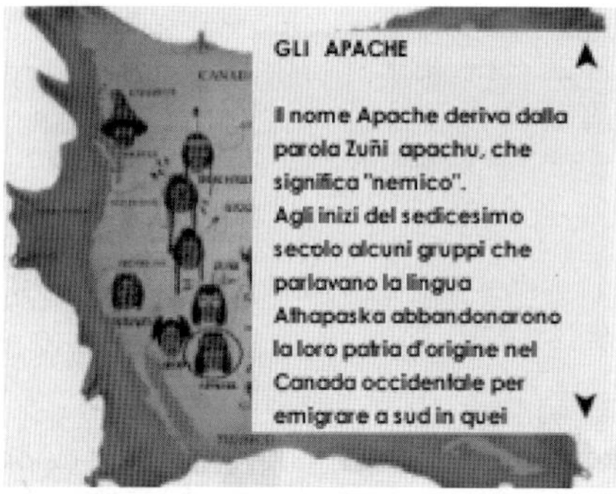

Figure 3: Map of the American Indians tribes (description of the Apache tribe)

Figure 4: Apache woman's full screen photo

There are, for example, legends of the American Indians that express the ideal of harmony between man and nature; Afro-American poems that take their inspiration from spirituals and Gospel song; African fables rich in simplicity and wisdom; poems of the Kurdish people that speak of freedom and humanity.

In this performance, the reciting voice is accompanied by music composed for the show and performed by musicians specialised in ethnic music. The hope is that such shows will be able to bring out the humanity shared by all peoples, the common forms of tradition, the universal feelings that unite us all.

Design of hypertexts for the theatre

The hypertexts for the theatre were conceived in such a way as to contain a vast range and variety of texts, sounds and pictures that can be grouped together and set in different sequences according to the reading one wants to make of the material - clearly this provides the basis for numerous potential performances. The shows become more dynamic, evolve through time, are never repeated in exactly the same form - a single performance takes shape as on the spot decisions are made as to what is to be shown and seen. Nothing is automatic, rather it is the presence and the interaction of the "actors" - the narrator, the musicians and the hypertext navigator/operator - that is crucial for the development of a show. This does not mean that there is always a perfect synchronism of the various media: at times, relatively extraneous elements are introduced - the projection of scenes regarding with general themes, history or art rather than the specific cultural item being dealt with - so as to create a degree of counterpoint that sets the particular into relief.

One technique is to project different pictures onto two separate screens (Figure 5): one projector, connected to a video-recorder, shows a collage of video-clips of scenes related to the cultural aspect in question. These clips are not in fact synchronised with the running order of the performance but in a sense provide a backdrop against which chosen elements and the narration itself interact. On the other hand, the pictures reproduced from the hypertext, projected on the other screen, are shown in close correlation with the narration as it unfolds and the music as it is played (Figure 6).

Figure 5: Otranto castle, 06/09/00 – Otranto, Italy

Figure 6: Carlo V castle, 25/11/99 – Lecce, Italy

The hypertext contains the texts of the poems and the tales, as well as the biographies of the writers together with photographs of these and also pictures of locations and of whatever else might better illustrate the central concern of a given text. Now, as pictures play a role of great importance in the shows, great care was taken both with their size and with their resolution when the hypertext was being put together. The resolution is, indeed, very high, and often the pictures which the audience sees are the result of a meticulous work of processing and retouching with appropriate software which, together with an effective play of colours, succeeds in giving the impression of the moment.

In traditional multivision shows, the fade-out effect of images is achieved by control panels or through projector-control computer software; in these multimedia performances, however, the same effects of fade-out or fade-in are already programmed inside the hypertext.

In addition to this, there are instances of visual transformation, when animation is used to transform one picture into another. In this case, a morphing program was utilised to create the transitional photograms between the initial and the successive picture. During the second phase, all the pictures thus created were then edited for the definitive animation.

Video films of people and places alternate with computer-created sequences assembled using video editing software, such as Adobe Premiere.

The fact that the editing procedure of a film sequence in Adobe Premiere is not linear has made it possible to insert, substitute, copy, transform and eliminate clips whenever necessary.

Through the assembly window we have been able to organise, separate or insert clips and arrange them on different levels, to join clips together and carry out the mixing of the clips' audio. The filters for film sequences, still pictures and audio make it possible to achieve effects of distortion, and of blurring; they also can be used to apply textures and colours to pictures, as well as bring about modifications to the sound.

CONCLUSIONS

There have been many experiments in the theatre with multimedia technologies; most often people have tried to use these technologies to create particular staging effects which would integrate and bring out aspects of a given work. Sometimes micro-telecameras follow the actors, these pictures are then mixed by the director and projected onto a number of screens: this effect is obtained through the use of interactive software, and at the same time acoustic effects are created all around the spectators so that they feel immersed in the sound texture of the show.

The effect is without doubt extremely suggestive, yet how much of the work's cultural message is really being communicated to the audience? To what extent can technologies used in this way, almost overpowering the presence of human actors, truly be seen as providing channels for cultural communication?

The crucial factors are the harmony and the inter-relations that come into being between all the component parts of a theatrical experience: - pictures, sounds, the speaking voice, the direction of the show, and the technical expertise involved - by means of well-made hypertext. In addition to these, the various technological instruments help intensify and bring as close as possible to the spectators the message that the cultural creations end up transmitting, acting at the same time at the levels of intellect, senses and emotions.

REFERENCES

1. Gentile A., "*Ritmo e Poesia Afro-Americana*" - FIREWORKS, University of Bari, Italy, Levante Editore, Bari, Italy, 1996/97, 82-91
2. Imax, *http://www.imax.com*
3. Firstimage, *http://www.firstimage.co.uk*
4. The Virtual Reality Society, *http://www.vrs.org.uk*
5. Multimedia show by Anna Gentile, *http://www.annagentile.it*
6. Tecnopolis Csata Novus Ortus, *http://www.csata.it/tecno_it/indici_it/jazz7_6.htm*
7. Serafino P., "*Jazz e megabyte*" - Il Corsivo, Lecce, Italy, October 2000, 30
8. Serafino P., "*Dove finisce l'arcobaleno*"- Il Corsivo, Lecce, Italy, September 1999, 24

ABOUT THE AUTHORS

Anna Gentile is a professor of "Telematics" at the University of Lecce, Italy. Her research interests are in Tele-Education and Networks Application in the area of cultural assets. She participated to the European Project ISIS, designing and co-ordinating Tele-Education Courses via satellite. A special attention she devotes to exploration of the use of hypermedial technics for definition of new cultural paradigms, such as shows with embedded hypetexts and antique books revived as hypertexts.
E-mail: *anna.gentile@unile.it*

G. Ilenia Paladini received the Laurea degree in Computer Science from the University of Bari, Italy, in 1997. She is currently a Ph.D. student in "Innovative Materials and Technologies" of ISUFI of Lecce, Italy.

Her research interests focus on stand-alone and on-line multimedia services. The focus of her current work is on implementation of multimedia services and design of user interfaces for Wireless Ad Hoc Networks.
E-mail: *ilenia.paladini@unile.it*

Broadcast Archives: Preserving the Future

Richard Wright (*)
(*) BBC Information & Archives
E-mail: Richard.wright@bbc.co.uk

ABSTRACT
A survey has been made of the holdings and preservation requirements of ten major European public service broadcast archives. The survey also covered 'the business they do and how they do it': the contribution made by the archives to the business of broadcasting. The archives gave their plans for future services as media moves from 'tapes on shelves' to mass storage and electronic distribution. This paper links the urgent requirements for preservation to a strategy for building the digital archives needed for the future.

KEYWORDS: multimedia, archive, preservation, digitization, digital library, funding, preservation technology

INTRODUCTION

The Problem: Millions of hours of valuable multimedia content - and irreplaceable European heritage - is at risk because of ageing media and technology. The content of European public service broadcast archives is the social and cultural history of the 20th century, on a European scale. Largely paid for by the citizens, it should be preserved for their benefit.

Broadcasting is a twentieth century industry, and television a post-war industry. Videotape came into general use thirty to thirty-five years ago. Ageing of audio recordings, film and videotape, plus obsolescence of videotape equipment, combine to produce a problem not previously encountered in broadcasting: all broadcast archive material from the beginning of broadcasting to roughly the end of the 1980's is now at risk. The "preservation requirement" is approaching broadcasting like a tidal wave, bigger and more potentially destructive each year.

An estimate ten million hours of material of national and European significance are at risk. The cost of preserving such material is between 100 Euro per hour for audio and videotapes and 2000 Euro per hour for film. The total cost of preserving this material using present methods and technology is well over one billion Euros. Unless new, more cost-effective preservation methods and technology can be found, the preservation price may simply be too high and significant portions of the audiovisual memory of the last half century will disappear.

PROJECT PRESTO

Broadcast archives and technologists have joined forces to better understand the issues and problems, and develop solutions and joint standards. This joint effort has been made possible by funding from the European Commission Fifth Framework Programme in Information Society Technologies: Cultural Heritage Applications. PRESTO is a 21-month, 4.8 M € project to develop broadcast archive preservation technology.

But - preservation is also transformation: to digital media of some sort, opening new possibilities for storage and access. Therefore this project has also considered future methods of archive usage (and revenue generation) in order to achieve the true 'best cost' when considering the possibilities, over the next 20 years or more, for usage of the preserved content.

Principal partners:
The project is led by the BBC (British Broadcasting Corporation), and full partners are two of the other largest European national archives: INA = Institut National de l'Audiovisuel and RAI = Radiotelevisione Italiana.

Technical partners:
The technical partners are all specialists in various aspects of media preservation, digital media and multimedia technology: Advanced Computer Systems, Italy; e-vod, France; ITC/IRST, Italy; JOANNEUM RESEARCH, Austria; NTEC, Germany; Snell & Wilcox, UK; Vectracom, France.

User Group:
In addition to BBC, INA and RAI, seven more national European broadcast archives are included as a user group, providing full data about their contents and operations, and evaluating the progress and outcome of the project. These partners are NAA, the Netherlands; NRK, Norway; ORF, Austria; TRT, Turkey; SVT, Sweden; SWR, Germany; YLE, Finland.

FINDINGS OF THE ARCHIVE SURVEY
A survey was made of the holdings and preservation status of ten major broadcast archives. These archives represent a significant portion of total European broadcast archives, including some of the largest individual collections.

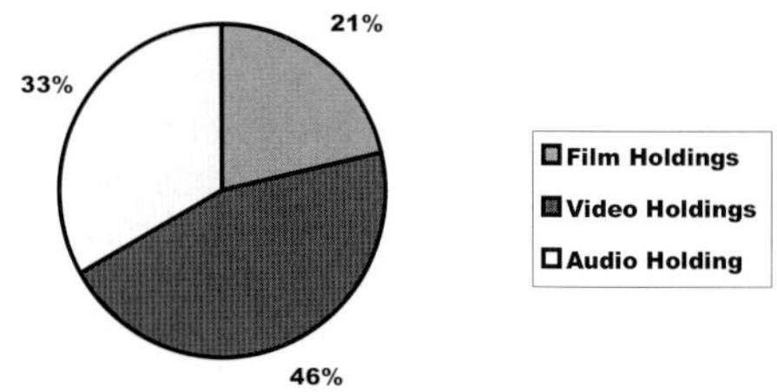

Holdings:
The survey found about 1 million hours of film, 1.6 million hours of video recordings, and 2 million hours of audio recordings in the ten archives. **Total European holdings of broadcast material are probably ten times larger**:

- **10 million hours of film**
- **20 million hours of video**
- **20 million hours of audio**

Access:
The content is unavailable to the general public and often unavailable even to national archives and educational institutions. Much of the content is unique, master material that cannot be allowed to circulate generally, and all of the content has rights issues. **A major goal of preservation work for broadcast archives must be to find joint solutions to preservation and access problems: preservation for access.**

Preservation Status:

- **Obsolescence:** At least 2/3 of the material in archives cannot easily be used in its existing form
- **Deterioration:** Approximately 1/3 of the material has one form or another of deterioration
- **Fragile media:** Roughly ¼ of the material cannot be released for access because the media are too easily damaged

Cost per use:
True cost of an asset is total lifecycle cost. True benefit is related to the number of times that asset is used over the lifecycle. Archive preservation strategy should aim at the "lowest cost per use" over the life cycle of the new media, NOT at the lowest transfer cost.

Bulk Preservation:
It costs money to select material, and to develop statistics for archive usage. The value of archive material very high – up to 100 times the cost of preservation. **In general, it makes economic sense to preserve the bulk of current content, rather than taking a piecemeal approach.** As a rough rule-of-thumb, the high value of material when actually used means that **one minute of sold or re-used archive material will pay for one hour of preservation.**

Security Copy:
Much of the material in broadcast archives is unique, and exists on ONLY a single master carrier. One of the major benefits of moving to digital technology, especially datatape, is the decrease in cost for backup master copies once media leaves the analogue domain (this is especially true for film). The costs and benefits of datatape masters, over at least a 20-year item lifetime, need to be assessed within an overall context of moving from media-based to mass-storage based processes during that time period.

Actual Costs:
Information from all ten archives gave the following overall media transfer costs per hour of media, in Euros.

	Like-for-like preservation cost/hr	Usage over life-cycle of material	Cost per use
Film	2000	5	400
Video	200	5	40
Audio	120	5	24
	Mass storage preservation cost/hr	Usage over life-cycle of material	Cost per use
Film	3000	10	300
Video	300	10	30
Audio	180	10	18

These costs, for the year 2000 in Euros, exclude much significant detail, and do not represent the actual costs for any particular archive or project

Digitisation and mass storage is about 50% more expensive, but is expected to double the usage of an asset.

Funding:
The aim of preservation work is to retain for the future, as cost effectively as possible, that portion of existing broadcast archives that will contribute most to future usage. The conclusion from current overall archive usage figures is that **the value of an item must be more than four times the preservation cost** in order to be financially justified on a commercial basis.

Commercial basis: For most broadcast archive material, this condition can easily be met – because one minute of sold or re-used archive material will pay for preservation of one hour of archive material.

Heritage basis: For material that cannot pass the 'commercial economics' criterion as outlined above, **there should be a safety net of assessment for cultural/historical value, and a separate funding mechanism.**

PRESERVATION WORK

A considerable amount of preservation work is in progress across European broadcast archives. The ten archives in the survey had the following amounts of material undergoing preservation:

Total film preservation hours per year: 8 000, plus an additional 10 000 hours of 'sepmag' work. Sepmags are magnetic soundtracks for film. Many of these are on acetate material, and suffering from 'vinegar syndrome' – the conversion of acetate to acetic acid, which can then spread across an archive, attacking all archive contents.

Total Video preservation hours per year: 36 000, mainly work on the formats of the 1970's and 80's, specifically 1" videotape and U-matic.

Total Audio preservation hours per year: 20 000 from BBC, INA, NAA, ORF, and an additional 25 000 at RAI, which has committed to digitisation not only of its entire audio archive, but of the entire radio production process.

However, **preservation work is not in any way keeping up with new archive intake**. Total intake of new material per year as reported in the survey was:

- 4 000 hours of film
- 170 000 hours of video = 4x rate of preservation
- 80 000 hours of audio = 4x rate of preservation

REQUIREMENTS FOR LOWEST COST PER USE

Process

The essential requirement is an overall perspective of the preservation work. Efficient preservation work needs a defined, well-engineered process, not just lots of piecemeal transfers. Attention must be given to the following elements

- **End-to-end process** – a view of all the work involved, to capture all possible efficiencies (to identify unnecessary steps, and find when information needed at one point could be created without manual intervention at another point)
- **Identification of costs** – to identify cost-attractors: steps which would benefit from investment (in technology, physical plant, or moving of staff or equipment) in order to reduce costs
- **Development of cost-effective approaches to batch processing** – investment in systems or equipment which more than pays for itself for volume work. Examples are
  - **automatic quality-checking equipment**
  - **software for overall control of a preservation end-to-end process**
  - **purpose-built preservation**

areas to optimise equipment and streamline movement of material,

- **investment in in-house rather than contracted facilities.**

- **Links to the wider business process:**
  - **programme documentation,**
  - **subtitle information,**
  - **rights clearance,**
  - **creation of new metadata to improve the catalogue**
  - **extraction of key frames**
  - **automatic speech recognition.**

These issues are not considered part of preservation work, but need examining at the end-to-end perspective of the entire business. At the business level, it may become cost effective to include work within a preservation project which pays off to another part ot the business. An example would be digitisation of programme documentation (scripts, running orders, contracts) as part of preservation work, in order to increase usability of the new media, and decrease costs of activities such as rights clearance.

Link to future business – **the most important single issue which does not fit into a like-for-like, item-based piecemeal approach to preservation is using preservation work to create the future archive.** An archive needs to move into the technology that will be needed in the next decade, rather than simply producing the particular tape format needed in the current year. It is hard to do strict cost-benefit evaluations of future requirements, and hence hard to develop a business case for justifying transforming archive technology.

The decision making requires managerial vision, a commitment to the eventual development of "tapeless working" in TV and Radio. At the minimum level, this vision consists of accepting that electronic access would double the use of archive content. At a wider level, the vision may involve a broadcaster abandoning the perspective of lowest cost, greatest return on investment, and accept that with the demise of ¼" tape in Radio, and the certain growth in server-based operations in TV, a broadcaster could be motivated to conversion of the archive on grounds of "necessary technology" rather than "lowest cost". **It is vital to the future of broadcasting and broadcast archives to have an increased appreciation of the value, not just the cost, of new technology.**

Key-link Technology

While the overall process is the most significant factor in the cost effectiveness – and future success – of preservation work, there are significant individual 'key links' in the process chain that are at present labour intensive or otherwise ineffective. PRESTO has identified the following Key Links, and is developing innovative replacement technolology

Videotape Transport / Playback - It is particularly important to have the best possible electronic processing, to ensure reliable and stable signal detection, to the greatest degree possible using modern methods. This will minimise the dropout, and hence minimise the transfer errors. This approach is better than relying entirely upon dropout correction, which is not perfect and is essentially a restoration technique, not a preservation technique.

It is also possible to improve the mechanics of old transport, and improved handling is being developed for 1" and U-Matic machines, particularly addressing the problem of tape motion impairment (sticktion). Dropout correction will be re-examined with regard to digitisation, because if a video signal is being captured to a server there are opportunities for signal correction which are not possible in a standard transfer (eg using information from adjacent frames and from preceding as well as following lines)

Film Transport – A major problem in broadcast archives is news item films from the 60's and 70's, often spliced item-by-item onto 'day reels'. This material often falls apart when running through film transport, and manual cleaning and resplicing is 90% of the preservation cost of this material. A fully-automated method is being developed for this material. Alternative methods of film handling which do not require re-splicing is also being investigated.

Quality Control – This remains the main transfer cost, but with modern transfer operations requiring four or even five items being transferred simultaneously, full manual monitoring is not performed. Automatic methods to identify basic problems with audio signals will be developed, as a 'software toolkit', to back up the spot checks performed by the operator. This will allow the operator to concentrate on properly setting up a transfer, ensuring initial quality, and then relying upon automation to warn of signal dropout or gross problems during the course of the transfer.

Signal Capture – while video is easier than audio in terms of multi-signal monitoring, it has far more problems with signal dropout or transport errors (head clogging) during playback. The gross errors will appear on the monitors, and the trivial errors (single-line dropout) can be immediately compensated in hardware. In between these extremes, the operator will not know how much brief loss of signal has occurred. Improved monitoring of playback, with automatic logging, will be developed.

Error Logging – many aspects of signal playback my have problems. Signal restoration projects have identified dozens of artefacts in archive signal: blotches, scratches, flicker, grain, noise, fading. Many of these artefacts can be automatically detected, with varying degrees of difficulty and success. Although preservation work is not the same as restoration, it is desirable when doing preservation work to capture any useful information that may be needed for subsequent handling of the item. In audio, such a log is now an addition to the EBU Broadcast Wave Format [1] standard.

Other key links for audio and video are:

- **Lossless encoding (audio and film/video)**
- **Multi-level encoding for film** (creation of a master format which efficiently supports conversion into all required presentation formats for both cinema and TV)
- **Multi-level encoding for video** (efficient creation of various quality levels: keyframe, internet, browse, studio)

Efficient **metadata management** will be developed for handling multiple versions of the same underlying material (essence). Practical aspects will be incorporated in the encoding 'key links':

- ensuring exact time registration
- metadata maintenance across all versions
- time code
- other embedded metadata
- links to external metadata (catalogue)

Standard Metadata – The final aspect of metadata management developed by PRESTO will be a demonstration of the methodology for using a preservation process to standardize metadata. Data from multiple broadcast partners will be converted (mapped or restructured, as required) to form a common catalogue.

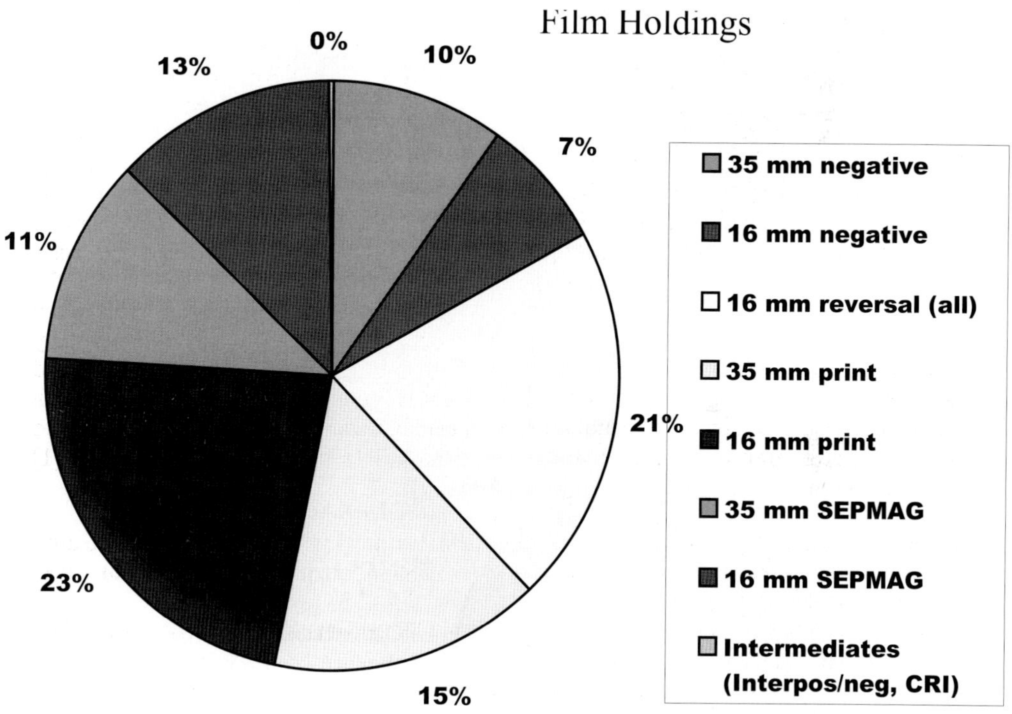

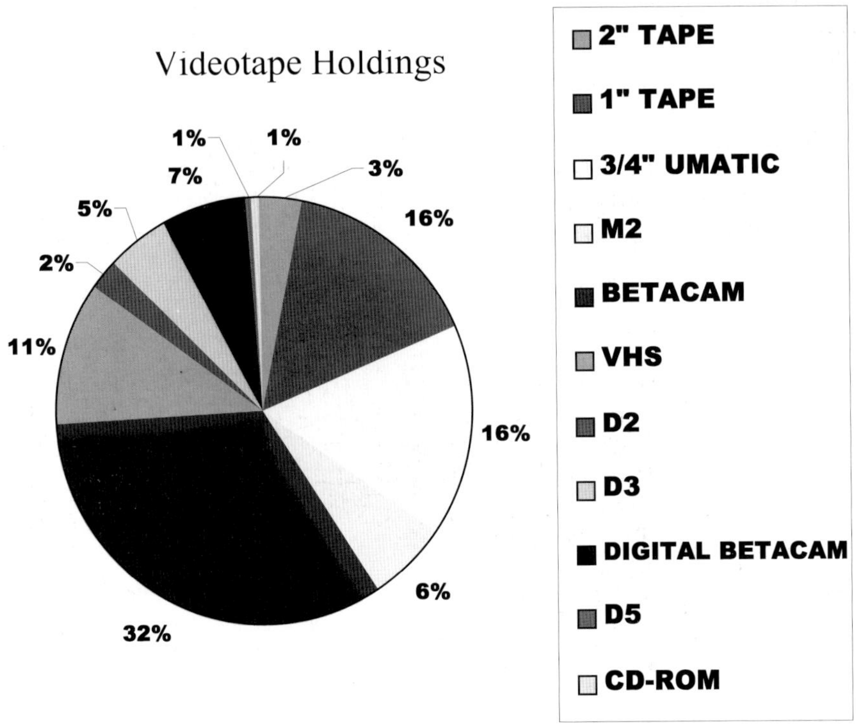
Videotape Holdings
1%
1%
3%
7%
5%
16%
2%
11%
16%
6%
32%
2" TAPE
1" TAPE
3/4" UMATIC
M2
BETACAM
VHS
D2
D3
DIGITAL BETACAM
D5
CD-ROM

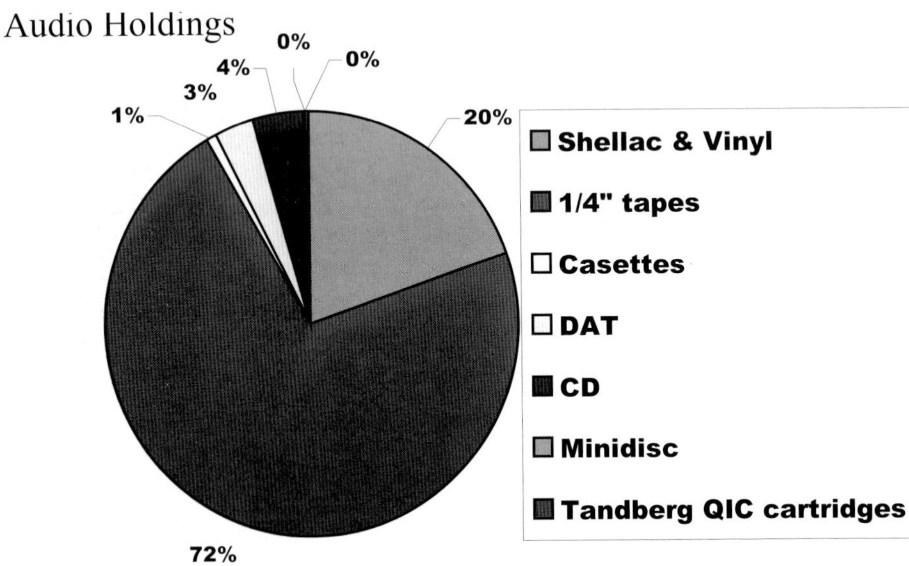
Audio Holdings
0%
0%
4%
3%
1%
20%
72%
Shellac & Vinyl
1/4" tapes
Casettes
DAT
CD
Minidisc
Tandberg QIC cartridges

PRESTO WEBSITE

More information and full deliverables covering all the above – and more – is on the website:
http://presto.joanneum.ac.at

REFERENCES

1. EBU document T-3285, July 1997: **Specification of the Broadcast Wave Format.**

http://www.ebu.ch/pmc_bwf_ug.html

ABOUT THE AUTHOR

Richard Wright is the Technology Manager for the BBC Information and Archives Department, and coordinator of EC Project Presto.
E-mail: *Richard.wright@bbc.co.uk*

Facing Challenges of New Media on the Crossroads of East-West and Past-Future (Lithuanian Case Study)

Nerutė Kligienė (*) *and Laimutis Telksnys* (*)

(*) The UNESCO Chair in Informatics for the Humanities at the Institute of Mathematics and Informatics, and Vilnius Gediminas Technical University, Vilnius, Lithuania
E-mail: nerute@ktl.mii.lt, telksnys@ktl.mii.lt

ABSTRACT
The presentation focuses on three aspects of applying advanced information communication technologies (ICT) in the field of cultural heritage: preservation, dissemination, and training on ICT in case of countries in transition - Lithuania is one of such countries. The aim is to specify economic and social conditions for ICT development in such countries and to show how advanced ICT provide equal opportunities to big or small countries, of prosperous or transient economics, as well as to ethnic minorities, illuminating their cultural or religious diversities and making all this a common Heritage of mankind and allowing for the preservation of cultural identity. Problems of cultural heritage accessibility and need of training in informatics for the humanities considered.

KEYWORDS: Multimedia, digitization, training on digital publishing

PRESERVATION AND SAFEGUARDING CULTURAL HERITAGE

Much of the cultural heritage is currently available only in paper form. Archives, libraries, museums are places where the majority of the scientific and cultural resources are stored and preserved. For centuries, incunabula, maps, music sheets, manuscripts, facsimile and old prints have been carefully archived and protected over times of wars and natural disasters, as a deposit for future generations. Unfortunately, the direct access to this huge deposit of valuable material for professional and personal use is rather complicated, expensive, and not safe for vulnerable and fragile documents. For example, the first Lithuanian book, prepared by Martynas Mažvydas, the Lutheran priest and rector of Ragaine, (Lithuania Minor), printed in Königsberg, 1597 has only two copies extant and they are preserved at the Vilnius and Torun (Poland) University libraries. Naturally, the access to this valuable monument of language and culture long time was very limited.

In the occasion of the commemoration of the 450th anniversary (1547-1997) a computer multimedia book *Year of the Lithuanian Book* (Fig. 1) was issued [24]. It contains pictures of pages from the first Lithuanian book, presents its historical background and cultural context. It provides with the opportunity to listen to Martynas Mažvydas' hymns and get acquainted with old maps as well as the place names of East Prussia, or Lithuania Minor. A reader can find information from all over the world about jubilee events and their coverage in the mass media, bibliography, documents, and chronicle of events. The 450th Jubilee of the first Lithuanian book by Martynas Mažvydas was included in the 1997 UNESCO Calendar of Jubilee Dates as one of eminent dates with which UNESCO has been associated in 1996-97. After the 450 years the modest original book of a small format (18 x 11cm, 79 pages) is represented as an interactive CD-ROM containing incomparably more information, pictures and sound records.

Figure 1: The first electronic book, CD-ROM (1997) on the first printed Lithuanian book –*The Catechism*, 1547

Preservation in ICT context means digitization and safeguarding of cultural values as superior quality digital copies for long-term loss-less data storage oriented to contemporary and future users. This involves storage and manipulating huge amounts of digital information representing images (digitized museum's treasures, medieval manuscripts), sound records (folklore or speech collections). For example, Lithuanian language, still rich in dialects, are interesting for linguistic studies world-wide as one of the most ancient Indo-European languages, therefore must be safeguarded and made accessible for scientific research everywhere. The first attempt in this direction was made by creating the Multimedia Dictionary of Lithuanian Dialects [10]. This interactive CD-ROM demonstrates multimedia dictionary possibilities: a reader can see a picture of an object, to listen to pronunciation of a word in different dialects and to see how it is written in Lithuanian language. Therefore it can be used without any a priori knowledge of a language. A scientific interest of Italian scholars in Lithuanian language ought to be mentioned separately: the exhaustive studies were published in Italy [5, 15], and later on translated into Lithuanian.

During a period of a several last years, some part (but rather small) of stored cultural heritage has been issued on CD-ROMs in Lithuania. A few of them has been already published or prepared to release on CD-ROMs (they are mentioned in the extended list of References [7, 10, 11, 13, 21, 24, 25]). This gives an insight into themes and content which was made accessible electronically upper-most.

CULTURAL HERITAGE OF ETHNIC MINORITIES IN NEW MEDIA

New opportunities for self-expression of cultural communities of ethnic minorities in digital space and spread their wealth were used extensively in Lithuania. Lithuania as a multi-cultural and multi-lingual country is known from the 14th century, the time of the Grand Duchy of Lithuania, which has been a realm and good home for many nations with different cultural and religious background, and which had cultural and trade relations with many European and non-European countries. Several

Lithuanian ethnic minorities, such as the Karaims [12] or the Tatars [14] celebrated the 600th anniversary of their settlement in Lithuania recently. In spite of a rather small number of those nationals (less than 0.2% of whole population; the smallest minority – Karaims have only 257 persons living now in Lithuania) they were not assimilated during ages and have safeguarded their customs, their religion and their identity. New media means such as WWW and CD-ROMs do open wide opportunities to those ethnic cultural minorities: their web sites were created in the framework of a Lithuania State program *A Virtual Exhibition of a Millennium of Lithuanian Cultural Heritage* [1]: Jews in Lithuania [8], (fig. 2), Lithuanian Karaims [12], Lithuanian Tatars [14], Russian Culture in Lithuania [20].

Figure 2: Starting page of the Web site http://litvakai.mch.mii.lt

The past, history, customs, traditions, language, religion, social movements are considered here in details and are illustrated with pictures and records of music, language, hymns. This project has to be developed adding new materials and issuing the CD-ROM on the Cultural Heritage of Ethnical Minorities in Lithuania [21]. It is interesting to notice that there is no analogous CD-ROM on the cultural heritage of the main population of Lithuania but instead there is a digital book-encyclopedia about Samogitia [25] - one of the ethnographic regions of Lithuania. It is a good proof that new media provides equal opportunities to small and big communities, fosters and stimulates their self-expression in a new media. This can be seen from the list of multimedia projects, given in the extended list of References at the end of this paper. Many of these projects are implemented at the UNESCO Chair in Informatics for the Humanities and the Multimedia Center for the Humanities at the Institute of Mathematics and Informatics, in collaboration with many other Lithuanian cultural institutions: Institute of Lithuanian Language, Lithuanian Art Museum, Vilnius Art Academy, Martynas Mažvydas National Library, Institute of History.

Figure 3: Home of the Lithuanian Art Museum [9]

PROBLEMS AND SOLUTIONS OF CULTURAL HERITAGE ACCESSIBILITY

Cultural heritage nowadays cannot be represented by text only, multimedia elements emerge inevitably. This requires extended computer memory resources and broadband communication channels in order to share with others digitized cultural values. Computer networks are still not ready for transmission of heavy flow of cultural information. This is alike anywhere but a problem is especially hard in countries in transition.

Lithuania as one of new independent Baltic countries, which achieved a political independence in 1990 by non-violent struggle and has to experience all ups and downs conditioned by economical and social difficulties of post-communist period in its cultural revival and rediscovering its spiritual values. More precisely focusing on the specific economic conditions for ICT development in Lithuania, the following factors should be mentioned. Firstly, the critical mass of users necessary to spark the market "explosions" can be expected when Internet penetration will reach 10%. Currently it reaches only 8% (as percentage of total population) and this is caused by the high cost of network access (in average evaluated as 10% of a middle-level salary per month). Lithuanian Internet surfers can only dream about low-cost, high-bandwidth, and ubiquitous connections to the network so important for cultural heritage presentation online.

Secondly, the high quality equipment and an experience necessary for digitization and storage of cultural values is still rarity in Lithuanian museums or libraries and no funding foreseen for this. Among negative factors mentioned above, a positive one should be mentioned also. That is availability of highly skilled in informatics personnel at the universities and firms ready to take up actions and to help in representing a cultural heritage.

The important goal is a development of Internet-based services in order to extend cultural activities, to promote the use of advanced ICT for access to, understanding, preservation and economic promotion of country's cultural heritage in European or world-wide context. But that is rather hampered by economical factors, mentioned above.

Even more important goal is to face a reality of country in transition and, using rather limited resources, to follow up world-wide process of so called digital revolution. The new era is the explosive development of the Internet and possibilities offered by the rapidly growing technical advances of the digitization, enabling easy access to various treasures of the cultural heritage sometimes is only an illusion. Say, a portal site *Museums of Lithuania* (see fig. 4), [17], rich in content is hardly accessible from other countries and miss many opportunities to represent cultural heritage stored in Lithuanian museums.

Figure 4: Starting page of the Web portal Museums of Lithuania [17]

Accessibility is heavily depending on transmission speed in information channels and costs of communications [2]. The last time a bottleneck effect is mentioned more often when communication concerns Central and Eastern European countries and not only there. The so-called the 'last mile problem' is pressing everywhere. The narrow-band network cannot be used effectively because of heavy traffic, while high-speed modern channel means high costs what are not affordable for country in transition or some single user. There is an alternative solution. Involved people should be trained how cultural heritage material, usually enormous digital files, should be prepared for dissemination using more elaborated presentation and advanced compression methods. The great advantage of information technologies is that ICT are not dependent on the geographical situation of a country or its earth' resources; ICT are dependent on mind, science and human factors. The countries such as Lithuania, although being temporally restricted economically, are not restricted in mind development. There is a great scientific potential and information technology specialists are able to use advanced ICT in technically and economically limited environment. Dissemination of cultural heritage via narrow-band communication channels of countries in transition needs specific development of WWW sites and elaborated compression techniques to save up channel's resources and make a heritage accessible online.

Figure 5. The cover of the CD-ROM Monasteries and Convents in Lithuania, 2000, [16]

This evokes also a need to train people [3, 6, 23] how such materials should be prepared for dissemination while using advanced ICT in limited possibilities environment. Training subject is discussed in the following section.

TRAINING PROJECTS ON INFORMATION TECHNOLOGIES

In previous section was mentioned a great impact of ICT people into dissemination of cultural heritage, but not everything can be done by the efforts of information technology's specialists only. The progress in the fields of hardware and software is nothing if a human factor, the ability of the community to understand, accept and use the advanced technologies is forgotten. Teaching and training users and trainers in the fields of cultural education become one of the most important challenges in creating Information Society, a popular slogan of last years in many countries.

Training on ICT for purposes mentioned above is implemented as intensive course informatics for the humanities. The latest example of a specialized ICT training is the project supported by UNESCO entitled *Courseware for Training of Trainers and Users on the Special Applications of Internet-Based Services in the Fields of Cultural Education* [3], created at the UNESCO Chair in Informatics. The aim of this project is to develop educational material completed with explanatory texts, examples, demonstration material on the special ways in which computers and, in particular, Internet-based techniques can be used in the fields of culture, such as museums, teaching of music, management of cultural sites and tourism, artistic creation and education, preservation of and improved access to historical documents, publishing. The training material is accessible worldwide online and on CD-ROM if requested from UNESCO. Other projects in education area are: *Training of Trainers on Digital Publishing* [23] supported by UNESCO, 1999 - 2000, and *Development of Electronic Publishing Program in Lithuania* [6], 2000, supported by the Open Society Fund-Lithuania (Soros Foundation).

Several years of development cultural objects in digital space had proved that it is extremely important to teach the content owners to represent their heritage in new media and to maintain their sites, when luck of funding is a common situation. Otherwise the number of forgotten or "dead" sites will increase dramatically and a culture of a small country will be diluted in a progressing process of globalization.

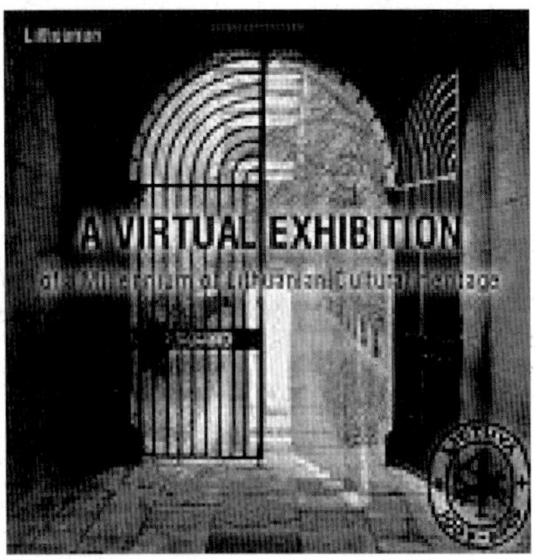

Figure 6: Starting page of a Virtual Exhibition of a Millennium of Lithuanian Cultural Heritage site [1]

CONCLUSIONS

When applying advanced information technologies at the field of cultural heritage concerning countries in transition, considered in this paper, it is concluded that:

1) ICT tools have a special importance for small countries and particularly for ethnic minorities to represent their culture about which the rest of the world does not know so much. This gives enormous opportunities for self-expression of cultural communities and resistance to a common process of cultural globalization;
2) There is a discrepancy between needs of multimedia representation of cultural objects and bottleneck in computer networks. Dissemination of cultural heritage via narrow-band or limited in costs communication channels evoke a need of development of new methods and special approaches to save up computer network's and channel's resources;
3) Training in informatics for the humanities has a special importance in current situation.

REFERENCES

1. A Virtual Exhibition Of A Millennium Of Lithuanian Cultural Heritage *http://alka.mch.mii.lt*, Lithuanian and English versions. A portal site represents Lithuanian cultural heritage from ancient time until nowadays. The Balts, Lithuanian State, society, ethnic minorities, folk and elite culture, landscape &architecture, art, religion, intellectual life are overviewed in depth by more than 25 project teams working on separate themes and having this joint connection in a Virtual Exhibition Site.

2. Content And Commerce Driven Strategies In Global Networks, Executive Summary, European Commission, Directorate General XIII/E Luxemburg, 1998.

3. Courseware For Training Of Trainers And Users On The Special Applications Of Internet-Based Services In Fields Of Cultural Education. English version, downloads. *http://daugenis.mch.mii.lt/UNESCOeducation.* Educational material completed with explanatory texts, demonstration material on the special ways in which computers and Internet-based techniques can be used in the fields of culture. preservation and improved access to historical documents.

4. Digital Culture For The Humanities. *http://sk.mch.mii.lt,* Lithuanian version. The aspects of Cyber culture, interesting and useful links, training courses and services related to software and hardware problems met by inexperienced user are represented here.

5. Dini, P. U., Le lingue baltiche, Firenze; La Nuova Italia, 1997.

6. D-Pub (Digital Publishing). *http://mm.mch.mii.lt/index.en.htm,* Lithuanian and English versions. Important information about digital publishing courses, outlines of main training modules, online registration, downloads of teaching material.

7. Historical Collections Of The Vilnius University Library, 1998, the CD-ROM prepared and released by Vilnius University Library, in the framework of the Memory of the World programme, UNESCO sponsored. [CD-ROM, ISBN 9986-9204-0-X].

8. Jews In Lithuania. *http://litvakai.mch.mii.lt,* Lithuanian and English versions. History, language, religion, outstanding persons, Vilnius - Jewish Jerusalem and many facts in Jewish culture in Lithuania are overviewed on this site. Joint Project with the Lithuanian Jews Community

9. Lithuanian Art Museum. *http://LDmuziejus.mch.mii.lt,* Lithuanian and English versions. Represents the Lithuanian Art Museum in Vilnius and all Branch museums and galleries spread in over Lithuania: their history, collections, educational activities, exhibits and the latest exhibitions, pictures, interesting texts, sound records are displayed. Joint Project with the Lithuanian Art Museum.

10. Lithuanian Dialects. Multimedia Dictionary, Volume 1, 1998. Published in cooperation with the Institute of Lithuanian Language, supported by UNESCO. [CD-ROM 9986-668-19-0].

11.Lithuanian Folk Textiles Patterns. Online *http://www.mch.mii.lt/AudiniuRastai* / This CD-ROM running project, 2001, provide an historical and regional perspective of Lithuanian textile patterns and a framework for interpretation and analysis of the designs. It dwell on the importance of hand-made weavings both within the family environment and as expression of the social fabric of the country. Joint Project with the Institute of History.

12. Lithuanian Karaims *http://protenis.mch.mii.lt/karaimai,* Lithuanian and English versions. Represents culture, history and customs of the smallest ethnic minority in Lithuania.

13. Lithuanian National Martynas Mažvydas Library Collection Of Parchments.
The oldest privileges, the Great Duchy of Lithuania parchments, Pope bules. The material available in English, Polish, Italic, and Old Slavonic languages, the CD-ROM to be issued 2001 and the web site is under construction. Joint Project with the National Library.

14. Lithuanian Tatars
http://wwww.gaumina.lt/totoriai
Lithuanian and English versions. Customs, culture of Tatars what are settled in Lithuania since 15th century.

15. Michelini, G. "Ancora sulla Margarita Theologica lituana: interferenze con il tedesco nella traduzione del lessico latino" AtSGM, 32, 1990-1991.

16. Monasteries And Convents In Lithuania. A multimedia book presenting monasteries and convents in Lithuania: the history and their impact on culture and art in their historical development with particular attention to the artistic values of objects, a lot of pictures, representing their architecture and arts. Several original pieces of music created in Lithuanian monasteries can be listened to, here having the first publication of newly discovered opus of music. [CD-ROM, ISBN 9986-571-58-8], 2000. Lithuanian version is also available online: *http://vienuolynai.mch.mii.lt.*

17. Museums Of Lithuania. *http://muziejai.mch.mii.lt,*
Lithuanian, English, German, and French versions. A Portal site were over 250 Lithuanian museums are represented. Museums in countryside and big towns, the collections and important information for visitor as well as the latest exhibitions at any museum described can be found there. The pictures of the most interesting items, some sound records are displayed. Joint Project with the Lithuanian Art Museum.

18. New Fine Art Editions *http://leidiniai.mch.mii.lt*, Lithuanian version, English under construction. Lithuanian art publishers and publication of art books and albums since 1997 are represented. Joint Project with Vilnius Art Academy.

19. Postilla - 400.
http://postilla.mch.mii.lt, Lithuanian and English versions. The first printed (1599) Collection of Sermons for Lithuanians. The events of the 400th anniversary celebration, conferences and exhibitions are presented. Joint Project with a Fund of Samogitian Culture.

20. Russian Culture In Lithuania
http://www.rkc.lt/paveldas
Lithuanian, Russian and English versions. Represents cultural activities and outstanding persons of 19th century.

21. The Cultural Heritage Of Ethnical Minorities In Lithuania. The CD-ROM running project 2001 including Russian, Russian Orthodox, Jewish, Tatar, Karaim cultural heritage.

22. The Samogitian Parks And Museums. *http://mpkelias.mch.mii.lt*
Lithuanian and English versions. Project deals with elite culture heritage (manors and museums, parks, their past and present). Joint Project with a Fund of Samogitian Culture.

23. Training Of Trainers On Digital Publishing. Educational material completed with explanatory texts, examples, and demonstration material. The training material is available on CD-ROM if requested from UNESCO.

24. Year Of The Lithuanian Book- Lietuviškos Knygos Metai - Jahr Des Litauishen Buches, 1997. All about the first Lithuanian Book - THE CATHECHISM of Martynas Mažvydas published in 1547, history, cultural context, the 450th anniversary. [CD-ROM, ISBN 9986-680-09-3]. Online http://PirmojiKnyga.mch.mii.lt Lithuanian German, and English versions created in the framework of the State Programme to Commemorate the 450th Anniversary of the publication of the first book in the Lithuanian language.

25. Žemaitija-Samogitia. *http://samogitia.mch.mii.lt*, Lithuanian and English versions. Unique ethnographic region of Lithuania with interesting history where the customs of ancestors are still strong. Joint Project with a Fund of Samogitian Culture. [CD-ROM, ISBN 9955-441-01-1], 2000.

ABOUT THE AUTHORS

Nerutė Kligienė (PhD in mathematics, Associate Professor) is graduated and obtained her degree from the Vilnius University. She is a senior research assistant at the Institute of Mathematics and Informatics (IMI), lecturing at the Vilnius Gediminas Technical University, the author of more than 70 papers and a textbook for students, the leader of multimedia projects and of an intensive Digital Publishing program for the humanities at the IMI, an expert on digital publishing.
E-mail: *nerute@ktl.mii.lt.*

Laimutis Telksnys (Professor, doctor habilitatis in informatics) is the head of UNESCO Chair in Informatics for the Humanities at the Institute of Mathematics and Informatics. He is the head of the Project Culture Heritage Multimedia Presentation, Lithuania, which demonstrate possibilities of advanced information technologies for the promotion and sharing of culture heritage in multimedia form via Internet and multimedia CD's, the author of 130 scientific publications, the member of the Lithuanian Academy of Sciences.
E-mail: *telksnys@ktl.mii.lt*

Electronic IPR (eIPR) Management Technologies

Franco Bartolini(*), *Roberto Caldelli*(*) *and Vito Cappellini*(*)

(*) Department of Electronics and Telecommunications, University of Firenze
Via S. Marta 3, 50139 - Firenze, Italy
E-mail: {barto, caldelli, cappellini}@lci.det.unifi.it

ABSTRACT

The progresses over the last few years of digital technologies, the rapid development of the Internet and of other communication means, have caused an ever increasing need for protecting Intellectual Property Rights (IPRs). The problem of the IPR protection for multimedia works exchanged through telematic networks have promoted industrial and research projects for the development of Electronic Copyright Management Systems (ECMSs), i.e. systems able to manage automatically all the issues related to trading works through telematic means by protecting their IPR. The goal of this paper is to exploit the main requirements that such systems should satisfy, the players that should be involved and the technologies that will be most probably employed.

KEYWORDS: IPR protection, ECMS, MPEG.

INTRODUCTION

The progresses over the last few years of digital technologies, the rapid development of the Internet and of other communication means, have caused an ever increasing need for protecting Intellectual Property Rights (IPRs). Traditional laws on IPR protection do not seem suitable to solve all the problems raised by this technological revolution. Just the fact that each work could be easily digitised, stored and transmitted without loss of quality an almost uncountable number of times, is causing, at the international level, the revision of the concept of IPR protection itself. On the other side the large diffusion of Internet, more than a menace of increasing copyright violations, can be, for the authors and the copyright owners, a good chance to diffuse their works and a big business opportunity 1. In the United States a White Book regarding the problems related to the protection of the Intellectual Property Rights in the Information Society has been published on 1995: it gives suggestions regarding the modifications needed by American laws on Copyright (*American Copyright Act*) to consider the new digital and network scenarios. In 1996 the WIPO (World Intellectual Property Organisation) supported the signing of an international treaty, whose main statement is that copyright protection must be granted also for works transmitted through electronic media. All these documents witness the interest for the problem of the IPR protection for multimedia works exchanged through telematic networks and have promoted industrial and research projects for the development of Electronic Copyright Management Systems (ECMSs), i.e. systems able to manage automatically all the issues related to trading works through telematic means by protecting their IPR.

REQUIREMENTS

An ECMS can be considered as an ensemble of services, connected through a network environment, co-operating together, to allow the protection of the IPR of multimedia data, on the basis of contracts agreed among the involved parties. In particular, two technological revolution have made the problem of IPR protection so new and challenging today. The first is the advent of digital techniques, and the second the explosion of telematic networks. The possibility to represent every kind of work (being text, picture, video, music, …) in a digital format has given birth to a new type of creation, i.e. multimedia creations, where different kind of data can be integrated to produce a new composite object. This possibility offer a big chance to authors to better express their creativity, but, on the other side, the IPR problem becomes quite complex: in order to compose the new object, all its components have to be licensed by respective authors; an ECMS should, thus, allow to easily track the information related to the IPR of each component. Furthermore works in digital format could be copied almost an infinite number of times, without any degradation, thus encouraging the trading of illegal (i.e. not licensed) copies of the creations. The development of open telematic networks (in practice Internet) makes highly easier the distribution of creations, thus increasing both their diffusion and their business value. This is again a big opportunity for authors. Nevertheless, some problems are also coming, together with the advantages: in particular illegal distribution becomes easier too, and privacy issues have to be dealt with. One of the most important requirements of an ECMS is that it should be able of efficiently playing in an open network environment, and consequently to deal with security problems. In particular, two aspects are implied: first integrity of the electronic transactions must be granted, in such a way that users can be sure that they obtained what they really asked for, and the rights are paid to the true copyright owners; second the users would like their privacy to be preserved. It will be very important to build up archives where the information on the IPR of a particular work could be stored and easily retrieved, in such a way to grant the correct distribution of the fees to the copyright owners. Such archives should also be able to make available this information at different level of detail, according to the type of user who is accessing it: a purchaser would only need superficial data (e.g. regarding authors, year of creation, an so on), while a right-holder should be able to access (and possibly modify) also confidential data (e.g. shares, prices, and so on). An ECMS should also be able to deal with the new type of transactions, typical of a digital environment, as for example licensing sub-parts of a creation. An ECMS should also offer means to track copyright violation, and allows to honest purchasers to demonstrate that they legally acquired a creation. Thus some way should exist to securely link the copyright information to the works that are to be traded. Another requirement which is strictly connected to the two just mentioned (i.e. the set up of IPR information archives and the need for linking copyright information to the creations), refers to the ability to be exhibited by an ECMS to manage some way to identify the creations. With this aim a big effort is currently being done by international standard organisations such as ISO and CISAC, to develop a Digital Object Identifier (DOI), that should be used for every kind of creation (being it a text, a picture, a video, a music piece, etc.) and that will represent for digital creations what are now ISBN and ISSN for books and periodicals. Most probably an ECMS will need to be able to work in a multi-code environment, given that it will be very difficult that an unique code will be used for every type of creation, at least in the next years. Furthermore an ECMS should be as easy as possible to be used by a purchaser who is interested to legally acquire copyrighted materials, in order not to deter such a kind of use. Besides to end users, ease of use should also be granted to those authors who are interested in developing composite creations, and thus need to retrieve the information related to the IPR of a work in an easy, and secure way. Indeed the problem of ease of use is very important: many today proposed solutions, in an attempt to grant IPR to be preserved in a strongly secure way, offer to the users a service which is complex to be used. In this field a trade-off will need to be found between the interest of IPR owners (for whom security is the main issue) and of end users (who will not use systems which are too complicated). In this area will be played the success of ECMS.

THE GENERAL ARCHITECTURE

For understanding all the issues related to *e*IPR management, it is convenient to refer to a so called, business model, that is to a model describing all the players involved in the exchange of works as well as the transactions occurring among them. In particular we will here refer to the Business Model developed in the ESPRIT 20676 IMPRIMATUR Project 4 whose scheme is depicted in Figure 1.

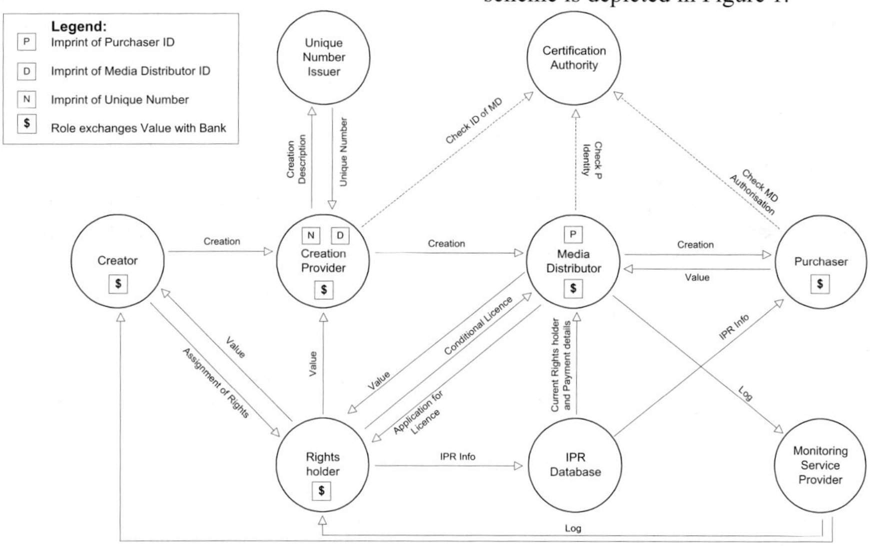

Figure 1 Scheme of the IMPRIMATUR Business Model

It is not always needed that all players are present and all transactions occur, simplified processes can be implemented according to the particular type of application and to the needs of the users, anyway the scheme of Figure 1 offer a comprehensive and general overview of the mechanisms that could be implied by the trading of IPR sensitive digital objects. By referring to

Figure 1 a possible simple example of business process can be hypothesised: a creator creates a multimedia work, assigns the right on it to a rights-holder he likes, and receive back some value (typically money or percentage of royalties). The multimedia work is the transferred to a creation provider who prepares it for being traded securely and forward the prepared work to the media distributor, for example an on-line shop; a purchaser accessing the media distributor site, can decide to buy the multimedia work, for which he exchanges some value (payment) with the media distributor; the media distributor forward part of the obtained value to the rights-holder. The purchaser can then get information about the IPR tied to the multimedia work he has acquired, by querying an IPR database where the rights-holder has previously stored all needed information about the IPR of the work. In this way, the purchaser can be granted that he has acquired the work in legal way, can know all the limitations of use that he has to undertake, and so on. More complicate business processes can easily be imagined by referring to the scheme of Figure 1 which is quite general. The value of such a scheme is also witnessed by the fact that it has been included as an "Example of a generic description of a model for content delivery and rights management" in the Technical Report describing the new standardisation effort of MPEG, that is MPEG-21 5.

Players

Let us now more deeply investigate the roles of the players involved in the ECMS 4. These players can be listed as:

- The author, who will be the creator of the copyrighted work.
- The rights-holder (or copyright owner). The author does not want, usually, to be involved with the administrative issues related to the management of his IPRs; he thus sells to another player the rights to exploit his creations. This player is thus mainly responsible for licensing the use of the creation: he defines the conditions of use, records the IPR information at a registry, and collects the royalties deriving from trading of the works which he is in charge to administer.
- The creation provider (or service producer), who will be in charge to prepare creations for being traded in the ECMS, and thus, for example, to embed into creations those mechanisms that will allow the tracking of copyright (watermarks). This player is the repository of the technical knowledge needed to grant that copyright can always be protected.
- The media distributor (or service provider), who will have the duty to distribute to purchasers the creations, and thus will have to satisfy the request of his clients, and to grant that the IPR on the distributed material is protected and the related fees correctly paid. This player will have to offer different simple ways to pay for the purchased works (Electronic Payment Systems), to access the databases where IPR information is stored, to be able to use all technological tools needed to protect the copyright of the creations he trades (watermarking, cryptography, secure protocols).
- The IPR register or database, which is the repository of all the information related to the IPR of the works to be protected, and that will have to be

accessible, at different levels of details and confidentiality, by all the other ECMS players.

- The Unique Number issuer, who will be responsible for assigning the unique identifier (e.g. the DOI) to each creation, for facilitating its tracking.

- The controller, who will be a Trusted Third Party (TTP) responsible for monitoring that all transactions have been carried on legally. The role of the controller is very important, and the establishment of TTPs having in charge this duty will be one of the key point for the successful diffusion of ECMSs.

- The certification authority, which is also a TTP, that is charged to authenticate all ECSM players, by means of electronic certification.

Transactions

Some of the single transactions that can occur during a generic business process related to the trading of digital objects can be listed as following:

- Transfer of the work: the work which is object of the trading can be transferred from the creator, to the creation provider, then to the media distributor and finally to the purchaser. These transfers usually occur in the just described order.

- Transfer of values: this is the exchange of money which is given for compensating the use of the multimedia work. This transfer usually originate from the purchaser, who pays the multimedia distributor, which transfer part of the value to the rights-holder which, based on previously defined agreements, will compensate the creator and the creation provider.

- Assignment of rights: the rights on the multimedia creation can be transferred from the original creator to a rights-holder (usually a licensing and collecting society such SIAE in Italy, SACD in France, GEMA in Germany, MPCS-PRS and ALCS in UK, etc.). Also in this case protection from misappropriation and forgery is needed.

Technological issues

In the realisation of really effective ECMSs some technologies will undoubtedly play a major role. In general two approaches can be distinguished to solve the problem of IPR protection: the first approach is aimed at developing systems able to prevent copyright violations (as for example the IBM proposal named Cryptolope), the second is on the contrary aimed at developing systems able to track copyright violations (as for example the ECMS developed in the framework of the EC funded project IMPRIMATUR). The ECMSs attempting to prevent copyright violations, strongly rely on cryptographic techniques: the traded work is embedded into a kind of cryptographic envelop, and only authorised users can have access to it. It seems that for tracking copyrighted material watermarking techniques will be of great help 23: they should allow to robustly embed IPR information just inside the multimedia creations, thus helping for example the controller to check the legal status of works traded through the network. Nevertheless it is today apparent that watermarking can not be considered a cure all solution, and it will be important to integrate such techniques with cryptographic tools]. In order to preserve privacy and grant integrity of transactions secure protocols, such as SSL for on-line transactions or S/MIME for off-line (email based) transactions, will also be largely used. Finally, a key role for the actual growth of creations trading over open networks, will be played by efficient and easy-to-use Electronic Payment Systems: if these will be effective, they will encourage authors to make available their works for selling on on-line shops, and purchasers to buy them, thus making ECMSs a big chance for increasing the business opportunities of copyright. In particular, that one of the Electronic Commerce is, today, a very active and rapidly changing field. Next few months should say if very simple but not very secure solutions (as for example that used by the Amazon On-Line Bookshop) will be preferred to more secure but also more complex to be established and to be used solutions (such as for example the SET standard).

MAIN APPROACHES FOR E*IPR* MANAGEMENT

In the design of effective ECMSs, two different approaches can be distinguished to solve the problem of IPR protection: the first approach is aimed at developing systems able to prevent copyright violations (as for example the IBM proposal named Cryptolope), the second is on the contrary aimed at developing systems able to track copyright violations (as for example the ECMS developed in the framework of the EC funded project IMPRIMATUR). Both categories of systems require that multimedia works are properly managed before their distribution.

Cryptography based

Based on this approach digital objects are wrapped in an encrypted system and integrated with an application (the reader) allowing to use the work only in a controlled manner; e.g. the images can just be displayed, but they can not be printed, or the audio files can be played but can not be stored in the hard disk of the user. The content can not be accessed without the proper application. The main disadvantages of this approach is that a standard for the embedding applications is difficult to be established; moreover when the works are enjoyed (for example, they are visualised in a PC screen, or they are played), it is still possible to capture and copy them without constraint. According to this approach, some systems have already been developed: CryptoLope from IBM. and LiquidAudio. In this kind of systems the fundamental role is played by the cryptographic technologies: as a matter of fact, the traded work is packed into a ciphered box, and only authorised users can access to it.

Watermarking based

Based on this approach digital objects are watermarked through digital watermarking techniques 23 whose aim is to tightly and robustly embed IPR related information into purchased creations (the hidden data can be the name of the copyright owner, or the unique code identifying the work); it us thus possible to check the legal status of the content exchanged through the network. Nowadays it is evident that watermarking techniques can not be considered as a cure-all solution, but it will be necessary to integrate watermarking with cryptography tools. An advantage demonstrated by this of approach is the unnecessary adoption of a particular format for the watermarked multimedia content, since the IPR information are directly embedded into the content itself.

MPEG-4 IPMP

The problem of intellectual property

management and protection (IPMP) 6 was deeply felt also in the sphere of MPEG-4 standard and peculiar attention was dedicated to this issue, trying to merge the need not to limit MPEG-4 characteristics and the demand to give some effective tools for copyright safeguard. Since MPEG-4's target applications range from low bit rate Internet telephones to high fidelity video and audio systems, the intellectual property management and protection systems required had to be as diverse as these applications were; furthermore the level and the type of protection strictly depends whether on the content or on complexity and sophistication of the associated transaction models. Despite these considerations, initially it was tried to conceive a MPEG-4 IPMP including facilities such as encryption and watermarking within the MPEG-4 standard itself, but later the need not to bind MPEG-4 application potentialities and to adequately answer to the widely different IPMP requirements, led to a standardisation of a generic interface to private and non-normative IPMP systems (Figure 2).

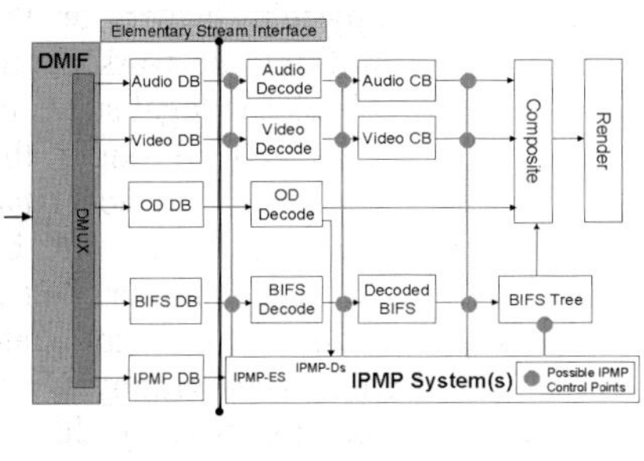

Figure 2 IPMP Framework in the ISO/IEC 14496TerminalArchitecture

This choice allows to optimise the domain-specific IPMP tools to the individual application. This interface is designed to be a simple extension of basic MPEG-4 systems constructs, it consists of IPMP-Descriptors (IPMP-Ds) and IPMP-Elementary Streams (IPMP-ES), as evidenced in Figure 2. Second ones are like any other MPEG-4 elementary stream and the first ones are extensions to the MPEG-4 object descriptors. Both provide a communication mechanism between IPMP systems and the MPEG-4 terminal; MPEG-4 objects requiring management and protection have IPMP-Ds associated with them and they indicate which IPMP systems are to be used and give information to these systems about how to manage and protect the content. In Figure 2 some possible IPMP control points (*hooks*) are indicated; these represent points, in the MPEG-4 terminal, where IPMP control might be desired and where a proprietary DRM (Digital Rights Management) system could be attached. Anyway, which has to be the watermarking approach to use, has not been standardised and which must be the peculiar features to be supplied, is not fixed, and only generic indications are to be followed. From a primary analysis, many of the existing watermarking techniques could be adopted to work in this scenario and their integration with the MPEG-4 standard does not appear so critical.

MPEG-21

The most recent standardisation effort of the MPEG, following this guideline, has been named MPEG21. Its goal is to achieve "an environment that is capable of supporting the delivery and use of all content types by different categories of users in multiple application domains" 5. Up to now 7 basic elements have been identified as constituting the framework:

A. Digital item declaration: it regards the establishment of an uniform and flexible abstraction and interoperable scheme for defining digital objects.

B. Content representation: it regards the technologies needed for representing in a suitable way the multimedia content (this has been until now the main objective of MPEG through MPEG1, MPEG2 and MPEG4).

C. Digital item identification and description: it regards the standard to identify and describe digital multimedia content.

D. Content management and usage: it regards the interfaces and protocols for creating, manipulating, searching, storing and delivering of digital multimedia content.

E. Intellectual property management and protection: it regards the possibility of granting that the rights on multimedia content are reliably managed and protected.

F. Terminals and networks: it regards the possibility to enable transparent use of multimedia resources across a wide range of networked devices.

G. Event reporting: it is related to the possibility to describe the interaction that occurred during the use of multimedia content.

At present, in particular, the goal of MPEG21 is to identify the issues that still need a standardisation effort by ISO. Among the elements listed above, those directly connected to the problem of *e*IPR are mainly element C and element E. In particular with reference to element C it seems that MPEG21 has the aim to work to establish a framework for common digital item identification and description by harmonising and integrating with the existing standards. With reference to element E, MPEG21 has the aim:

1. to define the technical, legal and financial attributes of a trusted environment for IPMP systems (by mainly extending the work that has been already done about IPMP on MPEG4 and MPEG7);

2. to define rights expression languages for describing contractual usage rules for digital items;

The objectives of MPEG21 seem indeed very ambitious, nevertheless the benefits that could be obtained by the successful development of such a framework would be really important and could push the effort towards its goal.

CONCLUSIONS

In this paper the need to develop ECMSs to deal with automatic management of all issues related to trading works through telematic means have been highlighted; furthermore the players involved in this kind of systems and the requirements to be satisfied have been indicated. Moreover the tight connection between this sort of infrastructure and whether the standards guidelines or the EC Projects has been pointed out.

REFERENCES

1. J. P. Barlow, "Selling Wine Without Bottles: the Economy of Mind on the Global Net", in Wired, March 1994, p.85.

2. M. Barni. F. Bartolini, V. Cappellini, A. Piva, “Interoperability issues in watermark-based electronics copyright management”, Proceedings of 4th Int. Conf. on Visual Systems and Multimedia, Gifu, Japan, November 18-20, 1998.

3. M. Barni, F. Bartolini, G. Bini, V. Cappellini, A. Fringuelli, G. Meucci, A. Piva, “Enforcement of copyright laws for multimedia through blind, detectable, reversible watermarking”, Proc. of IEEE ICMCS, Florence, Italy, June 7-11, 1999.

4. ESPRIT 20676 IMPRIMATUR Project, “The IMPRIMATUR Business Model, Version 2.1”, November 1997, *www.imprimatur.alcs.co.uk/.*

5. ISO/IEC JTC1/SC29/WG11, “TR 18034-1, Information Technology – Multimedia Framework (MPEG21) – Part 1”, September 2000.

6. J. Lacy and N. Rump and P. Kudumakis, "MPEG-4 Intellectual Property Management and Protection (IPMP) Overview and Applications Document, ISO-JTC1/SC29/WG11/N2614, December 1998.

ABOUT THE AUTHORS

Franco Bartolini graduated (cum laude) in electronic engineering from the University of Florence, Florence, Italy in 1991. In 1996, he received the Ph.D degree in Informatics and Telecommunications from the University of Florence. He is now Research Assistant with the University of Florence. His research interests include digital image sequence processing, still and moving image compression, image watermarking.
E-mail: *barto@lci.det.unifi.i)t*

Roberto Caldelli graduated (cum laude) in electronic engineering from the University of Florence, Florence, Italy in 1997. In 2001, he received the Ph.D degree in Computer Science and Telecommunications Engineering from the University of Florence. He is now a Postdoctoral Researcher with the University of Florence. His research interests include digital image sequence processing, still and moving image compression, digital watermarking.
E-mail: *caldelli@lci.det.unifi.it)*

Vito Cappellini obtained, in 1975, the Full Professor degree in Electrical Communications at Florence University where he was director of the Department of Electronic Engineering and Dean of the Faculty. His main research interests are: digital signal-image processing, digital communications, watermarking, multimedia systems and art-work analysis-restoration. He is "fellow" of the IEEE, member of EURASIP and AEI.
E-mail: *cappellini@lci.det.unifi.it*

Networking historical sources: A demand-side driven approach

Andrea Rosenbusch (*)

(*) Swiss Federal Archives, Bern, Switzerland
E-mail:andrea.rosenbusch@bar.admin.ch

ABSTRACT
Many archives have introduced online archival databases in the past few years. However, apart from being independent of place and time, hardly any of these databases make finding records any easier than it was before. This paper suggests to put more emphasis on the needs and behavior of users and to provide them with the context in which records were created, thus making it possible to link archives in distributed locations and cultural heritage in other organizations as well as enabling users to learn archival search strategies and make better use of the databases provided.

KEYWORDS: User focus, accessibility, network

INTRODUCTION

The Swiss Federal Archives is undertaking a project to create new spaces in the World Wide Web in order to make cultural heritage more accessible. Today, the structure of the Web largely reflects the organizations or individuals supplying information resources, not the contents they provide. Users of the Web looking for heritage related information and material (in our case, the focus is on modern history – political, social, economic) are not so much interested in the structure of the providing institutions as in certain topics. So far, there are very little subject-based entry points to the Web with the exception of search engines, and the possibilities of linking resources have hardly been fully exploited. The main aims of our project therefore are:

- To create entry points to our collection which are based on users' interests and abilities and which take different user groups into account;
- to provide the context of the sources: On the one hand, this makes it possible for users to ascertain the value of the information received; on the other, resources can be linked via their mutual context across institutional or other barriers.
- to initiate a database supplied by several partners independent of the type of resources provided on the basis of a common metadata set.
- to play an active role in building specialized networks which focus on subjects and thus create new entry points to several collections.

THE WEB AS AN INFORMATION RESOURCE

The complexity of the Web is an ideal prerequisite for the arrangement of information, and hyperlinks are ideal for exploring and learning. The complex structure seems to be ideal to represent archival search paths, which depend on vast knowledge of the context of records and connections between the elements in this context. Hyperlinks also to some extent offset the necessity of creating and adhering to common standards which are so important when sharing databases. However, the Web demands information be largely self-explanatory to a growing and increasingly diverse number of people. The challenge, in our opinion, therefore lies less in the use of technology itself than in the arrangement of knowledge to make the maximum use of technology.

A FOCUS ON USERS

If we want our users to be able to search our resources independently on the Web, we need find out more about their needs and behavior and tailor our applications accordingly. Up to now, very little work has been done in the archival field to find out more about our users' needs. An exception is the recent study by Wendy Duff and Catherine A. Johnson. (1) Even less work seems to have been conducted on making Web applications user friendly. (2) The first step in our project thus is an empirical study of our users with the aim of taking their perspective more into account.

The Federal Archives holds written inquiries dating back over half a century as well as e-mail requests from the last few years. The latter seem to reflect the changing public addressing its questions to the archives as well as a change in strategies for finding information in the age of the Internet. In order to be able to perceive trends, we will analyze samples of such inquiries over several years. We aim to find anwers to questions on the following topics:

- The background of the inquirers (e.g. are they trained in or new to historical research),
- the search paths so far pursued (e.g. is this the first place they are looking for information),
- the way their attention has been directed towards this archives (e.g. have they found it via the Internet or been referred to it by someone else),
- the range or specification of knowledge they have on the topic they are doing research on (– this has a strong tie to the search paths, as a well-informed person with experience in research has many more possible access points to information),
- the kind of answer they have in mind (e.g. are they looking for a specific document, a piece of background information or a range of records).
- The degree to which the questions asked can be answered with the current finding aids.

Based on the findings, we will need to decide on how best to guide our users to the sources they are looking for: Is it by categorizing user groups according to experience or by arranging sources according to topics of interest? Can we assist users by making them specify exactly what the are looking for and prompting them to give details which might lead to that information? Does the selection of the approach(es) influence which user groups are best served?

As the last dot point above implies, we also expect the results of our survey to possibly have an influence on the finding aids we provide. It might be necessary to digitize or even create additional finding aids. In any case, we will have to enable users to make the most of the complexity of our finding aids without getting lost and without having to read too many explanations. All of these questions and tasks show us we are embarking on an ambitious and lengthy project. In the meanwhile, however, we will try to simplify matters somewhat and carry out some projects which we believe from experience will enhance the usability of our archives.

CONTEXT

Most people who first use an archives do not know what it holds. The accessibility of information on the Internet adds to this problem. Enter a search term in an online archival database and see what you can make from the results! Users need to be able to judge the relevance and reliability of the information they are being given. Whilst the reliability must be assured by the provider of the information, i.e. the archival institution, the relevance of hits can only be estimated with much contextual knowledge, because archival finding aids are often very selective. This can lead to misleading results, showing many hits in irrelevant areas or no records at all because no detailed lists are available online. Therefore, users may have to pursue a variety of search strategies to find what they are looking for. For example, somebody is looking for biographical information on a Swiss school teacher who emigrated to South America in the 19th century. The Swiss finding aids will probably not list the name and maybe not even feature anything on schools, because these records were never shipped back to Switzerland. However, the diplomatic mission is bound to have some data like the date of arrival in South America, marriage, and death. The teacher might have been active in the local Swiss club which sent its records back home, and there might be some records relevant to the subject. To find documents, it is essential to know what kind of records might have been produced in the first place. Descriptions of record creators and their functions therefore are key access points for researchers, and even more so because they are not necessarily pertinent to an archival institution. Records which reflect the activities of a record creator or a function carried out by multiple record creators need not be held by the same institution, and custody might even change over time, but the context of the creation of the records will always stay the same. (3) Knowledge on record creators and their functions is also essential for the interpretation of records. No record creator is impartial, and the more records from different sources are scrutinized, the more differentiated research becomes. Finding distributed resources is the concern of many researchers. (4) To enhance the possibilities of the description of the context of records and to integrate descriptions of records themselves, we aim to participate in a shared database with other institutions of cultural heritage.

A SHARED DATABASE

Context is an essential element when a search is conducted by a machine on the Internet. Only if sufficient context is given can the user judge the usefulness and trustworthiness of information. The larger the range of resources used for the search, the bigger the problem. However, if only a sample of resources is searched through, is the information received complete or at least adequate? The compromise we suggest is that institutions, e.g. from the archives, museums and library sectors, create common databases to help users identify those resources. There is a difference between portals of institutions and common databases. A portal like that of UNESCO (5) is an improvement for researchers, but it still requires queries to be conducted in each resource separately. The common database should be a possibility to overcome barriers between different kinds of providers, e.g. archives and libraries. We expect our users to mainly look for subject-based information, independent of it being a record, an object or a published resource. Let's assume our user is not looking for teacher X. in the shared database, but for the emigration from Switzerland to South America. With participants in the network from Switzerland alone, the user should be able to find records from a number of State and some private archives, objects in museum collections as well as some publications.

Creating a common database however poses some problems. Semantic and syntactical standards for the exchange of machine-readable archival data and the architecture for harvesting and storing the data have to be agreed on. Because archival description can be conducted on multiple levels (from aggregations of records to single documents), it is also necessary to agree on which level of detail description should be shared. This question still needs some investigation, especially into the user-friendliness of high level and contextual description. The question of common standards was important before the Internet was widely used. In today's World Wide Web, things have become a bit more flexible. On the one hand, the Dublin Core metadata initiative which allows to describe resources independently of their format or type is being increasingly embraced by all kinds of providers of information on the Web. On the other hand, the relatively general character of the Dublin Core metadata set can be offset by linking retrieved data to the database of the institution which originally created them. The user can then search the original database which might fulfill more sophisticated requirements. An example for sharing metadata is an award-winning portal to pictorial resources, PictureAustralia. (6) The contributing organizations each map metadata from own online-accessible databases to the Dublin Core metadata set. These metadata are harvested on a regular basis and stored in a central database which is then queried by the portal's visitors. The users receive thumbnails of images and the pertinent Dublin Core metadata in the shared area, but are directed to the providing institution's online database if they want to see an enlargened version of the picture and learn more about its context. Contributing to the network therefore adds only a small amount of work and costs to the institutions, does not influence their internal standards and practices, and is popular because the holdings are much more readily accessible. Though it might not have all the advantages of a highly standardized central database, the simplicity of the model guarantees maximum flexibility to the contributors and allows them to cater best to the needs of their patrons. It is here that the competitive element between heritage institutions battling for user frequencies and public and private funding takes effect: The joint database increases the chance of their resources being found in the first place, but it is the organization's own collection as well as its website and/or database(s) with its informational, navigational or visual qualities which can constitute its advantage over others.

BUILDING SPECIALIZED NETWORKS

The approach taken above is an important step to link cultural resources to one another. However, it still has an institutional starting point, and the kind of resources linked to each other will be determined by the participants in the partnership. For popular topics, we believe a truly subject-based entry point would be preferable for users. This allows broader participation by giving everyone the opportunity to link their resources to the network. For example, individuals, families, libraries, archives and museums could contribute documents, photographs and objects (references to and/or digital representations of) Swiss emigrants to South America. Maybe some archives in Uruguay has records of the Swiss teacher we were looking for and is able to publish them through this network. This more informal approach might lead to less organized and maybe also less reliable information. But if it works, it can be much more than an information resource: It reflects what people are interested in, it is dynamic and diverse, giving records from non-institutional custodians equal weight to those of renowned organizations, and opening new access points to collections. Hopefully, it will animate debate and provide a wider perspective and more mutual respect for all resource providers involved.

CONCLUSIONS

In the case of networking historical sources, technology does not have such a great impact on *what* we are doing as on *how* we are doing it. The first generation of online archival databases has greatly simplified the previously tedious use of archival finding aids. Now we must begin to design applications which assist the users in actually finding what they are looking for. With the help of user studies, we hope to be able to guide our users in a better way according to their experience and the questions they ask, and to make archival finding paths more transparent. In order to achieve this, we must provide more information about the context of records and use this contextual information to link records in distributed locations.

ABOUT THE AUTHOR

Andrea Rosenbusch has been working as an archivist since 1993, mainly at the Swiss Federal Archives. In 2000, she spent six months at the National Archives of Australia to specialize in different aspects of electronic records. As an historian, she has published on innovation in the Swiss chemical industry in the 1920s and 30s and is regularly involved in making historical exhibitions.
E-mail:
andrea.rosenbusch@bar.admin.ch.

REFERENCES

1. Wendy Duff and Catherine A. Johnson: A Virtual Expression of Need: An Analysis of Email Reference Questions. Paper held at the Society of American Archivists' Annual Meeting in Denver, August 28 – September 3, 2000. In: *American Archivist* (forthcoming).

2. Andrea Rosenbusch, Are Our Users Being Served? A Report on Online Archival Databases. In: *Archives and Manuscripts*, May 2001.

3. Gavan McCarthy, Utilizing the Web to Build a Network of Archival Authority Records. In: *Janus* 1999.1, pp. 96 – 106, especially pp. 98f., and the exemplary site ‚Bright Sparcs' by the Australian Science and Technology Heritage Centre at *http://www.asap.unimelb.edu.au/bsparcs/.*

4. National Scholarly Communications Forum, Round Table No. 10, November 1999, Archives in the National Research Infrastructure, Communiqué and

Resolutions, resolution no. 6, *http://www.asap.unimelbedu.au/nscf/roundtables/r10/r10_resolutions.html.*

5. *http://www.unesco.org/webworld/portal_archives/.*

6. *http://www.pictureaustralia.org/.*

A View on the Design of Usable Museum Appliances

Piero Mussio(*) *and Augusto Celentano*(#)

(*) Dipartimento di Elettronica per l'Automazione, Università di Brescia, Italy
(#) Dipartimento di Informatica, Università Ca' Foscari, Venezia, Italy
E-mail: mussio@ing.unibs.it, auce@dsi.unive.it

ABSTRACT

This paper poses the problem of designing Museum Appliances (MA), which are usable, i.e. easy to learn and easy to use. A recently proposed model of multimedia Human Computer Interaction is used to explore the dimensions of the MA usability, identifying the features which characterize this type of appliances, and to outline the an approach to usable MA implementation. The reasoning is bottom up, in that it starts from the study of some existing MA within the frame offered by the model, to derive the features which characterize an MA and the usability requests to be satisfied.

INTRODUCTION

Museum Appliances (MA) are complex systems designed to support humans in their interactions with a virtual or real museum. An MA allows its users to navigate in the real or virtual museum space, presenting them data which document the current state of the interaction and how the interaction can be continued. However these data can be correctly interpreted only by users who recognize their nature and know the rules for their interpretation, i.e. by users who belong to a certain culture. But MA users have different cultures, skills and dexterity levels and interact with the system in different contexts to achieve different tasks.

The problem arises of designing *usable* MAs, that is, easy to learn and use [8]. This paper addresses this problem by capitalizing on a recently proposed model of Human Computer Interaction [2] and on several practical experiences[4]. Users–MA interaction is studied in the frame of this model identifying a set of communicational and operational features an MA has to posses to be usable. Indications are derived on the design of MAs which enjoy those features which are culture dependent. An architecture is thereafter presented which is able to customize the ways in which MA interacts with its users adapting itself to their culture.

MUSEUM APPLIANCE

Museum Appliances are software–hardware multimodal interactive systems designed to support users in their interactions with a virtual or real museum. They can be classified in several categories according to their primary purpose – e.g., supporting user orientation and movement, providing introductory or shallow information, providing specialized information for professional users, and so on –, to the complexity of their operation, to the completeness of information they can provide and to their mobility. Not surprisingly, the more they grow in complexity, the wider is the information they can furnish. We could adopt a simple classification scheme centered on three basic kinds of devices:

- Hand held devices, primarily audio

devices, whose purpose is to provide a shallow information for the novice user, and a help in following guided tours.

- Kiosks, i.e. multimedia devices able to give a variety of information for users of different skills. An extension if kiosks are CD-ROM based guides and electronic books. Both are based on the same technology, but they serve different purposes.
- World Wide Web sites. They could be considered extensions or technological variations of CD-ROM, which can provide dynamic and open information Many museums have set up Web sites that help the user to set up a tour, or provide practical information, or serve as an anticipation of what the visitor could see at the museum.

We base our discussion mainly on visual interfaces typical of CD-ROM and Web based appliances and address issues related to information access, navigation and orientation.

MA provide three main supports to their users, which allow them to perform their tasks without loosing themselves in the information space:

- maps, i.e. orientation devices that help the user to understand where he/she is, from where he/she came, and where can go further;
- guided tours, i.e. orientation devices that help the user in stepping through the information in a regular and motivated way, according to a knowledge goal;
- multisensory devices that supplement knowledge about the user movement (physical or conceptual) with easy to perceive *stimuli* that integrate rather than replace the information he/she is accessing at any moment.

The MA supports its users by sending them multimodal messages: these messages can only be understood if the user recognizes them as messages and is able to decode them. To make this point clear, let us frame the user–MA interaction in the Pictorial Computing Laboratory (PCL) model of WIMP Human Computer Interface [3].

Without loss of generality, we first consider a WIMP MA, in which the interaction is based on the exchange of images between two participants, namely the human user and the MA itself. Both human and MA interpret every event occurring during interaction with reference to the whole content presented in the MA screen display. The content is formed by text, graphs, pictures, icons, etc., therefore representing a multimedia message, and materializes and conveys the meaning intended by the sender. The message must be interpreted – i.e., associated with a possible different meaning – by the receiver. The screen is the communication surface between human and MA, thus acting as a bi-directional channel in the transmission of messages. The receiver interprets the message based on cognitive criteria which could be described as more or less formal rules (or conventions) which come from experience, habit, cultural background, etc.

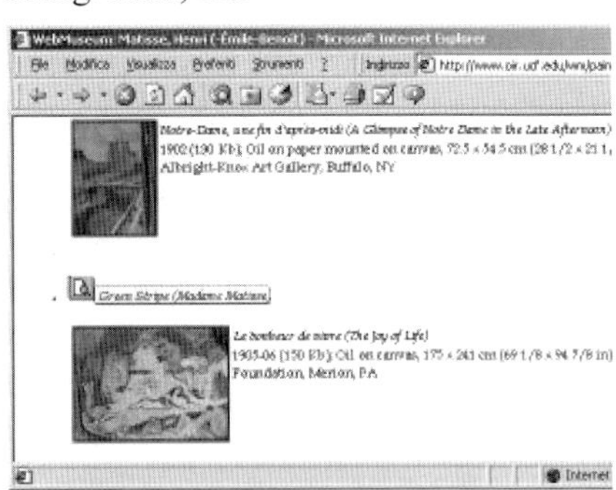

Figure 1. A list of paintings from WebMuseum

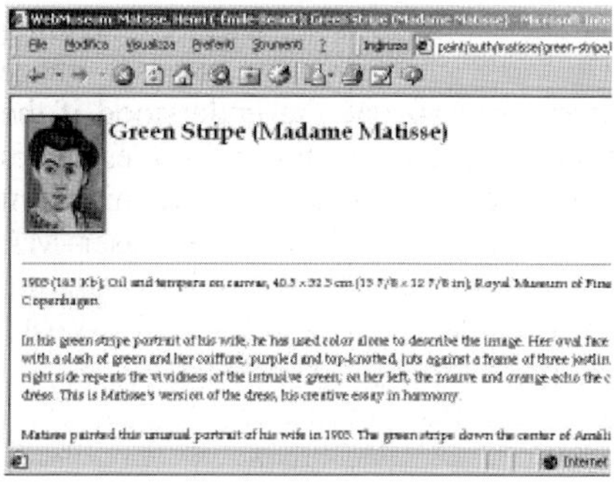

Figure 2. A painting description

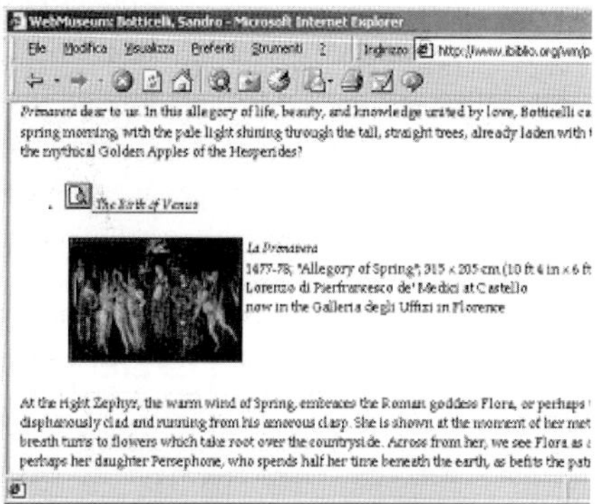

Figure 3. A comment on Botticelli's paintings

INTERPRETATION OF MUSEUM DOCUMENTS

Figures 1–3, taken from the WebMuseum [9], help to understand this process, showing how users can interpret museum documents based on visual properties and layout.

Figure 1 is a list of references to paintings, each of which is introduced by a small image and a technical description. The image is an index and has a descriptive function: it gives the user a glimpse of the artwork. The description is perceived as an introductory, objective, technical one due to its limited size and the presence of recurring text patterns (Title in italics and painting size being the most immediately perceivable ones). A second type of indexes is used: a symbolic icon (a magnifying glass) coupled with a string of characters suggesting that a more detailed description of a painting whose title is represented by the string can be reached, as in Figure 2, which shows a separate document with a simple visual layout, suggesting no further details. The absence of visual cues in the text asides the paintings in Figure 1 suggests that only some artworks are described more thoroughly.

In Figure 2 the small size of the image suggests that it is an index, as in Figure 1, while the blue frame which surrounds it suggest to whoever is a Web surfer that a larger image should be accessible by clicking over it, blue being the traditional color for links in WWW documents. In the large-sized image standing as a separate document it is evident that no other information is associated to it.

Some structures in each image on the screen act as cornerstones, i.e. they help the user in navigating through the virtual space: typically the image title, which identify the overall virtual space, i.e. the painter Matisse in the Web Museum, and the situation, i.e. the *Green Stripe* painting. The user is able to interpret the purpose of each document after a brief glance at its layout. A brief tour in the WebMuseum site shows that these patterns are recurring, even if sometimes deviations from this style can be found. An example of a different pattern is in Figure 3. This document is not a simple list of paintings: a criticism on Botticelli's artworks discusses two paintings, the *Birth of Venus* and *The Spring*. The former is described in a separate document (hence the icon), while the latter is described inline

(hence the small painting image). The choice of organizing the discussion in this way is an author's choice and is not motivated in the document.
The interaction process with an MA can thus be modeled according to the PCL model, which was however derived from the study of critical interaction processes, in which every mistake has a defined cost. This is not the case of MA interaction. This difference in the context of application will lead to different criteria for their design, even if the same model of interaction is adopted. The interaction process is in both cases modeled as the generation of a sequence of images appearing on the screen at successive instants of time t_1, ..., t_n, each one interpreted both by the human and by the MA. Each image derives from the transformation of the previous one according to user's actions and MA's computational activities.

Human and MA communicate by materializing and interpreting a sequence of images. Humans interpret and materialize the images using human natural cognitive criteria, the MA using the criteria programmed in it by the system designers and programmers. Two semantics are always implicitly defined in any interaction: one internal to the MA, in which each image is associated with a computational meaning, as defined by the designer and implemented in the MA; and one proper to the user performing the task, depending on his/her role in the task, as well as on his/her culture, experience and skill.
Ambiguous or equivocal situations may arise because of these two interpretations: *Ambiguity* arises when one of the two communicants – the human or the computer – associates two different meanings with a message, and flips from one meaning to the other during a same reasoning process. *Misunderstanding* arises when the two communicants associate two different meanings with a same message. Each communicant has no ambiguity in what he/she is doing, but different cultures or situational reasons determine different interpretation. Conflicts may arise between the two communicants, which make the user unhappy.

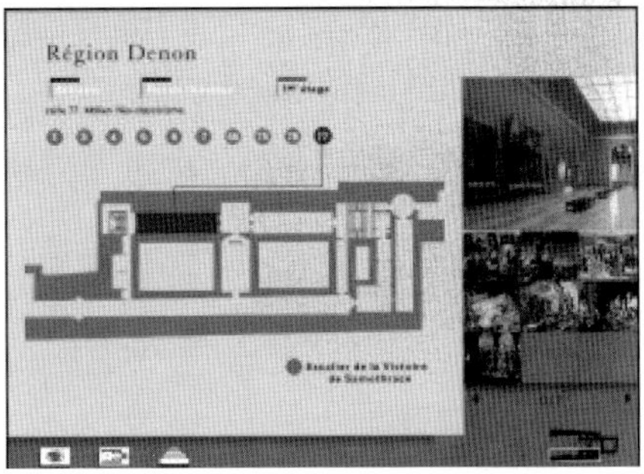

Figure 4. A map from *Le Louvre*

As an example let us consider guided tours and maps. The goal of a guided tour is defined by an author who could rely on a culture different from a visitor, therefore designing a tour not meaningful in the visitor cultural context, or difficult to understand. Maps should be more objective because they document a real or realistic layout, but ambiguity can arise also in this context. Figure 4 shows a floorplan of a region of the Louvre Museum, excerpt from the CD-ROM *Le Musée du Louvre* [6], with a room selected. The room is highlighted on the map in blue, and pictured on the upper right side of the screen. On the lower right side a mosaic of the paintings exhibited in the room and contained in the CD is shown. The floorplan and the mosaic are clearly maps. The room image is not perceived as a map in this context: it is only a preview of the real museum room. A click on this image shows the screen of

Figure 5 (frames are visible only when the cursor moves over a painting, so Figure 5 is actually an artifact obtained by several superimposed screen shots). Differently from the preview image this screen is a clickable map used as an intermediate step to reach the painting documents.

Figure 5: A room acting as an active map

As observed in [5], users can achieve their tasks if they associate with each structure in the image and with the whole image a meaning similar to the one associated by the MA, i.e., if an *adequate communication* is reached.

REQUIREMENTS ON DESIGN

According to the PCL model, an interaction process is based on two interpretation and two materialization processes. From this point of view, the goal of a successful design is to bring the system semantics to reflect the user's one, so that both the user and the system materializes messages which are properly understood by the user and adequately managed by the MA. Moreover users should be in the condition to perform actions, familiar to them, in a way that is correctly perceived by the system.

The MA designer has to take into account both the communication aspects of the interaction, regarding how the exchanged messages are interpreted, and the operational aspects which regards how the messages are materialized. At the communication level design must consider:

- variety: multiple facets of the same information (images, text, sounds, but also descriptions, criticism, comments) able to convey complementary meaning to a user in order to improve his/her knowledge about that information item;
- customization: a form of personalization targeted to the specific user needs and habits beyond the ones common to a wider cultural or geographic environment;
- deepening: the ability of navigating into an information item to discover new more specific information according to a progressive disclosure principle;
- accessibility: the ability of knowing what further information exists, and how it can be found.

At the operation level design must consider:

- navigation: the ability of moving among different information items effectively, i.e., without missing information due to the difficulty of understanding how to move;
- orientation: the ability of knowing at each stage of the visit where information is, what information has already been examined, what new information can be accessed;
- personalization: the ability of accessing the information in a way consistent with the cultural habit of the user;
- retrieval: the ability of finding an information item without knowing the path to reach him, or the knowledge that an information item is definitely missing.

A fundamental distinction between cultural heritage appliances and industrial systems appliances is in the use of metaphors in a user interface. Industrial systems make large use of metaphoric information for representing information and data about the real world. Metaphors must be easy to understand, therefore simple and related to a common understanding. Cultural heritage systems render real world information in such a way that it can be recognized as a representation of the real world, mediated only by the technology that displays it to the user. Therefore there is no need for common understanding or adoption of standard conventions, unless they are targeted to giving the user a sense of realism. For example, the use of the same scale for representing a set of paintings gives the user the knowledge about the relative sizes of the real worlds works; the use of a same colour resolution (for paintings) or lightning (for sculptures) makes easier for the user to compare different works. Metaphors are confined only to the operational part of the interface, and not to the informative part.

On the whole, an MA designer has to pursue user satisfaction, as an industrial appliance designer do, but in a different context: Museum Appliances and Industrial Appliances are designed with different goals, a mistake of the user as well as wrong behaviors have different costs and their success must be judged according to different usability metrics. Moreover MA users may belong to a many different cultures, which all are valid interpretation tools of the Museum content, while in an industrial environment the users' culture variability may be considered more restricted.

However at a high level of abstraction, the guidelines to correct design are similar: (1) The designer has to develop a *usable* MA: the system is easy to learn, to use, to navigate, efficiently and effectively. However, easy to learn bears now a peculiar meaning, due to the fact that an MA may be designed to be used just once by a user – think of a hand held device, which supports museum visitors. (2) The designer has to develop a *viable* MA: no human action result into a non-meaningful situation or in a system crash. (3) The designer has to develop a *deterministic* MA: in every situation an action has to produce only one (predictable) result.

In the interaction, no message can be ambiguous: the human (or the computer) never assigns to a message different meanings during the same interaction process and the communication must be adequate: i.e. both human and computer have to assign the same meaning to each message. These two features are difficult to be achieved, because of the variety of cultures to which MA users may belong to, and also to the difficulty to obtain a user profile. Last, the user navigating in the museum space (real + virtual) must always understand where he/she is and never gets lost in the virtual space.

FORMALIZING THE INTERACTION PROCESS

Let us now examine the interaction process in more details, following the PCL model. The images on the screen represents the state of the interaction, and their interpretation determines the next action of the human and the next computation of the MA. Humans interpret the images on the screen by recognizing *characteristic structures* (*cs* for short), i.e. sets of image pixels that users recognize as functional or perceptual units. The *cs* recognition results into the association of a meaning with a structure. Humans express the meaning attributed to the *cs* by a verbal description. In an image a human may recognize several *cs*: combining their meanings, humans derive the meaning of the whole image on the screen. Hence we can say that *cs* play different *roles* in the user understanding of the image.

For example, in Figure 1 the two paintings are recognized as *cs*. In this situation they play both the role of *indexes*, indicating to the user that the painting is present in the archive and that of *links*: the painting can be seen in full screen mode by clicking on it. A different *cs* is constituted by the rectangle with superimposed a file symbol and the magnifying glass icon. This last *cs* can be seen as a composed one, and its meaning can be derived only combining the meanings of its component *cs*. Several other *cs* can be recognized: every users, who knows Latin alphabet, is able to recognize the single letters appearing on the screen, while a user who also understands English recognizes the words and sentences, which therefore become complex *cs* for him/her. For this user the title in the top bar plays the role of a cornerstone, which help him/her in orienting in the virtual space. A user who uses a different alphabetic system can suspect that those structures may represent letters, while a user accustomed to different writing techniques, e.g. pictorial ones, may not even recognize the role of these *cs*. On the whole, the screen shown in Figure 1 can be properly understood by a user who knows English language and is familiar with the aspect of traditional desktop interfaces icons and web surfing.

On the other hand, the MA associates graphical entities on the screen with computational constructs. Each computational construct – here denoted by *u* – represents the meaning associated with a graphical entity on the screen, as intended by the system designer. It is exactly this association that makes the MA able to interpret the captured user actions such as clicking on a button, with respect to the image on the screen, possibly firing computational activities whose results are materialized on the screen, via creation, deletion, or modification of *cs*. The association between a *cs* and the corresponding *u* is called a *characteristic pattern* (*cp*).

These concepts can be formalized starting from the definition of a characteristic structure *cs* as a set of pixels perceivable on the screen and of a characteristic patters *cp* as a triple which specifies how the set of pixels (*cs*) is linked to a program *u* which describes its meaning.

In an image *i*, several *cs* can be identified. These *cs* became *cp* when a program *u* is associated to them. For example, in Figure 1, the two painting images are *cs*, whose description contains links to a program which changes the image on the screen, presenting the painting on a full screen. Moreover, *i* as a whole can be associated with a program *d* synthesizing the overall properties and the global meaning of the image *i*. We call *visual sentence* (*vs*), a triple whose elements are the image *i*, the program *d* and a specification of the relations among the *cs* in *i* and the programs which describe their meanings. A set of *vs* is a *Visual Language* (VL).

A formal theory that provides a finite definition of VLs by a special family of rewriting systems, the *Visual Conditional Attributed Rewriting Systems* (*vCARW*), has been described in [1, 2]. In this theory an interactive process can be specified by the possibly infinite set of all the sequences of *vs* that, starting from an initial *vs*, vs_0, are determined by the sequences of user actions and system computations in the process. Each sequence in the set describes a specific user-computer interaction session. The set of all sequences of *vs* that can be generated from vs_0 constitutes the *Interaction Visual Language (IVL)*.

On the basis of this definition, adequate communication occurs when in each *vs* of IVL, the graphical entities recognized by the MA are the *cs* recognized by humans, and the behavior of associated computational structures reflects the behavior expected by the users.

A STRATEGY TOWARDS REQUIREMENTS SATISFACTION

The recognition of the existence of two semantics is the starting point for reaching an adequate communication between human and computer. Indeed, adequate communication occurs when the human and the computer associate a similar meaning to a same message (or part of it) [5]. In order to identify the users' semantics, the PCL recommends to exploit the users' notation as the kernel of the definition of the IVL through which human and computer communicate. Users' notation embeds context, task and procedural knowledge possessed by the users. It embeds the knowledge explicitly and implicitly: explicitly in symbols and rules used for constructing the users' documents; implicitly, because the shape and the spatial arrangements of symbols allow users expert in the application domain to recognize structures and relations meaningful for their tasks, but too vaguely defined to be made explicit in a declarative format.

The adoption of users notations as the kernel of the IVL is a fundamental step to make the interaction process understandable and checkable by the users: users may justify results expressed in their notations on the basis of their experience and not on the basis of an algorithmic explanation. In this way, the adoption of user notation facilitates the reaching of the closeness of mapping between the real world, in which users operate, and the virtual world which supports their work. However, as pointed out by Mayhew [7], the mere adoption of this notation may bring to under-use the system. In fact, the users' notation has the advantage of being completely familiar to the user, but also the disadvantage of having been defined without taking into account the existence of computing systems. The PCL approach therefore proposes to *augment* and *adapt* the original notation to fully exploit the computing capability of the interactive systems. In the augmented language, symbols are able to show their state, for example, assuming a color to show that they have been selected, and can be associated to a specific functionality to favour the interaction with the user. For example, traditional supports exist also in real world museums (e.g., maps, arrows, colored pathways) and many studies exist on museum architecture.

Museums Appliances augment traditional tools, because they also support users in understanding and checking the information, mainly helping to understand the overall information organization. While in an industrial application the user interface for operating a complex apparatus must make evident what the industrial system does at any stage of operation, in a cultural heritage application the interface must make evident to the user what information he/she can find, in which form, and at which extent. Since MA users have different cultures, skills and dexterity levels and interact with the system in different contexts to achieve different tasks, different styles of augmentation and adaptation should be adopted to develop MAs for different users communities. What we are studying is to allow MAs to be adapted by outer actions or by self-evolution to the user according to some user profile.

One possible approach, which we are exploring, recognizes the importance of the role played by the *cs* in the user interpretation. *Cs* may have the role of cornerstones, which help the users in orienting himself in the virtual space; may indicate the possibility of executing an action; may convey information on the subject of investigation. *Cs* are distributed in the screen space according to their role, and often their appearance depends on it (an alarm or warning in west Europe is a red flag). Shape and position of a *cs* help the user in recognizing and in understanding it. Unluckily, these features depend on the user culture: users who read a text from right to left expect the title cornerstone on the right side of the screen. To take into account these features, the data related to a Museum Appliance are coded in an intermediate form, typically using an extension of some tagged language. Tag types are introduced which define the role that each bunch of data plays in the communication. For example, they specify if a text acts as a cornerstone, and should orient the user in the navigation within the Museum space (real + virtual); a tag related to an image specifies if the role of the image is to be an index, or a link, or both; in a sequence of images, tags specify their time dependency.

This intermediate form can thereafter be translated by an instance of MA, suited to a given user profile. Museal data are presented to the user according to his/her habits and needs. In other words, the MA not only translates the text, but also interprets the tags to adapt the layout, the symbols and the operational facilities to the specific user culture, skills and dexterity. Each MA receives the description of the museal space coded in the tagged language and interpret it for its users. The MA also accepts the users input in the user notation and interpret it to match with the Museum description. This organization can be achieved by designing the MA as a bi-directional translator among the user notations and the tagged language.

CONCLUSION

This paper stems from some practical experiences from which a classification of MA is derived as well as some usability problems emerged. MA are take-up-and-walk tools, which need to be understood at once by their users, who however have different goals and belong to different cultures. To face this problem, our proposal is to design MAs as bidirectional translators customized to a specific community of users, that is, users which have common goals and belong to a same culture. The museal space is described and maintained only once in a tagged language. Each MA customized for a specific user community translates the data of the museal space into the user notation and interprets the input from the users, expressed in their notation, with respect to the intermediate form.

REFERENCES

1. P. Bottoni, S.-K. Chang, M.F. Costabile, S. Levialdi, P. Mussio. Dynamic Visual Languages, *Proc. IEEE Symposium Visual Languages'98*, 14-21, 1998.

2. P. Bottoni, M.F. Costabile, P. Mussio. Specification and Dialogue Control of Visual Interaction through Visual Rewriting Systems. *ACM TOPLAS*, Vol. 21, N. 6, pp. 1077-1136, 1999.

3. P. Bottoni, M.F. Costabile, S. Levialdi, P. Mussio. A Visual Approach to HCI. *ACM SIGCHI Bulletin, Vol.28, No.3*, July 1996, pp.50-55.

   http://www.acm.org/sigchi/bulletin/1996.3/levialdi.html.

4. A. Celentano, F. Pittarello. Multi-sensory guided tours for cultural heritage: The Palazzo Grassi experience.
   Technical Report, Ca' Foscari University, Venezia, Italy

5. S.K. Chang, P. Mussio. Customized Visual Language Design. *International Conference on Software Engineering and Knowledge* Engineering - *SEKE'96*, pp. 553-562, 1996.

6. *Le Musée du Louvre, les collections.* Montparnasse Multimedia & Réunion des Museaux Nationaux, 1997.

7. D.J. Mayhew. *Principles and Guidelines in Software User Interface Design.* Prentice Hall, Englewood Cliff, 1992

8. Nielsen, J. (1993). *Usability Engineering*. Academic Press, 1993.

9. WebMuseum is available at several URLs: *http://www.oir.ucf.edu/wm/*, *www.ibiblio.org/wm/*; other URLs are referenced by these sites.

Initial thoughts for setting up a virtual exhibition

Cèsar Carreras Monfort (*)

(*) Department of Humanities, Universitat Oberta de Catalunya (UOC), Barcelona, Spain
E-mail: ccarreras@campus.uoc.es

ABSTRACT
In the last two years, the research **group Òliba**[1] has been collaborating with different Museums and institutions in Catalonia and the Balearic Islands (Spain) in order to introduce new technologies in Cultural Heritage Management. Most Spanish Cultural institutions have not introduced yet those technologies in their everyday life due to the lack of public funding as well as their scant experience in this field. Therefore our collaboration as University research group meant to these institutions a way to test computer applications in Heritage environment without investing a lot of economic and human resources.

KEYWORDS: exhibitions, computers, complement

INTRODUCTION
Museum collections offer on-line need to be digitalised in order both to preserve original documents and cultural objects as well as to make them accessible to a wide public as well as to specialists. The research of group **Òliba** therefore pays special attention to the long-term preservation of digital documents (image, video, and audio), through the development and implementation of standard formats. An on-line museum should be far from being simply a catalogue. New ways of presenting the information in the distributed collections will be approached, having in mind a versatile simulated physical lay-out. The research group **Òliba**, word which means "owl"[2] in English, was with the aims to evaluate the potential of new technologies for the dissemination and documentation of Museums cultural heritage. Our initial proposal was simply to test how virtual exhibitions could enhance the potential of "real" exhibitions in terms of more information access, public awareness and visits to the Museums.

From the academic point of view, the experience allows us to evaluate the potentials of Internet in the dissemination and documentation of

[1] The Group Òliba is directed by Glòria Munilla and Cèsar Carreras, both lecturers in Heritage management and Computer applications and Archaeology respectively in the Department of Humanities at the Universitat Oberta de Catalunya. Besides, the group includes to Multimedia specialists such as Piero Berni and José Maria Jansana.

[2] The owl was the pet of Atenea, the goddess of Arts and Culture in the ancient Greece, so we thought it was an appropriate name for our research group with few classical specialists. Olive-grooves that surrendered the Parthenon kept some owls in the late times, and the bird became an Athenian symbol struck in their famous coins, tetradrachmae.

Cultural Heritage, testing new technical tools (3D VRML, XML, Java EAI) and the people reaction towards this idea. Compared to other European countries, Spain still have a low percentage of Internet users amongst its population (around 10,5%)[3], so the development of new network applications will take some time to build up. Our first goal was to evaluate how virtual exhibitions may improve the diffusion of real exhibitions as well as their potential museography. This simple aim not only required a change of custom for the public, but also teaching Museum's staff the advantages of this new media and working together with them.

The present paper attempts to discuss all the problems we faced in the design of our last four virtual exhibitions and how our experience may help others not to make the same mistakes again.

Our present research project covers four stages, beginning with an analysis of the theories behind web Museums, leading on to methodological and practical analysis of previous experiences of digitalisation in Museums in Europe and States. The third and more original part involves developing pilot experiences of virtual exhibitions together with local and regional Museums from Barcelona and Catalonia. The fourth and final stage pretends to evaluate the experience in order to define standards for Museum webs, as a complementary platform for Museum didactics and dissemination, how people interact with different elements (3-D reconstructions, quizzes...), technical advantages and potential for the real Museums.

Virtual exhibitions should be understood in this context as **complements** of the real exhibitions, offering only a small part of what people can enjoy visiting the real physical Museum. The virtual platforms will show a taste of what people may see there, and on the contrary the virtual exhibitions will focus on displaying objects, information and images, which cannot be accessed in the Museum. Virtual exhibitions, as we understand them, should never compete with the real Museums but collaborate with them hand to hand.

CO-OPERATION WITH MUSEUM CURATORS: PEDAGOGIC ASPECTS

The experience of this two years has been rewarding since we have detected some pitfalls in the way Museum's staff approach to new technologies. In general, they believed Internet was only an alternative of paper publication, so they understood virtual exhibition as either Museum brochures or digital catalogues. Personally, we think a virtual exhibition may be an ideal complement to a "real exhibition".

How this ideal can be achieved? Well, we should foresee from the beginning that an exhibition must have to complementary parts: a real and a virtual. None of them should be the

3 The most recent report (December 1998 – April 2000) on uses of Internet by Catalan citizens (http://www.gencat.es/csi/cat/documents/estadistiques/documents/estadistiques_2000.htm) shows that there are usually only 23,7 % of population taking advantage of it, though at least 32,2 % have been connected at least once. This percentage is similar to a country such as United Kingdom, however Spain as a whole only reaches a 10,5% of users (http://www.c-i-a.com/199911iu.htm). Most people become users due to professional or educational reasons, only few as part of their entertainment.

same as the other, but they may contain common sections that will give a feeling of continuum between the two parts. Besides, the virtual exhibition should be addressing a wider variety of people than the real as well as an initial step before going to the Museum. The real exhibition ought to be design for the local people, who will visit it, and it will be central part of the whole exhibition. Once, people have visited the display they may come back to the virtual part in order to complement their information and experience.

Collaborating with Heritage institutions meant to explain in detail our approach to the Museum's curators in order to get their fully involvement in the project. However, it must be said that this was not the common case. On the contrary, most Museum staff does not understand why they should create a virtual exhibition, and the reasons they argued are the following ones:

- Virtual exhibitions get visitors away from real exhibitions, so they can be a **threat** for the own institution
- If visitors wish to remember an exhibition, they can always buy the catalogue, therefore some people identifies virtual displays as **catalogues on-line**
- The virtual exhibition can be only useful when the **real display is dismantled**[4], because they do not compete with each other
- The Museum's should focus only on **local visitors**, therefore there is no need to translate contents on-line for a global public
- Internet is only seen as media of dissemination of Museum's activities, in other words an **electronic booklet**

As you may imagine, one of our most important tasks was simply to convince them of the real advantages of the virtual world. Most our efforts, when we arranged to design a virtual exhibition, involved persuading curators and Museum staff that Internet could help in their curatorship task, since they were normally afraid of it. They were willing to use technology inside the Museum building, but they were scare of having an open resource outside in the net.

Of course, Museum curators were afraid of something new that could change their jobs and an environment, with which they were not used. Our pedagogical task was a key issue in the project development. The group Òliba initial aims when designing a virtual display were:

- Create virtual exhibitions that **complement** real ones, therefore the material and information included in each one should not be exactly the same, despite having some common traits (exhibition discourse, common objects…)
- Virtual displays should in fact **encourage people to visit** the real exhibition
- Virtual displays should provide **resources unavailable** of more difficult in the real world (interactive virtual reality environments, other objects and features not in the real display)
- Virtual exhibitions should be addressed to a **wider variety of**

[4] Exhibition of Palau i Fabre l'Alquimista by Lletra (*http://www.uoc.es/web/cat*).

public with diverse itineraries (from specialists to occasional visitors)
- Virtual displays should be a previous contact with the exhibition, so it will give **basic background information** about the exhibition topic to potential users in order to fully enjoy the visit
- Virtual exhibitions should be **an after visit resource** to consult any questions arisen in the tour through the Museum

Once, all these initial objectives were proposed to curators, they agreed with the initiative and lost their early fears. However, this ambitious approach involved an extra effort in setting up an exhibition, for instance which materials and texts will appear in the real or virtual display. Notwithstanding the acceptance of our principles, most Museum curators did not supply all the information required to fulfil these aims since it represented a new work that was not planned at the exhibition start.

FIRST EXPERIENCES: THREE EXHIBITIONS

This year 2000, the group Òliba created three virtual exhibitions in collaboration with different local institutions (the fourth one – "TREASURES OF NATURE" (*http://www.uoc.es/humfil/natura*) has not been analyzed in depth. Notwithstanding the fact that is too early for a full evaluation of these experiences, an initial account is described below.

"ALWAYS EIVISSA AND FORMENTERA"
(*http://www.uoc.es/humfil/eivifor*)

This itinerant exhibition funded by La Caixa Sa Nostra began on the 9th December 1999 and travelled for all the Balearic islands (Mallorca, Menorca, Eivissa and Formentera) as well as a short stay in Barcelona. The exhibition aimed to display the way Balearic people wish to preserve their signs of identity, maintaining some traditions and natural landscape without renouncing the economic development, which is mainly based in tourism. The virtual platform was ordered only one month before the real exhibition, so it was quite difficult to implement the web page in such a short span of time. Nevertheless, the final virtual exhibition fulfilled the aims of the organizers, who wished a complete dissemination of the exhibition contents to other Spanish people and foreigners.

With regards to design, the virtual exhibition reproduces exactly the real one, except from a major number of illustrations as well as a new virtual reality section. In this case, the virtual exhibition is not really complementary to the real one, therefore it does not represent our desirable model. The navigation structure was kept as simple as possible, allowing a direct access to any section from any point in the site. However, some minor problems were detected due to the hierarchical structure of the exhibition (i.e. architecture, natural environment), which deny this expected direct entry.

According to comments by the own users, the virtual exhibition was welcome as alternative to people who could not visit the real one as well as after visit resource. There were positive opinions in general, and also some suggestions for improvement.

For instance, clearer icons were desired to mark every single section, and also they complaint about the secondary level where the virtual reality was located. One of the problems we did not expect was font configurations depending on visitors, something easily solved with dynamic HTML (.css files). The virtual reality has an aesthetic effect on users, but it failed in this didactic potential, which was attempted in other projects. Besides, more animations were expected to make some screens more attractive. Visitors did not answer very often to our virtual questionnaires, at least we expected more participation than it really was.

With reference to the content, no extra information was created apart from the virtual reality. Perhaps, a series of transversal sections may have fostered the hypertext potential, making available alternative navigation through the diverse pages. Interactive sections such as Archduke and Casa Schmela, which combine clickcable maps, were highly rated by visitors. The virtual reality reproduces how the temporary exhibition may become a definitive collection in one public space such as the known octagonal market of the city of Eivissa.

The dissemination of this exhibition provides quite interesting data. First of all, there has been 1788 accesses to this page until the 15th June, despite the scarce announcement of this web. The virtual exhibition was announced in the local Balearic press, as well as the local Internet gateways such as Terra and Vilaweb. While the new was in the local press, we accounted the major number of visits, afterwards we detected a downshift in the visits.

People complain about how to access to this virtual display, since there are no specific gateways for update on Museums and temporary exhibitions, except the Spanish version of Museums WWW (*http://www.icom.org*). The only way to be aware of this is by the traditional press, the own Museum propaganda, our University home page (*http://www.uoc.es*) or by word of mouth. Hopefully in the future, the matter will be overcome by creating an up-to-date Museum gateway. Most people who visited our web site were, of course, Spanish (more than 60%), though there were users from other countries. Amongst them, United Kingdom stands out with 7%, USA with a 5% and Germany 3%, due to the fact that the site was in three languages (English, Spanish, and Catalan[5]).

The organizers after this year are planning to have a fixed venue in the town of Eivissa to set a permanent display in the future, combined with a number of information points in other location in the Pitïuses islands. However, they also wish to keep an updated virtual platform that will coordinate all these interpretations centers. Therefore, we believe the website has fulfilled its initial objectives.

---

5 The presence of a Catalan version in all the exhibitions is an attempt to preserve our minority language in a global medium such as Internet. The European Community is full aware of the difficulties of the minotarian cultures to survive in the digital age. Therefore it fosters a number of projects aimed to promote their cultural heritage as well as their language. Euromosaic project attempts to disseminate the less-spoken languages in the European Community (*http://campus.uoc.es/eurommosaic*).

"AUREUM OPUS: FIVE CENTURIES OF BOOKS' ILLUSTRATION"
(*http://www.uoc.es/humfil/aureum*)

The exhibition started on the 10th April and closed on the 6th June 2000, but the virtual display is still available on-line. The request of the Museum Frederic Marès provided ideal work conditions since it was ordered 3 months before the opening. Therefore, the group Òliba was able to create a complementary exhibition, which is so far our best model. The exhibition shows a selection of Catalan books of collectionists, some of them very old, with wonderful illustrations, which are not even available in permanent Museums collections.

From the beginning, it was arranged with the Museum staff which material could not be in the real display and then, only available on-line. Moreover, we thought about any potential public, even visitors who did not have any background in illustration techniques. That is why a preliminary section introducing those techniques was developed as a complement to the exhibition. Besides, two other transversal sections were created to provided information on illustrators and other one with recommended Internet links referring to the some subject.

The navigation layout was also kept as simple as possible with two frames for different menu levels. At the first level, the virtual reality was incorporated, which benefited its use; while at the second level, there were the diverse sections of the exhibition as well as the new transversal sections. Furthermore, the iconography used seemed to have pleased users and problems of fonts and screen configuration were overcome.

One of the most useful things in the virtual exhibition was the possibility of digitalising all the books' illustrations, since the displayed books were in close cases showing only one of the illustrations. Finally, it was decided to digitalise all the illustrations of a 10 books sample in two sizes (i.e. small and full screen), and then, one or two images from the remaining books. Therefore, this section plays a complementary role in the exhibition, whose function was appreciated not only for the users but the organizers and curators.

Another complementary and didactic element was the virtual reality, which in this case was a reconstruction of a XIXth century illustration workshop. The reconstruction on-line shows how the diverse machines (i.e. presses) work and was link to the techniques section. After the virtual exhibition came out, a second version of this virtual reality was set up including human beings in order to explain the illustration techniques in more detail (the second version can be accessed through (*http://www.dortoka.com/ang/frameset_culture.htm*).

In spite of the good comments on this virtual exhibition, people demanded more dynamic and interactive elements in the web site. Besides, they expected some animations in some sections of the display.

With reference to the content, the virtual replica is an excellent complement to the real exhibition with new sections, which provide a minimum background previous to the visit as well as a resource for the after visit. The navigation structure allows an easy access to any book illustration, and also users can jump to other transversal sections. It is believed that this web site took full advantage of the hypertextual format for no-linear access to information.

The exhibition had a wide dissemination thanks to the local press as well as the University, ICUB (*http://www.bcn.es/icub*) and Instituto Cervantes (*http://www.cervantes.es*) gateways. Unfortunately, the international dissemination was scarce since again, the specialized Museum gateways do not seem to attract many visitors. From the 10th April to 20th June, 1202 people had consulted the web site, which means a daily average of 20 people.

M'ESCRIURÀS UNA CARTA? (WILL YOU WRITE ME A LETTER?)
(*http://www.uoc.es/humfil/carta*)

This exhibition organized by the Museum d'Art de Girona is focused on the history of letters, the way people communicates by mail. It started on the 27th May and it will close in October this year. The work conditions of this display were also quite hard, since the textual and image material arrived only two weeks before the opening, which become hindrance for the whole project. Neither the design nor content was fully elaborated, since there was never enough time.

A series of complementary sections were created, which include biographies of characters mentioned in the text as well as some recommended links on the subject. Besides, a few virtual reality clips were defined as introduction to the diverse exhibition sections. In this case, the virtual reality did not have the didactic function, but only a illustration and alternative navigation menu. One innovative aspect of this web site is the employment of animations (created with Dreamweaver) for the main entrances of the diverse sections even the main page. These dynamic images have been highly valued by the users, though they only fulfill an aesthetic function.

With regards to dissemination, the web site has a serious handicap since it is only available in Catalan for an expressed desire of the organizers. Therefore, it is chiefly destined to local public. The group Òliba believes that this was a mistake, because Internet allows to access to any public even from a long distance. Now, the unique limits we put to this global access are basically languages and technical obstacles. So far more than 700 people have accessed to the web site, however 95% of them are Catalan.

THINGS WE LEARNED AND "NEW YEAR" GOOD PROPOSALS

The group **Òliba** still has new on-line exhibitions and projects under way[6], in

[6] At the time of writing this paper we are fully involved in developing a couple of virtual exhibitions with local an national institutions. One of them is a virtual tour through Barcelona Botanic Gardens that shows 8 Mediterranean climates all over the world from Chile to California, which is due to be on-line in July 2000. The exhibition complements the presential visit of these amazing gardens. A second project pretends to complement an itinerant exhibition about the Spanish exile after the Civil War (1936-9), focusing on how these children saw the war and the exile and how it affected their lives (it will be on-line in October 2000). Finally, the third project is a gateway more than an exhibition and will show the natural, ethnographic and

which we expect to apply all things that were learned from these previous experiences. First of all, the elaboration of a web site as complement of a presential display is quite complex, so it requires time of preparation and design, at least three months in advance. Simplicity in design and structure facilitates that people may navigate without extra assistance. As a rule of thumb, virtual exhibitions should be kept as simple as possible in terms of navigation structure. Animated images appear always attractive to the users even if they do not have special use, whereas interactive sections are not only appreciated but also quite useful.

Taking full advantage of the hypertextual format of a web site means creating new transversal sections, which complement the real display as well as permit alternative navigation to the resource. Probably, the most rewarding aspects of the virtual exhibitions are those complementary sections that provide a background to the users, increase the exhibition documentation (i.e. text, images, video, audio) overcoming all the problems that real display may have to face and deal in more detail secondary aspects. Those alternative navigations may fulfill the different needs of the myriad of potential users. Sometimes, they supply the contextual information required to fully understand the meaning of the real exhibition, so it is a quite useful background.

---

artistic heritage of a very small Pyrenees valley in Spain called vall de Boi. This valley contains more than 12 well preserved Romanesque churches (XI-XII centuries) with extraordinary frescoes, that is why it has been recently declared World Heritage by the UNESCO.

Finally, we would like to highlight the didactic potential of virtual reality, when the environment is properly chosen. It simplifies the communication between Museum curators and visitors, chiefly if it includes human beings, which make more understandable techniques or how machines work. More emphasis should be put on these virtual reality contexts, which can be of great help to reconstruct lost worlds and environments.

REFERENCES

1. Carreras, C. (2000a) "Nuevas tecnologías en Internet para la difusión del Patrimonio". En *Actas de TecnoArt 2000*, Barcelona.

2. Carreras, C. (2000b) "Virtual exhibitions: extending Museum potential". *Cultivate Interactive* 2 *<http://www.cultivate-int.org/issue2/virtual>*

3. Keene, S. (1998) *Digital collections. Museums and the Information age*. London.

4. Keene, S. (2000) "Museums in the Digital Space". *Cultivate Interactive* 2 *<http://www.cultivate-int.org/issue2/space>*

ABOUT THE AUTHORS

Cèsar Carreras Monfort is Lecturer in Archaeology and Computer applications from the Department of Humanities at the Universitat Oberta de Catalunya (UOC). He is DPh in Archaeology from the Universities of Southampton (UK) and Barcelona (E) and MSc in Computer applications from Southampton University. He is specialist in Roman archaeology (three books and few papers) and Heritage dissemination through Internet (leading group Òliba and other projects).

Dept. of Humanities
Universitat Oberta de Catalunya
Avda. Tibidabo 39-43 08035 Barcelona (SPAIN)
E-mail: *ccarreras@campus.uoc.es*

Cultural Heritage, New Media and Education

Discovernet©: building an education gateway

Lee Adendorff (*)

(*) Website Coordinator, Australian Museums OnLine
E-mail: leea@amol.org.au

ABSTRACT
Australian Museums and Galleries OnLine (AMOL) recently launched Discovernet©, a learning gateway to Australian cultural institutions and the quality educational resources they produce. Research for the development of Discovernet© showed that while children under the age of 14 used the Internet regularly, few online education products were designed specifically for this younger age group. The interface and schema for Discovernet© have been greatly influenced by the idea that young children are active users of the Internet and a portal such as Discovernet© has enormous potential to promote museums and museum objects as places and tools of learning and discovery.

Discovernet© comprises four sections: an interactive game called *Make your own exhibition* that uses real museum objects to introduce children to the curatorial process; a national database of museum education resources called *Study Booster* that incorporates specialist education metadata; a museum guide for students and teachers; and an *Australian Tales* section that presents historical or social contextualisation of the museum objects in AMOL's distributed national database, tied to curriculum objectives.

This paper looks at the advantages of the gateway model for museums' educational content on the web and the issues encountered in the building of Discovernet©, including the importance of active learning strategies in online education and the application of specialist education metadata.

Discovernet© was launched in conjunction with a web cast from a megafauna fossil dig in the Australian desert.

KEYWORDS: education, gateway, discovernet, active learning, education metadata, curriculum objective, children, students, interface design

INTRODUCTION

Australian Museums and Galleries OnLine (AMOL) is an established Internet gateway to Australia's cultural institutions. AMOL is a collaboration between government and the museum sector, and is currently based at the Powerhouse Museum in Sydney. AMOL serves the museum community through a series of professional online resources and tools, and the general public through a comprehensive online guide to Australian museums and galleries. Access to many of the nation's collections is also provided through a distributed national database of collection records that is continually growing.

AMOL's attention has turned recently to the education sector - one of the biggest visitor groups to museums and a consistent consumer of online resources. Almost all Australian schools, libraries and universities are now online and the Internet is widely accessed by children at school and at home. 95 percent of Australian children aged 5 – 14 have used a computer in the last twelve months and 46 per cent have accessed the Internet in the last twelve months (ABS, 2000). 46 per cent of Australian households with children under the age of 18 have access to the Internet at home (ABS, 2000).

Museums produce a wealth of dedicated educational products apart from their core exhibitions. These include publicly accessible databases, websites, online exhibitions, interactive material, travelling exhibitions, 'exhibitions in a box' and teacher and student packs. (Museums also importantly have human education resources, such as education officers, curators and researchers.) Even though most Australian museums have their own website, cross-institutional searching for complementary educational resources is problematic as metadata is in many cases sketchy and not particularly consistent with emerging Australian and international standards.

Discovernet© was conceived as a means to provide standardised, national access to the quality educational resources Australian museums produce and to promote museums and museum objects as places and tools of learning and discovery. A baseline user group of 8-12 year olds was drawn from research indicating that although young children regularly access the Internet, few museums are developing online products targeting children in younger age groups (Sumption, 2000).

We did not have the resources to build a gateway that could be all things to all children, and we have used this mid-range age group more than anything as a departure point for the exploration of an education gateway model.

Together with this target age group, three important features have informed the development of Discovernet©: active learning strategies; access and interface design, and quality data linked to curriculum objectives.

ACTIVE LEARNING – BALANCING EDUCATION AND ENTERTAINMENT ONLINE

The shift in recent years from didactic learning experiences to active learning in museums, where visitors can make their own meaning through interaction with objects, has had a significant effect on online museum education (Sumption, 2000). Undoubtedly the most important aspect of this shift for those of us designing products like Discovernet© has been the focus on interactivity, where children can make their own choices and possibly engage in creative play at the same time as retrieving information.

One of the most fundamental challenges of Discovernet© was not only to create an effective 'portal within a portal' (sitting as it does within AMOL's larger framework as an online gateway to museums and galleries) but also to achieve a balance between its educational and entertainment values.

Figure 1: Discovernet© homepage

We made extensive reference to existing active learning 'typologies' such as:

- Education meta-centres – where data-rich resources such as large databases are made available for comparison, investigation and evaluation;
- Expository teaching resources – highly structured computer-directed activities presenting a set of principles, laws or theories (or in our case, curriculum objectives); and
- Creative play – encouraging children to explore and test ideas through creative activities and the act of making. (Sumption, 2000)

While a combination of these typologies features in all four sections of Discovernet©, the last of these - creative play - features most prominently in the interactive game *Make your own exhibition*.

Make your own exhibition – creative play for learning

Make your own exhibition is an introduction to the curatorial process and the making of meaning in museums, designed for 8 – 12 year olds. The game is explicitly tied to curriculum objectives such as interpreting text and understanding context, writing and investigation. Using real museum objects drawn from one or several museums, *Make your own exhibition* asks students to choose artefacts, read fact sheets and write labels to go with each object as if they were compiling a real exhibition. At the end of the game the exhibition is displayed to an animated crowd who pass in front of the exhibits making positive and sometimes humorous comments about the exhibition.

A Kookaburra character (a Kookaburra is a native Australian bird) plays the role of the senior curator, giving information and instruction and guiding the child through the game. While *Make your own exhibition* allows certain creative freedom in the choosing of objects and the writing of labels, the parameters of the game are also moderately prescriptive, addressing metacognition problems that can occur with younger users who are offered *too many* choices (Sumption, 2000). Flash animation and cartoon-like illustration have been used to enhance the idea of fun and an accessible interaction with museums.

Figure 2: A cartoon character of a native Australian bird is the children's guide through 'Make your own exhibition'

Like several other online games of this type, Wildlife art tales (http://www.wildlifeart.org/arttales/) and MOMA's art safari (http://artsafari.moma.org/safari_menu.html) to name a few, *Make your own exhibition* seeks to enhance an active learning experience with creative play, based on the idea that museum learning is created through the experience of real objects (Clarke, 2001) as well as by the choice and control of the experience (Falk and Dierking, 2000).

ACCESS AND INTERFACE – APPEAL VS. SPEED

The Internet provides a country as big as Australia with many opportunities for overcoming its geographical barriers to communication. A student living in Western Australia can access information held in a Tasmanian museum thousands of kilometres away via the Internet. Living in rural areas no longer means the only access to museums is through a physical visit to an urban centre. Although the Internet does not replace these real museum visits, there is considerable educational worth in this remote access both for the student and for the museum that has a responsibility to provide the best possible access to its collection and knowledge.

There is however a wide variation in technical infrastructure across Australia, and our use of multimedia had to be carefully weighed against issues associated with download speed.

A second fundamental access issue was the interface design. Although the site had to be quick to download, it also had to have considerable graphic appeal for children, limited amounts of text and text instructions, and, in many cases, use graphics to explain or navigate the site. After looking at hundreds of portfolios, we found an illustrator with diverse experience who produced a series of engaging characters and a 'look-and-feel' to accompany the site's activities and content. In many cases we have used these characters as a structural device – playing the part of peer-guides who give navigation or explanatory instructions.

QUALITY DATA LINKED TO CURRICULUM OBJECTIVES

Museums are uniquely positioned as authorities on the subjects of their collections and many already produce tailor-made education products linked to curriculum objectives. In an online world crowded with as much information garbage as treasure, the authentic content of museums' educational resources provides a critical mass of quality data for an information portal like Discovernet©.

Discussions with teachers in the developmental stage indicated that although they were very interested in museums' educational resources, many felt they did not have the time to seek appropriate resources for their class activities. Many saw an Internet gateway to these resources as a valuable, time-saving solution.

Constant reference to curriculum objectives was thus vital to the planning process and within the gateway model of Discovernet© this has shaped both the choice and the classification of resources.

Study Booster – Dublin Core extensions and specialist education metadata

The standardisation of complex information in large databases has thankfully become a global trend, driven by the development of the now widely applied Dublin Core (DC) metadata standards.

Like several other areas dealing with large databases of archived specialist material, the education sector has been active in developing extensions to the DC fields to cater for their specialist search requirements. (Dublin Core project, 2001) In Australia, the education portal EdNA, (Education Network Australia – a collaboration between government and the education sector), has produced its own DC-compatible standards for metadata. Version 1.1 of EdNA metadata was ratified in December 2000, and extends the core DC set with fields for education sectors, institutional levels and reviews. (EdNA, 2001) This allows resources to be catalogued down to a very specific keyword set for each school year, as well as levels of academic qualification, for example Certificate level I according to the Australian Qualifications Framework. Review and reviewer indicate the presence and identity of a third party review of the resource.

Study Booster incorporates the EdNA fields together with the standard DC set and an additional set of Australian Government Locator Service (AGLS) metadata standards. AGLS was developed in late 1997 as the resource discovery metadata standard to describe both online and off-line Australian government resources. This of course includes resources produced by Australian public cultural institutions. Although it seems cumbersome to have nearly 35 metadata fields filled for each resource, the use of all of these fields in *Study Booster* is important for a long-term compatibility with national and international standards as well as efficient retrieval of resources from a database that will eventually contain thousands of records.

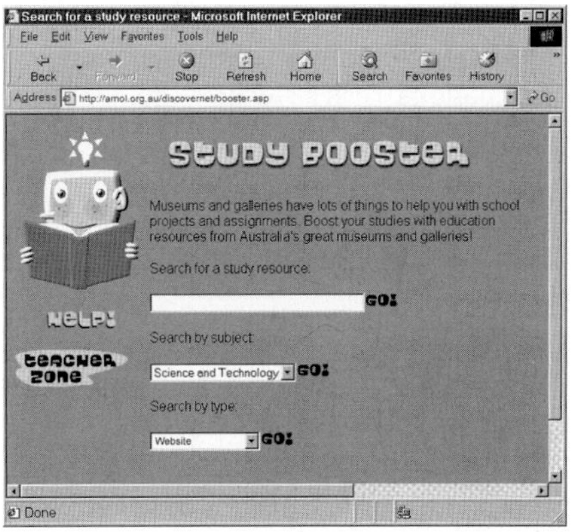

Figure 3: The *Study Booster* interface for students

Study Booster has much in common with databases of museum educational resources like the 24 Hour Museum's Curriculum Navigator (http://www.24hourmuseum. org.uk) and the Virtual Museum of Canada's Teachers' Center (http://www.virtualmuseum.ca/English/Teacher/). The target audience of school children for Discovernet© has however dictated the focus and design of *Study Booster,* which seeks to present a simplified and intuitive search interface for children.

From the first screen, children can use a text search, search by subject or search by type of resource (website, audio, video, print etc.). A secondary interface for teachers allows searching by curriculum links, educational objectives and relevance by school year. A classification schema for each resource was developed with the dual search requirements of both groups in mind.

Museum Locator and Australian Tales – 're-purposing' quality content for an education audience

The value of re-utilising quality content and data sets in different museum applications is well documented (Garzotto and Paolini, 1999). We found that this approach has many benefits when applied to the education sector and to databases that AMOL maintains and develops.

Re-utilising the content of our existing Guide to Australian Museums and Galleries, we developed a *Museum Locator* with an interface that resembles an alien spaceship console. Children can locate the nearest museum through an interplanetary search or search for a museum by type or collection. By the use of cartoon-like graphics to illustrate the idea that finding a museum could be fun and an adventure, we again hoped to add an element of play to an otherwise dry yet useful information resource.

Figure 4: Museum locator

As the Kookaburra in *Make your own exhibition* guides the child through the making of an exhibition, an alien character accompanies the child's search through the *Museum Locator.* The structural device of a cartoon character as informative guide has also broadened the education scope of a database such as this one. For example, when a result field appears after a search, so does the alien with a pop quiz, such as 'Do you know what a maritime museum collects? Do a search by museum type to find out!'

Australian Tales also employs the idea of 're-purposed' content, making use of a rich collection of museum stories developed by AMOL. These stories from museums and galleries are generated periodically when institutions place their collection records in AMOL's distributed national database. Their purpose is essentially value-adding to data sets, placing real objects from these databases in an historical or social context. The stories function as mini online exhibitions and often make use of some form of interactive multimedia.

Figure 5: Examples of AMOL's stories from museums – Golden Threads: the history of the Chinese in regional New South Wales, the Pandora marine excavation, and Indigenous oral histories from Far North Queensland

Conceptual and contextual frameworks are crucial to cognitive learning in museums (Ham, 1994). For Discovernet© to be truly valuable as a teaching tool, it was important that contextual frameworks such as those of the stories be relevant and related to curriculum objectives. *Australian Tales* seeks to do this through the addition of 'encyclopaedic' stories to the existing stories. These are brief contextual histories or stories linked specifically to syllabus topics that present relationships between objects in AMOL's distributed national database.

CONCLUSION

Discovernet© is very much a work in progress. While its attributes of active learning strategies, access, and quality content linked to curriculum objectives have been instrumental in shaping this portal, there will no doubt be modifications and improvements made to Discovernet© over the coming months and years. Hopefully its development and evaluation will add to the growing body of work into education gateways on the web, and appropriate online resources for children in younger age groups who we know are accessing the web both at school and at home on a regular basis.

Postscript: at the time of going to print, Discovernet© is being evaluated and tested in focus groups. This information will be presented at the ichim01 conference as a part of this paper and will be made available on the AMOL web site: http://amol.org.au.

REFERENCES

1. Australian Bureau of Statistics, *Use of the Internet by Householders, Australia*, August 2000

2. Clarke, Peter, *Museums Learning OnLine*, The Council for Museums, Archives and Libraries, 2001

3. Dublin Core web site, projects consulted May 2001. *http://www.dublincore.org/project/*

4. Education Network Australia (EdNA) consulted May 2001, *http://www.edna.edu.au/metadata*

5. Falk, J and Dierking L, *Learning from Museums: Visitor experience and the making of meaning*, Altamira, 2000

6. Garzotto, F and Paolini P, 'Using and Reusing archive information for multimedia applications', in *Selected Papers from ichim99*, 1999

7. Ham, Sam H, 'Cognitive psychology and interpretation: synthesis and application', in *The Educational Role of the Museum*,

Routledge, 1994

8. Sumption, Kevin 'Beyond Museum Walls: A critical analysis of emerging approaches to museum web-based education', *Museums and the Web conference proceedings, 2001*

ABOUT THE AUTHOR

Lee Adendorff is the Website Coordinator of Australian Museum OnLine. She has been working with digital media since 1994 in design and editorial roles.
E-mail: *leea@amol.org.au*

Zoom Media Lab & The Virtual Collection of Children's Everyday Culture

Sasha Pirker (*) *and Stefan Wagner* (#)

(*) Uma Information Technology, Vienna, Austria
E-mail: sp@uma.at
(#) Uma Information Technology, Vienna, Austria
E-mail: sw@uma.at

ABSTRACT
The Virtual Collection is a project by ZOOM Children's Museum, Vienna. It is a part of the new Media Lab, which will be opened in September 2000.
The collection comprises objects of popular children's culture - a field, which has, to date, been neglected in digital archiving. International cooperations with other Institutions are being targeted to achieve most possible variety. The collection will be accessible in the internet and serves research purposes.

KEYWORDS: children's museum, collection, new media, internet, research

UMA
UMA is a company specialised in interactive media environments and internet applications with widely recognised experience in the fields of conception, planning and application of large-scale 3D web applications. The company was founded in 1994.
UMA is represented in numerous public institutions of the federal government and the province of Vienna as well as in committees of the EU. In addition, UMA is recognized as a competent consultant for the internet technology sector. One of UMA's chief tasks is the development of digital, interactive environments for modern knowledge sharing. Novel concepts in knowledge transfer are thus developed in close cooperation with museums, universities, ministries and governmental institutions.

ZOOM Children's Museum
The ZOOM Children's Museum is to be the first children's museum located in one of the world's largest cultural centres. In autumn of 2001, the new museum quarter will be complete, and the children's museum will move into the new facilities comprising an area of 1,600 square meters.
One hundred thousand visitors are expected yearly.

The ZOOM Children's Museum and the company UMA have been working together in the implementation of new media in exhibits and workshops for quite some time.

INTRODUCTION
The Media Lab in the ZOOM Children's Museum in Vienna/Austria is to be the first permanent facility in a children's museum that confronts the topic "children and new media". Thus, the museum is setting leading steps in the field of education & new media, assuming an essential position as compared on an international level.

Parents, teachers and children discover a place which offers the possibility of taking part in playful activities and of contributing their own creative input achieved through the implementation of the most recent technology in connection with specially developed hardware and software.
Designed for school classes visiting the museum during the week, the Media Lab is divided into stations, in which the children reach a common outcome using teamwork.
Independent work, team spirit, respect and cooperation are promoted in the Media Lab in a playful atmosphere which encourages experimentation. The resultant works are then displayed on the web page of the ZOOM Children's Museum – the ZOOM World.

Part of the Media Lab is the Virtual Collection of the Children's Museum currently under expansion: with the reopening of the Children's Museum in autumn 2001, the collection will focus on popular children's culture, accessible in digitised form.
This archive represents the first basis worldwide for scientific observation of popular children's culture.
The Media Lab's specially developed stations along with the connection of the children's creations with the internet form an exceptional basis for cooperation with partner institutions.

THE NEW MEDIA LAB

Resulting from successful cooperation between ZOOM and UMA in the past, the idea of creating a permanent facility in the Children's Museum, which confronts the topic "Children and New Media" and "Virtual Collection" arose, becoming the Media Lab.
The Media Lab offers children the opportunity of entering a new world of experience and of discovering and living out their creativity with the newest media and technology.
Designed for school classes during the week and individual visitors on the weekend, the Media Lab appeals to parents, teachers and children, where innovative ways of learning are pursued in cooperation with scientists and artists.

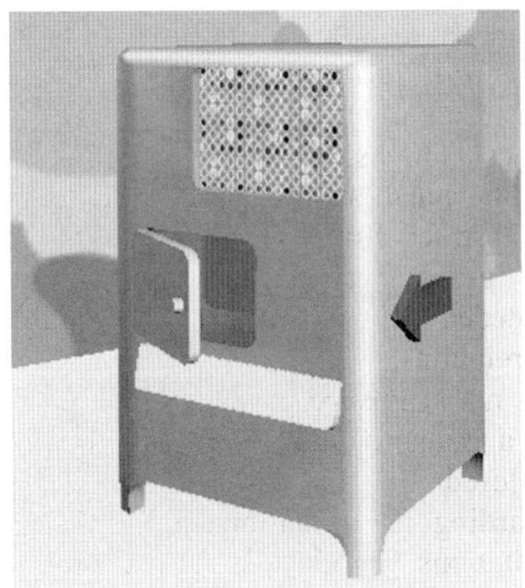

Trouser Pocket Scanner, model (input device for the Virtual Collection)

The Media Lab was designed as an interface between the real and the virtual world - in two aspects:
on the one hand, it offers the possibility of entering the world of virtual reality with actual objects produced in the centre itself and the technology offered therein, and, on the other hand, the results of the various classes are displayed on the web site of the children's museum, where they are accessible as a dynamic tool for further use.
The Media Lab is thus the production centre for and starting point of an exciting experience on the web.

Didactic Parameters and Reception

The goal of the Media Lab is to offer children the opportunity of becoming acquainted with levels of abstraction essential in meeting future demands, in a playful environment. Along with perception of the new "interface to the world", the children are confronted with the interface between man and computer or new technology.

The understanding of technical interrelations in the Media Lab not conveyed primarily through abstract ideas but through the active, that is constructive, encountering of cause and effect. This step consequently allows for the much-simplified comprehension of abstract structures, this means that the use of abstract terminology is preceded by the playful acquisition of correlations.

It is not a finished product which results but a technology potential, which can be explored by the children themselves regarding its limitations and qualities: limitations produced by the two-dimensional character of a surface or of the rigidity of predetermined forms and signs, as well as qualities, resulting from crossing the boundaries of this two-dimensional or linear form etc.

Children have the task of generating imaginative processes using curiosity and concentration and of tackling the challenges of interactive technology in an experimental environment.

THE VIRTUAL COLLECTION

The virtual collection comprises objects of popular children's culture - a field, which has, to date, been neglected in digital archiving. The collection serves research purposes and distinguishes ZOOM as a committed and professional institution internationally.

The Archiving of Objects With the Trouser Pocket Scanner

The first encounter with the Media Lab takes place in the foyer of the ZOOM Children's Museum with the "Trouser Pocket Scanner" (provisional title). The Trouser Pocket Scanner, specially designed to digitise objects to be added to the Virtual Collection, will be located in the foyer. In the future, further international museums are to follow as partner: they will also be provided with a trouser pocket scanner and make contributions to the Virtual Collection, thus creating a world-encompassing partner network.

The Trouser Pocket Scanner consists of an appealing casing, which disguises the scanner at first glance.

Implementation Scenario

By means of display and sound, the visitors are encouraged to open the door of the Trouser Pocket Scanner and to place their favourite objects on the rotating tray inside the device. The visitor then presses a start button and the scanning process begins. Simultaneously, questions appear on the display, which must then be answered (name of object, use of object, ...). Once the scanning process is complete, the device thanks the visitor, and a print out of the pictures taken in series emerges from a slot in the machine. The individual pictures can be glued together to produce a flip book. The Virtual Collection is presented on a large flat screen in the lobby and on the web itself. New objects are continuously uploaded and equipped with meta data by the museum staff. This is realised with the use of a database comprising a web interface.

Foyer ZOOM with Trouser Pocket Scanner (model)

Display of Objects in the Collection

In the collection each object is presented as single image.
With the object appears the relevant information the child has assigned to the object when scanning it: name of the owner, title and function of the object (e.g. good-luck charm). Also mentioned are the archiving criteria (index), the administrator has assigned to the object. Interested in a certain object the user has the option to switch from the 2d single image view to the animation mode in order to view the object from all sides.

All objects of the collection are administered in a database and thus can be presented in the internet in various structured ways. By ways of many different search critera the collection can be viewed: origin, intended purpose, etc.

THE MEDIA LAB

The Assembly Table

In order to allow a group of children to work with objects in parallel with one another, UMA have designed a new product, which allows teamwork via a digital input device, namely the Assembly Table.
The hardware consists of a table, onto whose surface screen contents will be projected from below. There are then Tracker Target Blocks on the table, each of which represents a digital object created by a child. If a child moves one of these Tracker Target Blocks, the existing picture will change in accordance with the movement. The special merit of this system is that users are not restricted to moving one object at a time (like on a monitor) but can simultaneously move multiple objects.

Individual Stations:

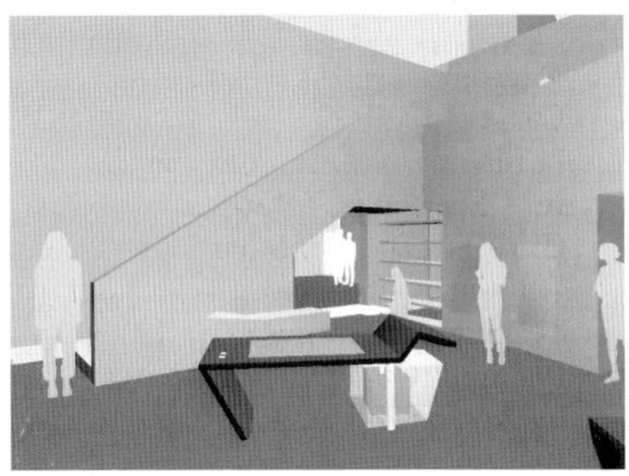

The ZOOM Media Lab (Rendering)

2D-Snapshot Animation

Cartoon-like 2D animations from hand-drawn and digitised objects and drawings are to be created. By moving the Tracker Targets children can put together their scene, and using multiple recordings can produce an animation. The end result is a moving picture animation, which can, for example, then have sound added to it.

The Virtual Classroom

A 3D class photograph of children in a virtual fantasy environment will be displayed. Individual pictures will be produced in groups, and digitally photographed in front of a blue screen. The results are then displayed at the Assembly Table. The end results are:

- An interactive 3D Stereo projection
- An interactive 3D world in web suitable format
- An interactive presentation in the foyer.

REFERENCES

1. Glos, J. and M. Umaschi. (1997) "Once Upon an Object: Computationally-Augmented Toys for Storytelling." Proceedings of ICCIMA '97 (International Conference on Computational Intelligence and Multimedia Applications), February 10-12, Gold Coast, Australia.

2. Negroponte, N., Resnick, M., and Cassell, J. (1997). "Creating a Learning Revolution." In Learning Without Frontiers (UNESCO) Technology and Learning Portfolio. Reprinted in Teaching, Learning and Technology Planning Guide CD-ROM published by Apple Computer.

3. Goldman-Segall, R. (1992). Collaborative virtual communities: Using Learning Constellations, a multimedia ethnographic research tool. In E. Barrett (ed.), Sociomedia: Multimedia, Hypermedia, and the Social Contruction of Knowledge(pp.257-296). Cambridge, MA: The MIT Press.

4. Ishii, H., Kobayashi, M. and Arita, K., "Iterative Design of Seamless Collaboration Media," Communications of the ACM (CACM), Special Issue on Internet Technology, ACM, Vol. 37, No. 8, August 1994, pp. 83-97.

5. Kafai, Y., and Resnick, M., eds. (1996). "Constructionism in Practice: Designing, Thinking, and Learning in a Digital World". Mahwah, NJ: Lawrence Erlbaum.

The Museu da República Data Base – a creative tool to educate and disseminate information

Ana Cristina Léo Barcellos Evres (*)
and Karla Estelita Godoy (#)
(*) Museu da República, Rio de Janeiro (RJ), Brazil
E-mail: aevres@hotmail.com / tecnica@museudarepublica.org.br
(#) Museu da República, Rio de Janeiro (RJ), Brazil
E-mail: godoy_karla@hotmail.com / bdr@museudarepublica.org.br

ABSTRACT
The strategu used to show the Museu da República's collection in a simple and attractive way, where collaborative work is incentivated and new proposals for heritage education are offered, includes basically experimantation and creativity. The process of changing established mentalities took a long time. Today the result of this system can be seen in www.republicaonline.org.br/

KEYWORDS: Brazilian history; Cultural heritage; Educational process; Information system

INTRODUCTION

The Museu da República, located in the Palácio do Catete, was head office of the Brazilian republican government for 63 years. It was created in 1960 to preserve the Brazilian republican memory, and has become through the years an important technological center of knowledge and education.

Figure 1: Museu da República (by Alex Ferro)

BRAZIL – A COUNTRY IN SEARCH OF DEVELOPMENT

Brazil has entered the 21st century with a population of 169 million people distributed over an area of 8.511.965km^2. More than two thirds of the population is living in the cities and 90,5% of these people in the southeast region, while the rest of the population is spread disproportionately among the other regions. Education is one of the greatest problem that brazilians are having to face in order to progress. Illiterates comprise a large part of the population and lots of people are still unable to have a job or even attend school. Only a few people have real access to opportunities, but even amongst these fortunate people one could find big differences which reside in the cultural aspects of the north and south part of Brazil. Whereas in the south the European and North American values predominate, in the north one can still find ways of living primitive.

Figure 2: Children of the Brazilian rural zone (Museu da República´s Collection)

Nonetheless, the difference of values and in the ways of expression of the cultures of the northern part and the southern part disappear when we take into account their richness and diversity, expressing not only one identity, but several. Therefore, one could say that Brazil has many cultures, instead of only one.

And these many cultures provide a large amount of artifacts, objects, documents and collections gathered as time passes, comprising a valuable cultural heritage kept in many museums. Brazil has over one thousand museums responsible for the memory of those cultures, but unfortunately disproportion between regions results in not preserving and diffusing them in the most appropriate way.

THE TECHNOLOGY ABOLISHING FRONTIERS

In Brazil, the number of PC's users has been rising over the years and there are today 10,7 million internet users. Most schools have added computer classes to their curriculums and this has helped immensely to abolish frontiers between the wealthy and the poor sectors of Brazil's society.

Museum collections are the primary sources of information with which elementary and high school students are not yet familiarized in Brazil. In order to overcome this handicap, the Museu da República has developed an educational tool that presents them the collections, documents and books, all inserted within a pertinent context. Information technology has become the most important ally able to democratize and extend the access to the collections spread all over Brazilian territory.

CHALLENGES

The Data Base was conceived as a management system that gathers three types of collections and provides access to information regarding more than 8.000 museologic objects, 80.000 documents from the Historical Archive and 20 thousand bibliographic records. The Data Base enables a joined recovery of informations. Other crucial aspects considered in the creation of the Data Base were: flexibility, simplicity and consistency in retrieving information.

The data input – Museu da Republica Data Base

The first great challenge was to gather in only one system distinct collections. There were many disputes concerning subjects and it was necessary to confront the library rules against new organizational methods for historical documents and objects.

Rules for data input, such as titles and author's names, were adapted to the reality of the Museum collections. Nevertheless, the specificity of each area was kept. The system allows an assured data input for historical archives, library and museology that keeps their integrity and respects the international rules.

Interdisciplinary

The second great challenge was to find the common information among the Museum collections. The accuracy and speed of the recovery of information depended on the correct inter-relation of the data. It was defined that the common points among the three areas would be subjects, authors, collections and places. From these common points records were created and used for the three areas (Historical Archive, Library and Museology) to guarantee the reliability and integrity of the data. The records are: Geographic Areas, Collections, Institutions, People and Subjects. Each of them allows the inclusion of biographies and entries providing a fuller vision to the users.

A simple, attractive and friendly language

The third great challenge the project faced was the adoption of a simple and attractive language able to be used by the largest variety of users. A manual was developed to help, step by step, the professionals in charge to feed the data.

An open system

The system is open, flexible and modular. It works on simple PC's (with a minimum of 450 or 500MHZ speed). The operating system uses Microsoft SQL Server and Microsoft Windows NT.

Because of its adaptabilities – new data fields can be included if necessary – so that the system can be used by the others Brazilian museums with similar collections.

The CD-ROM that describes the methodology used is being concluded until the end of the year.

CULTURAL HERITAGE: PRESERVATION THROUGH EDUCATION

From the Data Base to the Website

The Website *Getting to Know Museu da República* works as an interface to provide the access to the Data Base. Its construction brought out new challenges. After creating a system able to retrieve information from three different collections – historical documents, publications and objects – it was necessary to face problems of division and fragmented collections.

Even though the Museu da República was created to gather and preserve the Brazilian Republican memory, due to political reasons, historical documents and objects are sometimes located in other institutions. This could spoil the way of understanding the collection as a whole. But more than offer large lists, the Website was conceived to combine context, information and a friendly interface. In order to make this possible, the museum collections have been strategically classified according to parameters defined for elementary and secondary school, making identification of subjects easier to teachers and students.

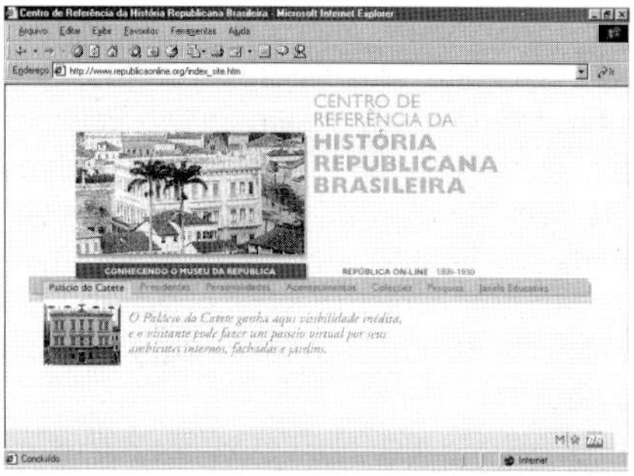

Figure 3: Access screen to the Website - Getting to Know - with the presentation of subjects

Defining themes

Palácio do Catete

The Palácio do Catete and its garden, which constitute the Museu da República, is one of the largest Website navigating themes. Constructed about 1860 by the richest coffee farmer during the imperial time, the Baron of Nova Friburgo, the complex is considered an exponent of eclectic Brazilian architecture, mixing artistic european style – as Neoclassicism, Manneirism and Baroque – and elements from other historical times. From its wealthy decorated rooms, came decisions that defined the destiny of nation, such Brazilian participation in Second World War and the creation of labour laws.

When the Palácio do Catete was selected as a Website navigating theme, the main idea was to show how important this construction is to retrieve information through objects and documents on display and its own architecture.

Presidents' Collections

As the Head Office of the Brazilian Republican government from 1897 to 1960, the Palácio do Catete was also a dwelling place for most of the 18 presidents that resided there. Thus, Palácio do Catete was known as "the presidents' home".

Therefore, there was an expressive collection from the public and the private life of the presidents that was big enough to constitute one of the Website navigating theme.

Events

Another important peculiarity of the Museum is its current subjects. As we live in a Republican regime, through the especific events like civil wars or popular and political movements we can recognize facts from ancient and modern history of the country.

Public figures and other collections

These themes follow the same principle, mixing ancient and modern collections. Some of the outstanding historical characters and other significant collections were chosen to show how relevant, distinct and comprehensive the heritage is from the Republican regime.

The website structure

The division in themes makes it simple and attractive to navigate in the Website. Users can find a sort of helpful options that liberate them from having a large knowledge about the Museum collection to get information.

From a technical view, all of these themes became queries that define the system of retrieval.

The system is dynamic, updated and whenever the Museum acquires new collections they can be input in the Website.

For experts users, the Website provides a section for free queries as well. The introduction screen presents basic information about the system's operation and there are also suggestions in orther to help the user find more about the subject chosen.

Get to Know in order to preserve

The Website has also a section called "Educative Window", specially designed for the young ones. This window is an educational tool developed by a multidisciplinary team with the purpose of creating educational activities using technical and historic resources. Students and teachers will find interdisciplinary activities based on Museu da Repúbica Data Base.

A combination of texts, images, colors and sounds excite senses and the imagination of the youngsters taking into account the multiple knowledge of the student, and providig the teacher with a tool to work subjects thar were thought as boring in the old times. Students and teachers are invited to interact and exchange experiences through the site.

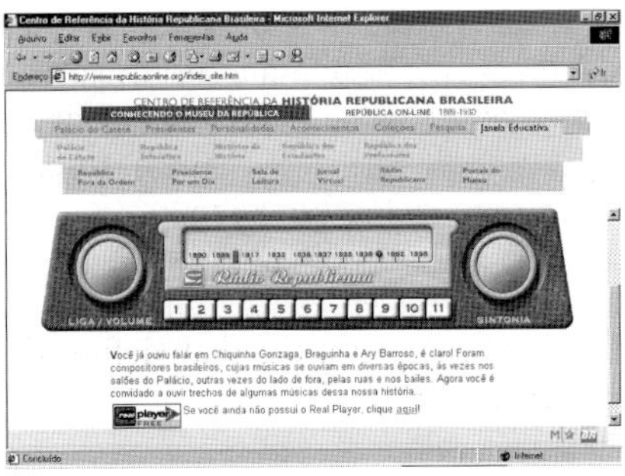

Figure 4: Screen of the activity *Republican Radio*

The Educative Window's performance has revealed multiple possibilities on interchanging informamation between the Museum and the schools, and specially among teachers who use it as a virtual board on classes. This relationship allows another view about the collections in general and a better use of it.

Paulo Freire [1], one the most prestigious brasilian educator formulated a theory taking into account the mode of local life of the student, raising the value of the culture he lived in, its singularity and identity. As far as Freire was concerned, the main role of education was liberate people, making them able to take their own decisions about their actions and play an active role in world. According to him, the access to information was the only possible way to get a complete education so that students could become able to change the present and the future.

At Museu da República, we are doing our best to perpetuate his way of thinking.

Staff

The creation of the Data Base and the Website was possible thanks to an efficient interdiciplinary staff that never gave up the dream of building a better for Brazil, bringing equitu and democracy to people on general.

CONLUSIONS

Cultural heritage: using to preserve

The responsibility of preserving cultural heritage becomes stronger when it is shared with society, that understands it better and begins to respect it. According to the philosopher Michel Serres, we can only attain equity among people, inventy a social tissue that minimizes violence, pacify the world and attain freedom throughout education. And he believes we should never give up the project of sharing knowledge and information. In his own words, "in order to save cultures from annihilation we have only used up to now museological preservation, protections that kill instead of promote them by reinforcing a touristic mass use of them". [2]

The concept of the Website *Getting to Know Museu da República* is based on three principles: using to preserve, giving access to information and developing creative educational activities.

But more than this, its ambition was to help to form citizens more conscient and prepared to share knowledge in this globalized society.

REFERENCES

1. Freire, Paulo. Pedagogia da Autonomia. Rio de Janeiro: Paz e Terra, 6ª ed., 1994.

2. Serres, Michel. Atlas. Lisboa: Instituto Piaget, 1994, 172-173.

3. Johnson, Steven. Interface culture: how new technology transforms the way we create and communicate. 1st ed. San Francisco: Basic Books, 1997.

4. Lévy, Pierre. As tecnologias da inteligência: o futuro do pensamento na era da informática. Rio de Janeiro: Ed. 34, 1993.

5. ________. As árvores de conhecimento. São Paulo: Editora Escuta, 1995.

6. ________. O que é o virtual? São Paulo: Ed. 34, 1996.

7. ________. Cibercultura. São Paulo: Ed. 34, 1999.

8. Negroponte, Nicholas. A vida digital. São Paulo: Companhia das Letras, 1995.

9. Minnesota Territory. *http://www.mnhs.org/places/historycenter/exhibits/territory/* June, 1st, 2000.

10. Museo del Prado *http://museoprado.mcu.es/* June, 1st, 2000.

ABOUT THE AUTHORS

Ana Cristina Léo Barcellos Evres: Curator and manageress of the Museu da República Data Base and the Museum Website, MSc in Museology and assistant to the Direction of Museu da República for the last three years. Co-ordination of Museum's editions and manageress of the documentations, research and preservation of the cultural heritage.
E-mail:
aevres@hotmail.com,
tecnica@museudarepublica.org.br

Karla Estelita Godoy: Curator, MSc in Museology and Co-ordinator of the Museu da República Data Base and the Museum Website.
E-mail: *godoy_karla@hotmail.com*
bdr@museudarepublica.org.br

Integrating Cultural Heritage Activities in Education Process: Experiences and Perspectives through New Technologies

Eugenio Morreale (*)

(*)Dipartimento di Informatica, Università di Pisa, Italy
E-mail: morreale@di.unipi.it

ABSTRACT
In this paper we first outline why pupils should be given the opportunity to visit and have direct contact with the cultural heritage. We then propose some 'didactic itineraries' organized by museums and other cultural institutes which are aimed at specific subjects and ages of pupils. We illustrate the main organizational aspects and the people involved in the various stages of the project. In particular, we refer to the Didactic Communicative Intermediator (DCI) who is the main person who guides the communicative, psycho-pedagogical and cognitive needs of the pupils. Through the DCI, pupils can acquire the target concepts and notions of the itinerary and can integrate these with those developed in the normal classroom activity of the pupils.

KEYWORDS: Internet based educational systems, computer assisted learning and instruction, authoring tools and methodology, collaborative knowledge construction using networks and the WWW.

INTRODUCTION
Nowadays schoolchildren are bombarded with all sorts of information in all kinds of forms by all types of media. In such a context, it is interesting to consider how the new media and technological supports may favour direct contact with museums and other cultural institutes, in the sense that these are real as opposed to virtual environments, which contain vast amounts of documents and examples through which (and through the musem staff) we can teach pupils a wealth of notions.

Involved in all this, is the emotional aspect of what the pupils actually feel when face to face with new environments, new objects and subjects, and new people. With the right preparation by the teachers, this leads to new experiences and a new way of learning; and this becomes a particularly enriching experience if well integrated into the daily classroom activities. One way to ensure this integration is the use of well designed and realized 'didactic itineraries'. These are specifically designed and realized for particular subjects and age groups and made available via the museums computer/electronic channels. They act as a base for the preparation, carrying out and 'conclusion' by the teachers and the pupils of the visits and a meaningful integration of the whole experience.

Using our recent experiences in this field, this paper will attempt to highlight:

- An example of design of a

dedicated didactic itinerary ("Mammals") formulated by a natural sciences museum for primary school pupils, in order to evidence the complexity, richness and importance of well conceived direct contact for/with the pupils on that basis,
- the overall potential offered by the latest technologies in the development and production of didactic itineraries in addition to various technical / organizational aspects. We will also focus on the people in charge of conserving the cultural heritage in terms of the design, development, production, documentation and maintenance, etc. of such didactic itineraries;
- The didactic aspects and the "didactic communicative intermediator" (DCI), whose role is to coordinate the comunicative psychological and cognitive impact towards the pupils and involve teachers in integrating everything into the scholastic activity,
- Some further considerations and conclusions.

DIDACTIC ITINERARIES: A CASE STUDY

Some of the main points why school visits and direct contract with museum and other promoters of the cultural heritage help to enhance the didactic potential are:
- such experiences can be *integrated* into the everyday school activities, and thus
- characteristics according to which the *visit is prepared and carried out*, and thus also
- *communicative, psycho-pedagogical and didactic bases*, according to which the didactic itinerary was conceived, designed and carried out.

If these factors are not fully considered then the visit merely becomes an escape from the usual school routine, which only superficially touches the pupils and leaves them nothing concrete to take away. This may be the result even if experts are on hand, who although having the best of intentions may use too technical language or refer to complex ideas that go beyond what the pupils can deal with, or other psycho-pedagogical factors (see following point) that the expert may ignore and neglect.
As an example of a possible approach toward the design, structuring and offering of didactic itineraries, below we summarize the fundamental psycho-pedagogical and the main structural aspects of the **"Mammals"** itinerary [1, 5] in its original version, which was conceived and designed for primary schoolchildren (~8-10 years) by Berni and Zuffi, who have primary school teaching experience (as Didactic Communicative Intermediary DCI, see next Sect.) alongside research in zoology (as Expert in Specific Fields ESF, see next Sect.) respectively, at the Natural History Museum of the University of Pisa (henceforth, the Museum) under the direction of Prof. M. Franzini.
For a detailed description of the design and implementation of the "Mammals" didactic itinerary, see [1, 5]. Here we summarize the main elements of its psycho-pedagogical and didactic structure and the activities suggested for (teachers and) students.

Psycho-pedagogical bases of the itinerary

The main aim was to help children to understand, in this case through observing the skeletons of 10 species of mammals, the relationship between the variances found and the functional adaptation needs to the environment. The starting and reference point is a human skeleton, ie the skeleton of the children themselves. This is used to show the children how they perceive it and how they schematize such perceptions. The children then compare this human skeleton with the skeletons of the mammals, again on a perception/sensory level. This enables them to hypothesize on why changes have come about in the structure of the various mammals.

Understanding how and why mammals diversify and are transformed as a reaction to the environment, helps children to focus on natural processes (above all in the animal world), and offers a dynamic perspective on the interactions between life environments and the physical characteristics of the organisms.

The itinerary is based on three main factors: (i) establish what the children already know; (ii) limit the work to mammals, specifically those which can be compared to a basic skeleton structure (10 animals); (iii) get children to give plausible explanations for these variations in terms of function, and adaptation to the environment, without any quantified time references.

This rationale entails radically altering the traditional approach to museums, where visiting classes are accompanied by an 'expert' who imparts knowledge, often in overly technical language, with questions being left to the end. Instead in our case, it is the children who ask the questions from the outset, the role of the expert being to guide the children in their discovery of the distinctive properties of the various mammals.

The underlying concepts of this didactic itinerary derive (i) from Piaget's ideas [8] on the formation of thought in children, and later studies on its 'magic' tendency [2], (ii) from research [9] into conceptual maps in children's thought, and (iii) from studies on preconceived ideas that may condition learning. A child's preconceived ideas on any subject covered in a syllabus must be known, so that teachers will ask the 'right' questions. In answering, the child will objectify his/her knowledge and will thus interact with the subject of learning. This interaction should modify any erroneous preconceptions, thus leading to a correct, stable, and lasting knowledge. When children are free to test out their own ideas and to compare these ideas with their companions they will be able to build their own knowledge through this social interaction. In this case the teacher takes on a new role as stimulator and mediator.

Didactic structure

The planned main Objectives and Activities and their detailed stages can be summarized as follows.

***Objective**: distinguish between mammals and non mammals*

Activity: children (i) divide pictures of animals into sets of mammals and non mammals using their own reasoning; (ii) compare criteria underlying their selection; (iii) if necessary, integrate their knowledge with simple information;

***Objective**: find out which animals were really recognised by the children*

Activity: children (i) write a free list of names of the animals (mammals and others) that they claim to know; after some time, (ii) teacher gives the list back to the children, and (iii) get them to write down what each animal eats and where it lives;

***Objective**: investigate what perception children have of their own skeleton within the body scheme*

Activity: children (i) do a

spontaneous and free drawing (without giving them any clues); (ii) get children to touch their body, telling them verbally which specific parts to touch, children now (iii) do a spontaneous and free drawing; (iv) use physical exercises so that children get a better perception of how their bodies are articulated; (v) children draw a skeleton, within the outline of a human body, on the basis of knowledge learned in the above physical exercises.

Main activities proposed to students

Bearing in mind the activities, the operating phases, and the worksheets outlined in previous Subsections, we will now describe only some of the pupil activities.

Maze game. In this game children find themselves in front of a maze in which there are animals - mammals and non mammals. The aim of the game is to follow a path along which there are only mammals. This is not that simple since when the child makes a mistake he/she is not warned of this immediately but only at the end.

Formation of sets The page on which the activity takes place shows two sets, one for the mammals and one for the non mammals, and a number of animal icons with the name of the animal underneath.

Skeleton colouring. On one side of the screen is a complete human skeleton and on the other side a dismantled skeleton. Pupils have to colour (using a colour palette) the skeleton parts using the same colours as in the complete skeleton.

Composing skeleton. The human skeleton is composed by dragging the individual elements to the outline of the human body. If the elements are not located correctly then they are automatically rejected. The pupil gets three chances, after which the positions are recorded. If the pupil gets the right result he or she is free to display the four main positions of the composed skeleton (from in front, behind, and from both sides) or to see a 3D animation which shows the rotation of a human skeleton.

TECHNICAL AND ORGANIZATIONAL ASPECTS: THE MAIN ACTORS

On the basis of the didactic itinerary outlined in previous Section, our subsequent experiences [6] represented a consistent evolution in fully exploiting technological potentialities in the field of education through cultural heritage activities, related to both the following points:

A) Whole System Conception

1. we broadend the overall consideration of the technical infrastructural tools with which to equip a Geopaleotology Museum realized in the Lerici castle, as for example with: robotics, seismic platform for earthquake simulation, etc. [4],
2. we structured the whole system architecture on a client-server basis,
3. we integrated the Hypermedia, Data Base for data management and storage (in order to give stability, continuity and permanence to the scheduled applications) and Communication Network technologies by introducing dynamic web pages construction for answering remote users requests,
4. use of sophisticated data, eg 3D interactive graphics, to enhance the communication of cultural heritage specimens (various graphic- 3D models, animations, etc, with consequent possibilities of:"navigation" in / "manipulation" of 3D models)

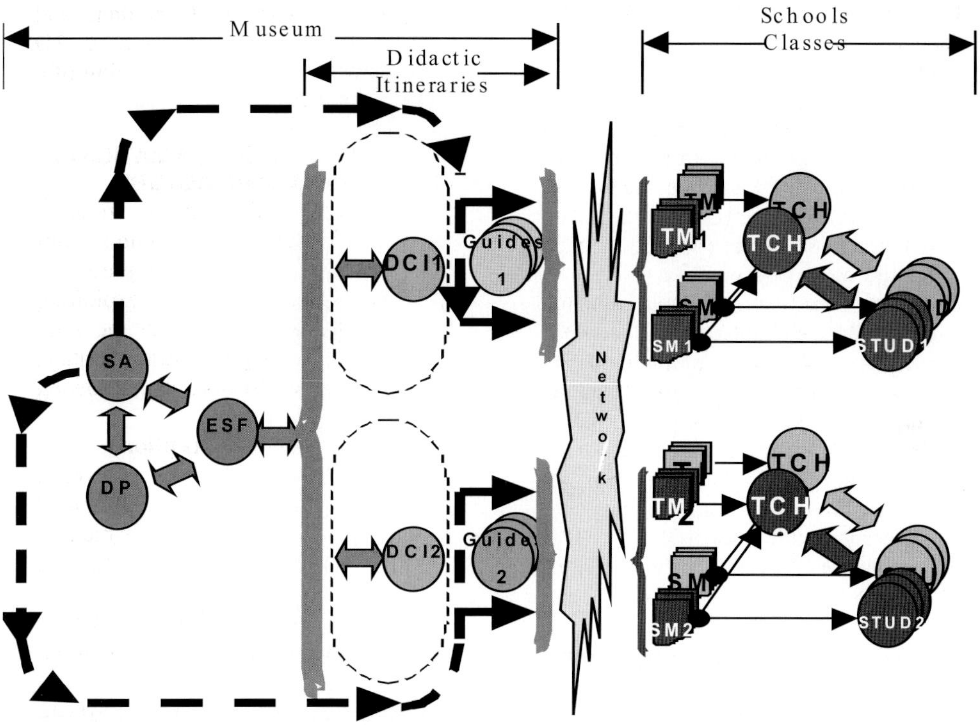

Fig. 1 People involved, Processes, and Interactions in Museum -School Applications

B) Specific Methodology on main Technical and Organizational aspects so that, thought their joint application, new paradigms and organizational structure of technologically aided activities for:

1. organization, conception, design, development, offer, documentation, update and maintenance, etc of didactic itineraries;
2. coordination, connection, maintenance and contacts with: (i) schools / classes, (ii) publishers / textbook authors;
3. new services allowed by the technology in the field of the

rich and meaningful innovations could be expressed and realized, as for instance,

cultural heritage as for example those concerning:

- contacts with other Museums, institutes dealing with cultural heritage,
- services for schools, teachers,
- coordination-Links offered to textbook authors and school publishers,
- coordination-Links with local organisation for administrative school functions, touristic promotion, etc.

This was further developed [6] by: integrating specific functionalities of Hypermedia, Data Base technology and Telematics, i.e. remote network communication, in order to:

1. give continuity and stability to the applications via an open approach to the gradual enriching and updating of the presentations through dynamic (most recent) data alignment within web page building;
2. allow for the accumulation and adaptability of the data;
3. enhance and make more involving the communicative and didactic potentialities of the informative materials via interactive access;
4. facilitate the identification of relevant data and means of presentation specifically for educational purposes, via various levels of data definition and organization;
5. widen the possibility of adapting the materials to various uses by different typologies of: i) users (students, teachers, guides, museum workers, visitors, specialists, etc); ii) application environments (schools, training, distance education, diffusion of culture, etc); iii) means of exploitation (local, remote, deferred access), etc, through a proper and rich schema and subschema data structure;
6. facilitate the re-use of the same information in different applications, in order to reduce production and maintenance costs, in particular in relation to expensive data, eg 3D interactive modelling.

The original version using 'traditional' data (texts, images, films and animations), was then extended to 3D interactive graphics to enable users to "work" with the models and examine them in detail [6]. The system was tested on widely different applications:

- on historic buildings and architectures, (historical aspects, materials used, building techniques, degradation, restoration, etc.) and taking down and rebuilding the "interior" medieval tower of the Lerici castle,
- on natural fauna and paleontological samples of an Ammonite (interactive atlas, shell sectioning, internal parts, reproduction of the functioning of the internal organs, simulation of behaviour etc.).

For a museum a similar approach can also facilitate the use of just one model (in particular for interactive 3D graphics) for many different applications thus allowing the overall costs in the development of Interactive 3D Network Applications to be reduced [6].

Let us refer to Fig. 1, where didactic itineraries are used to introduce and integrate scientific and cultural topics, in order to discuss main technical / organizational aspects and the main people involved (see circles in Fig. 1).

With reference to the left part of Fig. 1 note:

- the areas of the museum and school that the various people involved belong to,
- at the centre is the area related to didactic itineraries with the museums as producers and the schools as mediators to the end users, ie the pupils,
- people involved in scientific/ technical aspects of Museum activity for hypermedia preparation and offering; in the left part (Museum area) we notice:

ESF (Expert in Specific Fields), one or more experts

who will guarantee scientific validity of presentations;

SA (System Administrator), who ensures the integrity of the data, system and its functioning;

DP (Data Provider) who will propose and develop not only multimedia data, hypermedia and electronic solutions but also special effects (eg as in video games);

- and their relationships (see ⇔ in Fig.1) in the various technical phases of: acquisition of data, identifying the functionality of applications, design of interactive 3D applications, etc., until presentation,

and furthtermore, note also the processes:

P1 **design and production of support materials** by cooperation (⇔) of DCI, ESF, DP and SA,

P2 **make didactic materials available** to schools; two itineraries are shown (large dashed lines), each one accessed by two classes (different grey / colour shading)

P3 **interpretation and intermediation** (⇔) by school teachers with the pupils,

with the main people involved (see central and right part of Fig. 1). Specifically, with reference to any of the itineraries shown:

Within the Museum: People involved in the design and use of the didactic itinerary:

DCI (Didactic Communicative Intermediary)**:** the central figure who designs, defines and implements (with ESF and DP) the didactic itinerary (**P1**),

Guides: the central figures of the museum who exploit predefined didactic itineraries and who will be the live voice in exploiting the itinerary; they also help teachers to prepare and carry out the visit (**P2**);

Within the Schools: People involved in the use / exploitation of class work, in contact work for extra mural activities and in the integration between the two types of work and experience **(P3),** ie more specifically:

TCH Teacher, ie the main component for the proposal and the integration of the didactic itinerary with the didactics carried out in class, both in the preparation / conclusion of the visit and in enhancing the programme in institutional terms,

STUD Student, ie the "end user" of the proposal and presentation of the itinerary.

DIDACTIC ASPECTS: THE DIDACTIC COMMUNICATIVE INTERMEDIATOR (DCI)

Let us briefly summarize here:

- the central role of the DCI, by highlighting its main functions with reference to previous Example "Mammals" [1, 5],
- the reverberations on the teacher's activities that makes the itinerary possible for the pupils (P3).

The DCI should define:

- Psycho-pedagogical bases for the itinerary,
- Means for their use in the design and development of the itinerary.

In the case of the "Mammals" itinerary, the didactic itinerary should thus:

- directly and actively involve pupils who have to put into action and play the discovery of their own skeleton and those of their companions as a basis on which to compare those of animals (via the Museum),

- brainstorm their own knowledge and ideas on some animals, body parts and notions drawn from tales, pictures, animations, TV, etc.,
- offer tools that gain pupil's attention in motor and brainstorming activities (eg via fact sheets, drawings, materials to cut up / models / colours, eg The Maze Game, Formation of Sets, Skeleton Colouring, Composing Skeletons, etc.),
- offer feasible organization of the visits,
- offer feasible didactic verification of outcome of visits.

When carrying out these functions the DCI needs to:

within the museum collaborate with ESF, DP, and Guides in the design and development stage,

towards 'users' give attention / listening to:

teachers in order to be aware of their behaviour and any conditioning, their being more or less informed, willingness / resistance not only towards technological tools and support, but also in relation to didactic approaches inherent in the choice of subjects proposed and, above all, of the methods of presentation and discovery of these subjects (P3),

school children who in relation to their vision of the world – both technological and school – make a very useful source of ideas, behaviours, preferences and idiosyncrasies, which are often difficult to collect and understand appropriately by adults [3].

In relation to class work (P3), among the reverberations of the approach adopted in out example towards the teachers TCH, some may be connected to their willingness and ability to:

- manage: preparation, actual visit, verifications, etc,
- take on new subjects and approaches that they might be a little wary of,
- question their own methods,

with possible consequences on carrying out the stages of the itinerary and its conception.

Trying to summarize the main characteristics of a good DCI, let us say that s/he should be able not only to filter, choose, select and organically and harmonically compose inputs, proposals and possibilities coming from ESF, DP and school world, but s/he should be able mainly in the sensibility for inventing new solutions, propose and realize new ways of thinking and suggest activities which stimulate similar attitudes both in pupils and teachers.

Through the suggested approach and the related technological tools new form and help is given to:

- organization, conception, design, development, offer, documentation, update and maintenance, etc of didactic itineraries [7],
- coordination, connection, maintenance and contacts with: (i) schools / classes, (ii) publishers / textbook authors.

FURTHER CONSIDERATIONS AND CONCLUSIONS

The proposed approach facilitates, more than in the past, the contact and integration between:

- aspects of different disciplinary pertinence,
- extra mural activities carried out and supported by external institutions with internal school

activities.

Such possibilities of teaching/ learning in contact with realities outside the school should stimulate pupils:

1) in the perception of:

- aspects that they might otherwise miss,
- complexity of the web of interactions that might occur between these aspects, and specifically the size of the evolution processes and changes,
- multiplicity of the points of view in which such aspects can be considered;

2) to:

- elaborate and express their own interpretations, points of view, etc.,
- collaborate with others while reaching their interpretations, and assessing each other's points of view;

and at the same time encourage teachers in a more active and collaborative way to participate in carrying out their role as educators.

The organizational means that we have outlined can help make what would otherwise be one-off events organized by individual teachers and 'enlightened' heads of schools into something more frequent and structured.

Achieving all this has clear implications:

Museums, environmental and cultural institutes need to be able to organize what they have to offer and provide the relevant support. Moreover, the educational centres themselves need to prepare the staff who can be involved in such projects, particularly the DCI. Teachers, too, need to be prepared to be open not only to new technologies, but above all to the new related potentials, experiences and didactic approaches.

With reference to the applications of technological supports and their innovative use, let us summarize here the following aspects:

- use of sophisticated data, eg 3D interactive graphics, to enhance the communication of cultural heritage based on technology (various graphic- 3D models, animations, etc, with consequent possibilities of:"navigation" in / "manipulation" of 3D models);
- new and easier approaches to creative, technical and organizational cooperative activities mainly by DCI, ESF and DP, as recently proposed for didactic itineraries' design, implementation and dynamic updating and maintenance [7].

REFERENCES

1. Berni T., Carubbi F., Franzini M., Morreale E., Tagliaferri G., Zuffi M., Beni Culturali ed Ambientali e Scuola: un approccio per le Scienze Naturali nella Scuola Elementare, in: Andronico A., Casadei G., Sacerdoti G. (a c.d.), Atti Didamatica'96, Cesena 9- 11 maggio 96, Lavori Scientifici, pp. 210- 225, CILS, Cesena 1996, (in Italian).

2. Donaldson M., Come ragionano i bambini, Emma Ed, Milano, 1979, (in Italian).

3. Druin D. (Ed.), (1999), "The design of Children's Technology", Morgan Kaufmann Publishers, Inc. San Francisco, 1999.

4. Morreale E. , Landini W., Allotta B. (1996), "INTERO: infrastruttura INformatico - TElematico - RObotica per un Museo di Scienze Naturali e del

Territorio", Atti del Convegno "Sistemi multimediali intelligenti", SMI'96, Ravello (SA), 13- 15 Nov. 1996, pp. 228- 261, Arti Grafiche Gutenberg, Lancusi (SA), Nov. 1996, (in Italian).

5. Morreale E., Berni T., Carubbi F., Franzini M., Tagliaferri G., Zuffi M., Teaching Natural Sciences in Primary Schools: the Role of a Museum, in: Tomasz Müldner and Thomas C. Reeves (Eds.), Proceedings of ED-MEDIA/ ED-TELECOM 97 Conference, Calgary, Alberta, Canada, June 14-19th 1997, pp 745- 750, AACE, Charlottesville, Virginia, USA, 1997.

6. Morreale E., Gambino F., Guidi M., Savino P. (2000), An Open and Flexible Development Environment for Hypermedia Applications, Proceedings of the IASTED International Conference on Computers and Advanced Technology in Education - CATE 2000, held in Cancun, Mexico, on May 24-27, 2000, pp. 155- 160.

7. Morreale E., Savino P., Integrating Hypermedia, Database, and Network Technologies for Cultural and Environmental Heritage in Education, in press in Proceedings of 5th World Multiconference on Systemics, Cybernetics and Informatics (SCI 2001), July 22-25, 2001 ~ Orlando, Florida, U.S.A.

8. Piaget J., Le scienze dell' uomo, Ed Laterza, Bari, 1973, (in Italian).

9. Pontecorvo C. & M., Psicologia dell' educazione: Conoscere a scuola, Il Mulino, Bologna, 1986., (in Italian).

ABOUT THE AUTHOR

Eugenio Morreale is Associate Professor at the Informatics Dept. of the University of Pisa. He has made research work in many areas of Computer Science and related fields and applications (Switching Theory, Medical Records and Hospital Management, Natural Language Processing, Informatics in School and Education, Social and Cultural Aspects of Informatics) and in recent years works in the design of Hypermedia and related Applications in the fields of Education and Cultural and Environmental Heritage enhancing.
E-mail: *morreale@di.unipi.it*

[ars.edu] European network of Cultural Heritage Education Centres

Alfredo M. Ronchi (*)

(*) Politecnico di Milano, Department ISET – MEDICI Framework, Milano, Italy
E-mail: alfredo.ronchi@polimi.it

ABSTRACT

The aim of this project is to provide a strong link between Culture and Education through the establishment of a trans European network of Cultural Heritage Education Centres delivering digital educational objects specifically designed accordingly to different user profiles ranging between pupils, students, mentors and tourists. Education institutions and professionals should be taken into account as well. Spreading knowledge, popularising the Cultural Heritage (both tangible and intangible) is the duty of all humanity. Developing a system that will achieve this task will have a big social impact, it will preserve natural and cultural properties of outstanding universal value and will improve the quality of life. Digital educational objects should even help in preserving cultural diversities. In education, teachers can have all needed supports for their teaching tasks, on the other side students can found all the required help that will allow them to participate and to have more active role in the learning processes.

In order to achieve such result [ars.edu] will establish a Trans European Network of Cultural Heritage Education Centres. Such Centres should be both virtual and physical. The reference model inspiring the current project was originated two years ago on the occasion of the development of the educational web site named [louvre.edu]. The basic idea of this service, addressed to France market, was to provide copyright free customised content suitable for scholars and tutors. Comments, texts were expressly written for this purpose. Additional features such as the "Buro" concept were implemented as well. [ars.edu] will expand and extend the basic concept of [louvre.edu]

KEYWORDS: on line education, cultural heritage, digital educational objects

REFERENCE SCENARIO

The starting point of the [ars.edu] system is [Louvre.edu] and other similar experiences recently carried out at local level. [Louvre.edu] (http://www.louvre.edu) is an educational on line web, launched in 1999 from Louvre Museum, Ministère de l'Èducation Nationale de la Recherche et de la Technologie and Oda Èdition, which makes available exceptional cultural contents copyright free. The main features of [Louvre.edu] that will be kept and reviewed are:

- Museum and library guided tours
- Iconographic and textual resources access
- Guided and free search
- Creation of user "personal gallery" (Bureau)

The project is aimed to extend and upgrade at European
level local experiences. The main features to be develop should be:

- Definition of methodologies and solutions in order to exploit the application to different environments and contents
- Educational supports for developers and customises, pedagogical issues, learning paths
- Methods and tools addressed to select and exploit training and educational contents from content repositories
- Definition of authoring systems addressed to mentors (wizards, templates, etc)
- Management of different users profiles
- User tracking tools and techniques
- Adaptive multimedia (topics, format, level of detail, media, hyperlinks, etc
- Content repackaging on different media (video, paper, presentations, etc)
- Education and training results assessment
- Extension of the Bureau-personal gallery concept (authoring tools, intranet publication, content repackaging, etc)

WHICH OPERATIVE MODEL FOR ON LINE EDUCATIONAL SERVICES?

The efforts to be spent in order to set up valuable educational application are usually considerable. In addition access to relevant content, IPR clearance and, some time, *ex-novo* creation of additional content such as 3D reconstructions or computer-generated animations implies costs and efforts exceeding the aim of the application. On the other side, cultural community is interested in accessing some "objects" in order to further investigate or suggest new theories. Developing educational content we do not need to deal with interoperability of huge data sets spread worldwide, we just need a mentor who picks up or create reference objects (texts, images, 3D models, etc) addressed to the definition of "educational paths". Creation, publication and sharing of "digital educational objects" as a common platform for educational application in the field of cultural heritage are the scope of [ars.edu]. In order to achieve this goal [ars.edu] will create an on line "base of publication" for cultural services. The heart of the project is based on the presentation of the principal "foyers de création" which have marked the history of art and architecture in Europe. The concept of "foyers" at the same time usual and powerful is, in our opinion, the best to let people understand the peculiar nature of the European identity.

The full implementation of [ars.edu] project foresees the set up of collections of digital cultural objects related to a list of "Foyers Artistiques" selected as representative of a specific "learning path" in order to give a view on the European Culture.

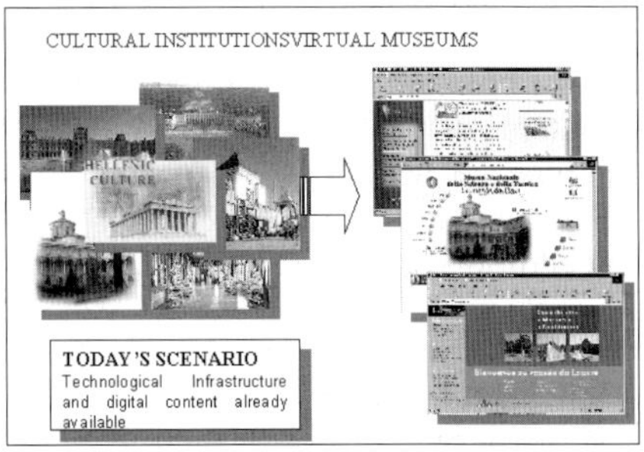

Today's scenario: from cultural sites to virtual museums - major part of cultural institutions has already a web site

DESCRIPTION OF SOME ADDITIONAL SERVICES PROVIDED BY [ARS.EDU]

One relevant concept is the development of tools & methods addressed to let the mentor choose relevant "digital cultural objects" and create an effective "learning path". There is no need to create or access huge digital archives in order to implement [ars.edu], in addition major content exchanges between content holders and the educational environment are performed once. Upgrade and integration, of already existent application, will mainly conceive user profiling, adaptive multimedia and content repackaging on one side and the creation of physical Education Centres in order to complement the network via specific facilities and human interaction.

So far the system will be exportable and compatible with several typologies of museums and archives.

Applications will be interoperable and based on accepted standards and protocols for digitisation and data exchange and allow a global and multilingual access to cultural heritage educational contents. Legal constraints affecting the exploitation of applications (copyright, licensing) will be considered as well and the final goal is, of course, to provide copyright free content on the web (for educational purposes only).

Typical implementations of the system, in addition to the usual one, should be:

Small / medium sized collections to be exploited or promoted even creating a network of related cultural contents. Eg. Museum of pottery in Germany chooses ars.edu solution in order to implement an educational Web site outlining historical production processes, crafts and skill. If there is any similar collection/tradition in Italy ars.edu should emphasize the relation, establish and manage information exchange;

Small collection of heterogeneous objects not so relevant as a whole to be visited frequently should be make available and shared using [ars.edu]. This will multiply the opportunities to be included in "learning paths" enhancing the visibility and public interest for the collection;

There are some "artefacts" not easy to exploit or even communicate, portions of the landscape such as "Les chateaux de la Loire" or the "Via Francigena". Using ars.edu it will be possible to create an educational service exploiting the content, the context and logical/spatial relations. So far it will be possible, for territorial cultural institutions (landers, regions, provinces, etc) to balance and distribute the number of visitors on several different sites enhancing the "cultural visibility" of hidden sites through a set of communication/education services provided by [ars.edu].

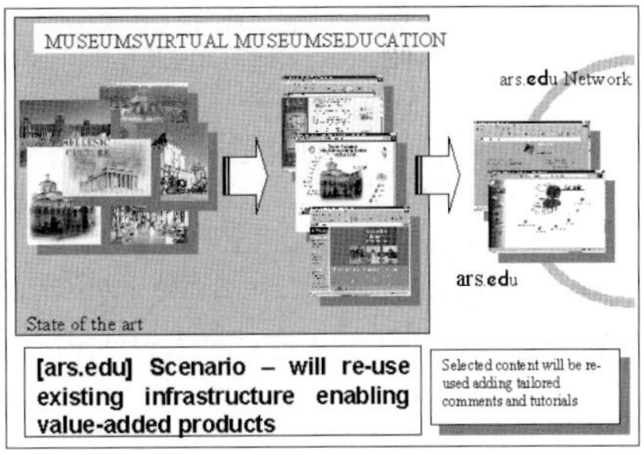

[ars.edu] scenario – will re-use existing infrastructure enabling value-added products

AN INSTRUMENT FOR PEDAGOGIC AND CULTURAL PROJECTS

As an educative and cultural site, [ars.edu], offers to colleges and schools, but also to private people, a collection of visual, textual and sound resources of great artistic foyers which marked the artistic history of Europe.

It will propose in six languages a detailed documentation about 1000 essential works and 400 important monuments: a pedagogic instrument rich of 6000 images together with a collection of documentaries which group together biographies of artists, information about the works and the presentations of great artistic foyers, of syntheses about different periods of history of art and architecture. [ars.edu] allows to develop an educative research giving to everyone the means to collect, choose, personalize and communicate freely.

Information may be accessible in multiple ways; you can search by theme, work, or starting from the artistic foyers and arte facts, monuments, which are related to it. At any moment you can relate the work to its historic context or, starting from a foyer, reach a particular object. Users can on the other hand personalize a selection that they will find every time they repeat the consultation to the site, or page and edit their research.

After inserting access code on the identification screen, the user arrives on a welcome page where lies the summary. The site is divided into four parts: Europe, Art and Architecture, Bibliotheca and Personal Gallery. A presentation text describes their content.

The part called Europe presents an historic and cartographic landscape of the main artistic foyers of Europe from the ancient times. From every foyer, we can assist to a presentation of the most important works and monuments related to this artistic centre.

The part called Art and Architecture, in which the most important action is the research of images, offers a visual thesaurus of works and spaces of architecture. This way a work becomes accessible beginning from its denomination, but also from its techniques, or also from its period and its author.

The selection of images is related to information about the work and historical presentations of artistic foyers.

On the other hand a direct access to the texts thanks to the space called Bibliotheca, which groups together all the texts of the site: information, biographies of the artists, synthesis of the different periods of the history of art, of the schools, of the great civilizations.

At the end, [ars.edu] offers a Forum, space of expression and exchange for the users. They can also be in direct connection with the educational centres and share the other their discoveries, their results and create a partnership of schools, which work on a common pedagogic project.

The site is, in addition, proposed to families and associations. Accessible by Internet subscription, [ars.edu] gives them a set of visual, textual and sound resources about main arts and monuments of the European culture, with that the user can:

- Discover the artistic history of Europe, the essential works, the principal foyers and creations centres, the artists;
- Plan a visit, a course, an exposition, a speech at school or at home;
- Create some documents in groups, take part to pedagogical exchanges among institutions;
- Plan cultural travels;
- To be informed and trained in the

field of arts and heritage for his own general culture, his curiosity or simply for his own pleasure.

The presentation of the works and monuments has many functions, allowing reading texts, to explore the details or to listen to audio documents. At any moment, you can go back from a work to the foyer it belongs. From this foyer, you can examine other works, read documents associated to this foyer, visit other creations centres... and see the different works of a period of a chosen country. The function of selection allows creating some diaporama that create an own museum of the user.

THE EUROPEAN MARKET FOR EDUCATION & CULTURE

Thanks to the work carried out care of the European Commission both in the field of ICT for cultural heritage (eg. DIGICULT framework) and educational multimedia (eg. Multimedia Educational Task Force) there is a good visibility on the overall scenario. In addition the European Information Technology Observatory (EITO) is to be considered as a reference for European market definition. The first condition for the development of such a market is a close and effective co-operation between those actors who are custodians of the cultural heritage - mainly of course museums, galleries and archives - and those actors who provide the means for ICT dissemination mainly commercial companies operating in the fields of information technology, telecommunications, broad casting, etc.

This is a major challenge, given the completely different business cultures in which these two sectors lie and their different knowledge background and expertise. Historically, museums have been suspicious of industry - fearful of a vulgarisation of the heritage and of being commercially exploited; while industry in turn accuses museums of being impractical and out of touch with harsh commercial reality.

Trying to summarise the potential educational market in Europe we can say that: 350 millions people; 4.7 millions teachers; 350000 schools; 67 millions pupils; in addition we have: 15 countries speaking 12 languages and three different cultural models (Mediterranean, German and Nordic).

On the Cultural Heritage side, in the last years computer graphics, hypermedia and telecommunications were applied in exploitation of museums, art galleries, architecture and other kind of works of art. Network access to museums and galleries seems to offer both easier access to cultural heritage and new revenue for its preservation and display. Many relevant players both in the Museums and ICT communities invested time and resources creating pilot projects and applications ranging between 3D reconstruction, image based rendering and virtual museums.

We are now in a position to consider if such investments are effectively useful and really increase and diffuse knowledge in the arts, sciences, and history and if they satisfy users' requirements.

Do virtual museums really provide added value to end-users? Museums, content providers and users are ready and willing new technologies for cultural heritage?

One of the usual tests is to evaluate the follow up of a visit to a museum (real or virtual), exiting the museum do we own "something" more than before the visit. One of the possible "catalysers" in order to ensure an added value is related to educational applications. Such services should be used both at home and on site. So far up until now there are many digital cultural data repositories, usually Web sites potentially suitable to join an aside Educational service. One of the key points is enjoy a set of methods and tools in order to select a limited amount of artefacts to be used in order to create a "learning path"

The communication process devoted to cultural heritage reflects a reasonable degree of complexity. We can present a work of art and propose an interpretation established in advance or try to offer to visitors all the necessary elements in order to evaluate autonomously the work of art. Some of the main difficulties inherent in the communication process regarding cultural heritage arise from the fact that usually the work of art pertains to a different historical and cultural context and it is nowadays probably situated in a different place and the original place is no more accessible. One of the main attentions that the communication manager must take back is the rebuilding of the original context, in such a way that it is possible to 'communicate' the work of art together with all such elements that may render an objective evaluation. Information science, specifically hypermedia and computer graphics should offer a fertile ground to develop such applications. In the rebuilding of the typical context of the work of art another important aspect specifically linked to computer graphics and 3D models is represented by 'space contextualization' which means the possibility to place the digital 3D object in the right location preserving the full set of spatial relations amongst the model and all other objects available on the scene. Advanced telecommunications applications provides some of the answers to the new education and training requirements: educational information (educational structure description, entrance requirements, curricula and syllabi, etc...), information on the use of education software materials, access for persons geographically, socially or professionally isolated, provision of "just in time" training which adapts itself to the needs of employment, improving employees flexibility, new teaching and learning methods which are more interactive and more customised to individual needs. In order to make possible the successful development of this market a number of conditions have to be fulfilled. Potential investors need to understand the potentiality of the return on investments. Companies should look at their investments in the field of art not only as philanthropic programmes but also as multiplier for potential market actions. While indeed many museums and galleries in Europe operate on a non-profit basis this is obviously not true for the companies which develop and market the technology necessary for the multimedia exploitation of European cultural heritage. These companies must be assured that a fair return on investment can be made.

Advanced distance education and training services should then be made accessible to individuals, schools, colleges and cultural institutions. Training centres with special multimedia and virtual reality equipment should be set up and made readily accessible across countries in order to provide courseware and training services for museums, education systems and public administrations.

TECHNOLOGICAL HIGHLIGHTS

Every day the population of the Internet community is expanding. On the other hand Cultural Organizations and Content Owners present a slow rate of adaptation to the new technologies mainly for two reasons:

- Scepticism for the case of electronic theft.
- Lack of Infrastructure for the appropriate presentation of the sensitive cultural topics.

This project's goal is to provide participants with the necessary common software infrastructure to:

- Maintain all content available to end-user and value-added applications / services. This will provide a uniform way (instructions and tools) for content providers to digitise / author, organize, maintain and make available content both for re-use and enhancement.
- Provide transparent access to content and content integration
- Provide for the management and protection of intellectual property rights, especially for digitised cultural content of high value made available on the Web
- Support the operation and collaboration within the human resources network established through the project.
- Perform the actual content acquisition.

Repositories

Applications that integrate all content provided by content providers in a format that can be reusable through value-added applications. Each repository functions autonomously. All content is in a presentation independent format – to the highest extent possible. Leading edge technologies (XML, XSL, RDF and so on) will be extensively employed. Main services that will be provided in the repositories are: querying (e.g. for HTMLs, via the Harvest systems, for XMLs, via Visual XML Query, for DBs, via standard SQL interfaces), data processing (e.g. Visual XML Tree, Visual DTD, HomeSite, native DB tools), other content conversions (e.g. Visual XML Creation, FOP, Xsplit, Versioning (e.g. RCS), Access Control (e.g. WWW server security rights)

The Metadata (MD) layer acts as the "glue" or the common denominator (abstraction layer) among our different underlying repositories, by having all repositories "export" or publish the metadata related to their resources. This use of metadata to enhance the discovery of electronic resources is widespread. Metadata can be navigational, schematic and associative (descriptive, restrictive, supportive, etc.) and are arranged within metadata sets. A well-known, domain-independent set, which is actually a standard, is the Dublin Core (DC), with 15 elements for abbreviated descriptive cataloguing. DC is, however, not enough to offer an integrated and more complete view into our resources. We further need our own metadata sets, with the following functionalities:

- application domain – specific content description
- more consistent content description (using auxiliary sets such as glossaries, thesauri)

The ontologies examined are pertinent to the domain of Culture, and included ICOM/CIDOC International Guidelines for Museum Object Information. Research on cultural metadata included the results of several cultural projects (e.g. Aquarelle). The metadata sets designed and implemented in a database repository include:

- Dublin Core
- Culture-specific (materials, artistic, glossaries of terms, keywords)
- Schematic (classes of XML content, elements of DTDs, database tables)
- Geographic-position related

Since metadata are stored within a relational database, it is easy to support features such as multiplicity of metadata fields and qualifiers (e.g. the LANG qualifier of Dublin Core). This also reduces data redundancy and improves integrity, as often-occurring text (e.g. keywords, locations) can be kept in separate sets and re-used by being linked to metadata fields, instead of being repeated in each metadata record.

Authoring systems addressed to mentors

The authoring system is designed to meet the need for a consistent and effective automation of the production of interactive multimedia titles of structured type, in which the behaviour of the user interfaces, the modes of navigation and interaction, the visual and media effects and the access to the information contained in the electronic book can be handled analytically and structured into a system of relational databases. The resulting data structure provides opportunity of sharing the project between several authors making in this sense possible the development of multimedia products in an enterprise context, which will rely itself on a common base of code sources and on a replication, on server or local, of the complete general project. The multimedia data, described in XML will be, as soon they'll be instanced, stored in the object oriented RDBMS. Highly innovative will be the definition of a DTD for the description of multimedia products considered as a set of media sounds, video images, text and the interactions ad actions that bind them together. The authoring system, developed on an object oriented RDBMS permits the guided insertion of data in several tables by the use of elaborate multi-component forms.

User tracking and profiling

The system is able to extract user's preferences and to catch his level of interest among the different contents, thanks to a careful analysis of his behaviour during the navigation. Users are classified on the bases of similar choices and expressed interests, in this way enabling, automatically or semi-automatically, the system administrator to propose information, matters and navigation personalized guided tours according to the needs and the level of experience demonstrated through the navigation modalities. Contents and proposed services adaptation to user's needs is a fundamental factor to favour an easy and personalized access to a complex information archive. On a parallel side there is the requirement to adapt pages information to user's navigation ability and to his knowledge level on the treated subject matters. The automatic ability of grouping and cataloguing users can suggest the implementation of services perfectly responding to the user needs in term of quality and quantity of information contained, so he won't be tired and leave the navigation. The project purpose is not only to realize a software platform perfectly corresponding to the above cited exigencies, but also flexible and customisable by the mentor himself.

The system must permit to analyse the navigation parameters (such as currently opened page identifier, session time, access time, banner utilization, stay on page time, search form...) which are contained in the web server log files and to relate them to a Data-base of proceeding registered data regarding each page. The realization techniques will be based on taxonomic trees describing the existing links among the several classes of contents assumed for user profiles definition.

The system must be able to extract, starting from incomplete indications, a range of indicators, which describe the overall behaviour of each user and characterize his interests' profile. The semantic hierarchy embedded in taxonomic trees permit to overcome a mere statistic approach, while the hierarchic link among contents classes permit to have an indicative general picture just starting from the first data analysis. The final user profile determination will use advanced fuzzy logic technologies enabling the identification of each user level of interest starting from the elementary parameters memorized in the logging database. Fuzzy engine is of an adaptive kind in fact the parameters which regulate its functioning are extracted from an automatic analysis of historical data to adapt the system to extremely differentiate contexts, reducing difficulties and insertion time in different operating environments.

Adaptive Multimodal Interaction

The adaptive multimedia can be activated by user profile module or by user choice. The Automatic content generator is the result of other pages abstract and result of user search. The user can set the level of detail of the proposed document: general, tutorial and detailed. General is only a survey of the topic, tutorial is a complete course, detailed shows a detailed information set. Adaptive Information Visualization comprehends all techniques, which manage visualisation settings of contents, i.e. graphic layout. The user can set the kind of page presentation and navigation model: Simple, Expert and Master. Simple is simple pages presenting only few relevant information, Expert is a rich abstract and Master is a portal-like model, containing a large mass of information from several sources. The navigation model set up can be defined into user profile or can be chosen from the user himself. Another kind of information visualization can include the creation of user-adaptive functional layouts i.e. tourism, learning, research, news, others: the user defines the objective of the navigation and gets the specific page. For example: if the user choose "tourism", in his page there will be a geographic map showing where he is; if the user wants to learn, the page will show a logical map of the contents. etc. The last sector of adaptive multimedia comprehends all techniques for the development or management of navigation-help tools.

ABOUT THE AUTHOR

Alfredo M. Ronchi Alfredo M. Ronchi is professor of Computer Aided Design at Politecnico di Milano - Engineering Faculty and Multimedia Publishing both at Politecnico - Industrial Design Faculty and DSI Information Science Dept. at State-owned University in Milan. Interests: Computer Graphics, IVR, Hypermedia and Networking. Alfredo Ronchi is actually Technical Secretary of the European Commission MEDICI Cooperation Framework. He is chairman of Eurographics Sessions, ACM MM and DEXA workshops, WWW Conferences "Culture Track". He is a European Commission expert both in the EU Telematics Applications Programme, IST programme, E-Content, and TEN Telecom. Alfredo M. Ronchi is a consultant of the Council of Europe, UNESCO Mediteranean Programme OCCAM, and member of the Scientific Committee of Fondazione Nuovi Media Comunicazione.

Looking for Content

Searching documentary films on line: the ECHO Digital Library

Giuseppe Amato [*], *Claudio Gennaro* [*], *Pasquale Savino* [*]

[*]Istituto di Elaborazione della Informazione, Pisa, Italy
E-mail: {amato, gennaro, savino}@iei.pi.cnr.it

ABSTRACT

Wide access to large information collections is of great potential importance in many aspects - economic, environmental, health, cultural, social, etc. - of everyday life. Historical video documentaries hold by national audiovisual archives, constitute one of the most precious - from a historical and cultural viewpoint - and less accessible cultural information.

This paper presents the ECHO (European CHronicles On line) Digital Library which aims at providing a reusable software infrastructure and new metadata models for old historical documentary films in order to support the development of interoperable audiovisual digital libraries. The ECHO Digital Library is being developed within the ECHO Project funded by the European Commission.

KEYWORDS: Audio/Video, Digital Library, Video Documentary, Information Retrieval, Metadata

INTRODUCTION

Wide access to large information collections is of great potential importance in many aspects - economic, environmental, health, cultural, social, etc. - of everyday life. However, limitations in information and communication technologies have, so far, prevented the average person from taking much advantage of existing resources. Humanity, in its continuous evolution, has accumulated an enormous quantity of information, knowledge, experience, art treasures, etc. One only has to think of the art treasures contained in our archives, libraries and museums, or of the immense and precious collections of observational data in the areas of space exploration, earth sciences, the environment, medicine, etc., accumulated during the last century. A huge amount of material has also been produced as video material. Many countries have national audiovisual archives, as well as there exist private collections of historical documentaries produced during the twentieth century. Such material is extremely precious from a historical and cultural viewpoint.

The ECHO project [1] aims at developing a Digital Library (DL) service for historical films belonging to large audiovisual archives. Actually being able to see and hear an account of a historical event, filmed in the original context, is very different from reading about it. The ECHO services will allow a user to search and access these documentary film collections. Users will be able, for example, to see an event which is documented in the country of origin and how the same event has been documented in other countries, or to investigate how different countries have documented a particular historical period of their life, etc. One effect of the emerging digital library environment, is that it frees users and collections from geographic constraints. This that we have to work across languages, cultures, international standards, etc.

The project involves a number of European institutions holding or managing unique collections of documentary films, dating from the beginning of the century until the seventies. These collections are of great value since they document the different aspects (social, cultural, political, economic) of life in European countries during this period of time. The set of services implemented by ECHO provide users with access to significant portions of their cultural heritage, which would otherwise be almost inaccessible.

This paper describes the ECHO system, looking at its functionality first and then considering the system architecture and the components used for A/V indexing and retrieval. All these functionality have been defined according to the requirements collected from a user needs analysis performed in the project.

The paper is organized as follows: next sections briefly report the results of the user requirement analysis, while section 3 presents the ECHO system functionality. Section 4 provides a description of the ECHO system architecture; section 5 provides the conclusions and illustrates the future research work.

THE USER NEEDS

The collection of the user requirements for an audio/video digital library is quite complex: indeed, the system should be usable by a large variety of users, which may have different needs. For this reason the people interviewed were selected from the following categories, all composed of potential ECHO users: (i) educational environment, (ii) researchers studying contemporary or film history, (iii) archivists, (iv) producers of A/V products, and (v) people marketing A/V products.

There are several ways for collecting user requirements; amongst all the possibilities, we decided to carefully select the users to be interviewed, to prepare a quite detailed questionnaire and to interview them in a face-to-face meeting. The questionnaire included questions on different areas, such as (a) how the user want the data entry to be managed (this section was mainly devoted to archivists), (b) what are the indexing features the user wants supported, (c) what are the required retrieval functionality, (d) how the query results should be presented, (e) what are the requirements for the reuse of the retrieved material, (f) what is the billing and accounting support needed, (g) which kind of material should be included in the digital library.

The analysis of the interviewees was quite extensive and the results covered many aspects of the ECHO system. However, in this paper we will briefly review only the most relevant topics; the interested reader may access the full project deliverable [2]. The requirements have been subdivided into the different areas of interest.

Data entry management. The documentalist consider relevant to have the possibility for an automatic insertion of A/V documentaries, plus the support of a metadata editor that allows one to insert other, more specific metadata and to correct the manually inserted metadata.

Indexing functionality. Indexing should be based on manually extracted features (such as speech transcripts, image features, faces, particular objects present in the video), as well as on a powerful metadata description that allows one to provide a detailed view of the documents.

Retrieval functionality. The retrieval should be based on all metadata used during the indexing, supporting free text and structured searching. Exact as well partial match retrieval (with ranking of the retrieved documents in decreasing relevance order with the query) is required. The access to single or multiple distributed archives is required. A support for cross-language retrieval is needed: the user should be able to formulate the query in one language and to retrieve documents that have been indexed using a different language.

Result's presentation. Query results should be presented in decreasing relevance order with the query. The presentation of the results should be personalized: the user should be able to specify the items of the metadata description he prefers to be presented. Furthermore, a visual abstract (having a length specified by the user) should be used for a fast browsing of query results.

Reuse of the Digital Library content. It should be supported the export of the entire video, of parts of the video, of the abstract, and of the video metadata.

ECHO SYSTEM FUNCTIONALITY

The requirements specified so far, led to the definition of a number of functionality that are briefly described in the following. A particularly relevant requirement consists in the possibility to support the *interoperability* of distributed, heterogeneous digital collections and services. Achieving interoperability among digital libraries is facilitated by conformance to an open architecture as well as agreement on items such as formats, data types, and metadata conventions.

In order to render a digital library of this type feasible, the project has to solve the numerous technical problems that currently bar the inclusion of film information in the digital environment. The aim is to make the film collections available to as broad as possible range of users. To achieve this goal, the following activities have been performed and the following modules have been included into the system:

Define an A/V metadata model

An important aspect of the project consists in the addition of a layer of metadata to the film archives. Metadata elements as presently defined do not describe film data well. A semi-automatic process was designed by which existing local catalogue records can be integrated with metadata elements, automatically extracted during the indexing/segmentation of the film material, into a common description, i.e., the common metadata model.

Design a multilingual retrieval user interface

Local site interfaces were implemented in the local languages; however, a common user interface in English will also be maintained on the project Web-site for external access. Online cross-language search tools were included. Cross-language interrogation is enabled by means of the employment of standard metadata formats, and via mechanisms which provide a mapping between the descriptive languages used by each partner.

Figure 1: ECHO system architecture.

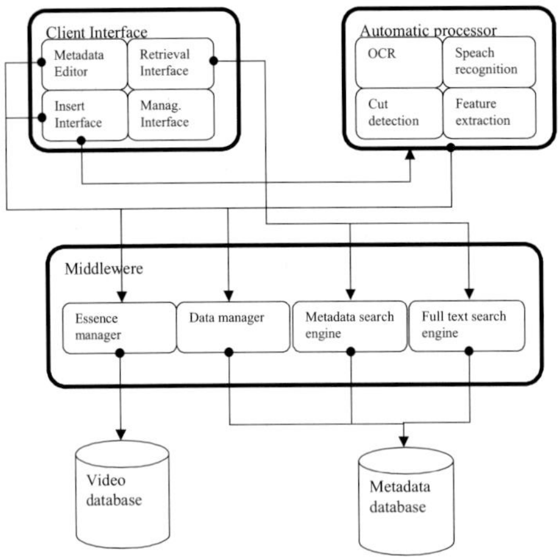

Provide an intelligent access to digital films

The utility of the digital film library can be judged by the ability of the users to retrieve information they need easily and efficiently. The project provides content-based searching and film-sequence retrieval. As the content is conveyed in both narrative (text and speech) and the image, a collaborative interaction of image, speech and language technology will be adopted in order to search the diverse film collections with satisfactory effectiveness. Speech recognisers for different languages are being integrated into the system.

Create visual summaries

The project is developing techniques to produce visual summaries. The aim is to capture the content and structure of the underlying documentary film in a brief visual abstracting process. The summary consists of a sequence of moving images, much shorter than the original film, but preserving the essence of the original message. It should provide a good overview of the entire film documentary.

THE ECHO SYSTEM

The ECHO system assists the user during the indexing and retrieval of A/V documentaries. The indexing is semi-automatic. Using a high-quality speech recogniser, the sound track of each video source is converted to a textual transcript. The transcript is then stored in a full-text information retrieval system. Multiple speech recognition modules, for different European languages will be included. Likewise, video and image analysis techniques are used for segmenting video sequences by automatically locating boundaries of shots, scenes, and conversations. Metadata is then manually associated with film documentaries in order to complete their classification.

Search and retrieval via desktop computer and wide area networks is performed by expressing queries on the audio transcript, on metadata or by image similarity retrieval. Retrieved documentaries or their abstract, are then presented to the user. By the collaborative interaction of image, speech and text retrieval technology, the system compensates for problems of interpretation and search in the error-full and ambiguous data sets. Exploration of the ECHO library is based on these same techniques.

Overall system architecture

The architecture of the ECHO system (Figure 1) is composed of three main components: *client interface*, *automatic processor* and *middlewere*. The client interface is the component directly employed by the users to interact with the system. The automatic processor component analyses multimedia documents, to automatically extract metadata. The middlewere component manages accesses to data stored in the video database and metadata database, on behalf of the other two components. In the following, we will give a more detailed description of each of these components.

Client interface. The Client interface is composed of four main modules, related to the corresponding activities that can be carried out by users of the system. The *metadata editor* allows users to manually edit and review metadata associated with multimedia documents. The user can either edit automatically generated metadata, as for instance scene boundaries, or he can add additional metadata manually. The *insert interface* is used when new documents are inserted. This module interacts with the metadata editor and the automatic processor components that automatically analyse the documents being inserted. The *retrieval interface* is used to search the system for interesting documents. Various possibilities are offered by this interface. Users can retrieve documents by performing full text retrieval, on the transcript or descriptions associated with documents, or selecting specific field of the metadata structure. Finally, the *management interface* can be used to configure and fine-tune the system.

Automatic processor. The Automatic processor is composed of four main modules, each one dedicated to a different automatic processing technique. The *OCR* module recognises textual video captions. The *speech recognition* module is able to generate a transcript in correspondence of an audio or audio/video document. The generated transcript is indexed and the corresponding document can be retrieved by performing full text retrieval. The *cut detection* module analyses a video document and automatically identifies scene changes. In this way, metadata can be associated with specific portions of the document, instead of the whole document. Finally, the *feature extraction* module analyses multimedia document in order to extract physical properties, which can be used to perform similarity retrieval. Typical features extracted are colour distribution, texture, shapes, and motion vectors [3].

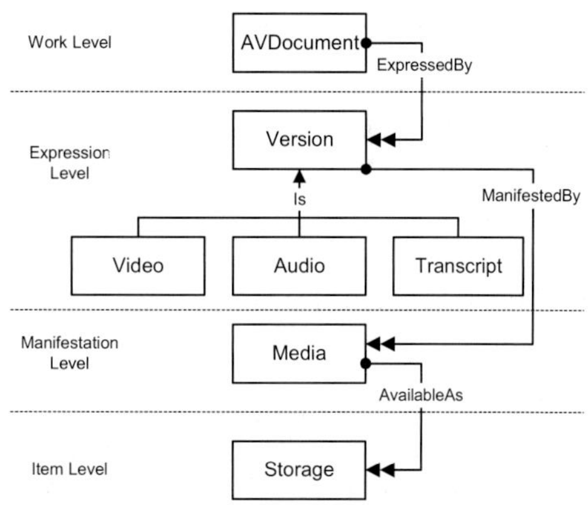

Figure 2. Schematic representation of the Echo Metadata Model.

The middlewere. This component manages the accesses to the underlying databases: the video database, that physically stores video documents managed by the system, and the metadata database, where all metadata associated with the documents are stored. The middlewere component is constituted of four modules. The *essence manager,* that handles the access to the video server, when other system modules, e.g. the metadata editor or the automatic processor, need to access to the video server. The *data manager,* that handles the access to the metadata database. Main operations supported are *read, insert, delete*, and *modify* metadata fields. The *metadata search engine* supports document search based on the full metadata content. The *full text search engine* can be used to search for documents by using textual parts of the metadata. For instance, the descriptions and the transcripts associated with documents are used to perform this type of search. The result of the two search engines can be combined to allow users to perform an integrated complex search.

Editing Metadata

The ECHO metadata model for audio visual documentaries [4] is based on the IFLA model, a general conceptual framework used to describe heterogeneous digital media resources [5]. This metadata model (as shown in Figure 2) is composed of four levels describing different aspects of intellectual or artistic endeavour: *work*, *expression*, *manifestation*, and *item*. The entities of the model are organized in a structure that reflects the hierarchical order of the entities from the top level (*work*) to the bottom (*item*).

The AVDocument entity is the most abstract one; it provides the general intellectual or artistic view of the document. For instance, let us suppose we want to describe a document about the Berlin Olympic Games in 1936. An AVDocument object will represent the abstract idea of the documentary film on the Games. A number of objects, of the abstract entity Version, could represent different language version of the same film, e.g., versions in Italian or in German. The Italian version could have three different objects that are specializations of the entity Version: a Video object, a Transcript object and an Audio object. Moreover, Version objects can be also part of other Version objects. For example, the Video object representing the Italian version of the whole documentary film can have other Video objects representing specific scenes of the film.

However, the Version entity does not represent any specific implementation of the film. This aspect can be represented by means of the manifestation level. For instance, a Media object could represent a digital realization of the document in MPEG format. More than one manifestation of the same Version, e.g. MPEG, AVI, etc., may exist. Nevertheless, the Media object does not refer to any physical implementation. For instance, the MPEG version of the Italian version of the Games can be available on different physical supports, each one represented by a different Storage object (e.g., videoserver, DVD, etc).

Since the metadata model is relatively complex, the design of the metadata editor is of primary importance. The editor is intended to be used by the cataloguers of the archive, that insert new audiovisual documents and that specify the metadata of the documents. The typical cataloguer workflow requires that a new audiovisual document is digitalized or transformed from one digital format into another. Then, the document is archived by the system in the videoserver. The document is processed for automatic indexing (extraction of scene cuts, speech recognition, etc.).When the automatic indexing has been completed, the user is informed by the system and the manual indexing can start. The user typically edits the textual description for typos or factual content, reviews or sets values of the metadata fields, adjusts the bounds of the document segments, removes unwonted segments and merges multiple documents. This phase is usually performed starting from the top level of the model (the AVDocument), and continuing by modifying/editing the lower-level objects connected to the AVDocument (i.e., Version, Media and Storage objects).

The interface of the editor is designed in such a way that it is possible to browse the tree structure of an audiovideo document. Figure 3 shows a screenshot of the interface: the window on left side displays a document like a folder navigation tool. On the top level of the tree, there is an icon representing an AVDocument object (the work of the "Olympic Games on 1936" in our example). Connected to the work object the editor presents the three main Versions

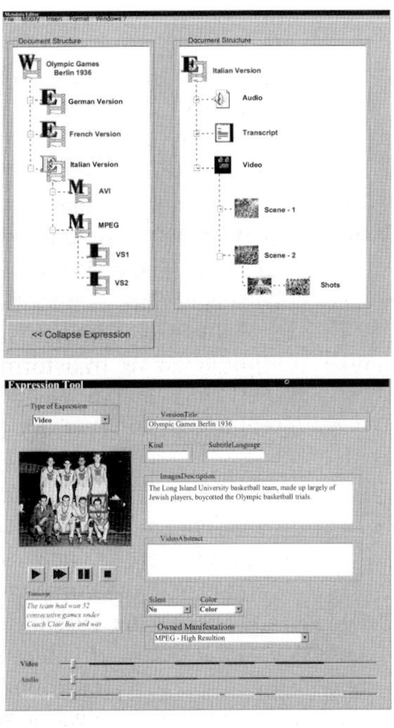

Figure3: A screenshot of the Metadata Editor. Document structure (up) and Expression Tool (down).

that belong to the AVDocument. Moreover, selecting an icon representing a Version (the Italian Version in Figure), it is possible to see the Media instances of the Version and, hence, the corresponding Storage objects.

The navigation tool on the left side of the window shows only the main expressions belonging to the documents (i.e., the expression which correspond to the whole audio/video document). The editor allows to browse a single Version one at a time by using a second frame on the right side of the window. In this way it is possible to see the possible Video, Audio and Transcript Versions (at least one of them must exist) of the document and, for each Version, to browse the video segmentations in scenes, shots, etc.

By clicking on the icon corresponding to a metadata object, it is possible to modify, in a separated window, the metadata fields of the object. A particular attention has been paid to the expression window design, i.e, the Expression Tool. Figure 3 gives an example of the Expression tool interface. Besides the textual fields, the Expression Tool allows the access to the metadata relative to the video segmentation, and allows one to modify them. More precisely, the user can view the video, hear the audio and read the transcript. The window shows also an overview of the video segmentation, by means of three slide tools (see the bottom of the Expression Tool window), which represent the video, the audio and the transcript (if any) of the whole expression. These slides are subdivided in partition that represent the media segmentation. By selecting a segment, the Expression Tool shows the Version corresponding to the subpart of the media (for instance, a scene or a shot).

CONCLUSIONS

The first ECHO prototype is under development, as well as the metadata editor. The first prototype (*Multiple Language Access to Digital Film Collections*) integrates the speech recognition engines with the infrastructure for content-based indexing. It will demonstrate an open system architecture for digital film libraries with automatically indexed film collections and intelligent access to them on a national language basis. This prototype will be refined and the system functionality will be enhanced in two successive prototypes. The second prototype (*Multilingual Access to Digital Film Collections*) will integrate the metadata editor, which will be used to index the film collections according to a common metadata model. Index terms, extracted automatically during the indexing/segmentation of the film material (first prototype), will be integrated with local metadata, extracted manually, in a common description (defined by the common metadata model). The second prototype will support the interoperability of the different video collections and content based searching and retrieval. The third prototype (*ECHO Digital Film Library*) will add summarization, authentication, privacy and charging functionalities in order to provide the system with full capabilities.

ACKNOWLEDGMENTS

This work has been funded by the 5th F.P. IST Research Program, Project No. 11994, ECHO (European CHronicles On line). We would like to thank the members of the ECHO Consortium for fruitful discussions which inspired the ideas presented in the paper.

REFERENCES

1. ECHO Web site: *http://pc-erato2.iei.pi.cnr.it/echo*

2. ECHO User Requirement Report, ECHO Project Deliverable D1.2.1, June 2000, *http://pc-erato2.iei.pi.cnr.it/echo/workpackages/wp1.html*

3. B. Furht, S.W. Smoliar, H. Zhang, "Video and Image Processing in Multimedia Systems", Kluwer Academic Publishers, 1996. ISBN 0-7923-9604-9

4. G. Amato, D. Castelli, S. Pisani, "A Metadata Model for Historical Documentary Films", Proc. of the 4th European Conference ECDL 2000, Lisbon, Sept. 2000

5. K.G. Saur München, "Functional Requirements for Bibliographic Records", Final Report, 1998, *http://www.ifla.org/VII/s13/frbr/frbr.pdf*

ABOUT THE AUTHORS

Giuseppe Amato graduated in Computer Science at the University of Pisa, Italy, in 1992. He is a member of the CNR-IEI in Pisa, working in the area of multimedia information systems. His main interests are model for content based retrieval of multimedia documents, access methods for similarity retrieval and approximate similarity search. He as published papers in many international conferences and journals in the areas of databes information systems.
E-mail: *G.Amato@iei.pi.cnr.it*

Claudio Gennaro is researcher at IEI an institute of National Research Council (CNR) situated in Pisa. His current main research interests are Similarity retrieval, Storage Structures for Multimedia Information Retrieval and Multimedia document modelling.
E-mail: *gennaro@iei.pi.cnr.it*

Pasquale Savino is a researcher at CNR-IEI in Pisa, working in the area of multimedia information systems. He has participated and coordinated several EU-funded research projects in the multimedia area. Currently, he is the director of the ECHO Project. He has published scientific papers in many international journals and conferences in the areas of multimedia retrieval. His research interests are Multimedia Information Retrieval, Multimedia Document Modelling, multimedia content addressability, and indexing.
E-mail: *savino@iei.pi.cnr.it*

Handling Sub-Image Queries In Content-Based Retrieval of High Resolution Art Images

Stephen Chan(*), *Kirk Martinez*(*), *Paul Lewis*(*), *C. Lahanier*(#), *J. Stevenson*(§)

(*) Intelligence, Agents and Multimedia Research Group, University of Southampton, UK
E-mail: scyc@ecs.soton.ac.uk, phl@ecs.soton.ac.uk, km@ecs.soton.ac.uk
(#) Centre de Recherche et de Restauration des Musées de France
E-mail: christian.lahanier@culture.fr
(§) Victoria and Albert Museum, London, UK
E-mail: j.stevenson@vam.ac.uk

ABSTRACT
The paper describes work which is part of the Artiste project to develop a distributed database of Art Images from four major European art galleries. One aspect of the project involves the development of content-based retrieval and navigation facilities. A particular objective is to provide a facility for retrieving an image from the collection or navigating to related information in the database, given a query image which may be only a part of a particular image in the collection. The query sub-image may be a poor quality reproduction of part of the original and may be digitised under significantly different conditions. The paper outlines one of the recent approaches we have developed to facilitate such modes of retrieval and navigation. It is based on the use of colour coherence vectors extracted from image patches for the query and target images at a range of scales with multiple vector matching in order to find the best sub-image matches. Some results of the application of the technique are described and its success at sub-image location from a collection of images, including many at very high resolution, is demonstrated.

KEYWORDS: Multi-scale, sub-image matching, colour coherence vectors, content-based retrieval.

INTRODUCTION

ARTISTE is an EC funded project under the IST work programme and aims to develop an Integrated Art Analysis and Navigation Environment hosted on a distributed database and accessed via the World Wide Web. The environment will include, amongst other things, facilities for searching and analysing digital images and will include tools, not only for retrieval using metadata but also for a wide range of content based retrieval activities. Partners in the collaboration include NCR Systems Engineering Copenhagen, Giunti Interactive Labs, Centre de Reserche et de Restoration des Musées de France, the National Gallery London, the Uffizi in Florence, the University of Southampton and the Victoria and Albert Museum London.

Content-based image retrieval is a challenging and active research area with the potential to provide powerful facilities for image searching. But although many techniques have been described in the research literature, it is probably not an exaggeration to maintain that the capabilities of content matching alone are still relatively basic as general purpose approaches although some powerful applications specific methods can be developed. General techniques based on such features as colour distribution, texture, outline shape and spatial colour distribution have been popular in the research literature and in content based retrieval systems but many techniques only work on complete images, will not allow the query image to be a sub-image of the image to be retrieved and require similarity in image resolution between query and target.

In this paper we describe in more detail some of the problems of content based retrieval using colour matching and present an approach to sub-image retrieval for collections which include large numbers of very high resolution art images.

CONTENT-BASED RETRIEVAL

Several extensive general reviews of content-based image retrieval techniques have appeared in recent years [4]. CBR has been applied to Art images since the 1980s with the Morelli project [6] and IBM's QBIC system for example has been applied to such images in a collaboration with UC-Davis [1]. Previous approaches usually applied a generic CBR system to a group of Art images to see how useful it would be. In Artiste our aim is to solve specific CBR problems in addition to producing generic solutions for queries.

Use of the colour histogram [5] for comparing images has been popular primarily because it is easy to compute, is fairly insensitive to small variations between images and is relatively efficient in terms of computation speed and storage requirements. It has the disadvantage that it does not capture any information about the distribution of the colours spatially within the image and various techniques to capture such information have been proposed including the colour coherence vector approach (CCV) [3]. In the multi-modal neighbourhood signature (MNS) method, developed by Matas *et al.* [2], signatures representing the colours in different bimodal sub-regions are used as a cue for image and sub-image retrieval.

SUB-IMAGE MATCHING AND THE M-CCV METHOD

In this part of the project, the aim is to develop a robust technique to retrieve database images of an artwork given a query image which represents all or part of the target artwork. The query image may have been captured at a different resolution from the database image, may only represent a fragment of that image and may be distorted or degraded in some way. For example, the query image may be a fragment from an image captured prior to restoration of an artwork and the target image may represent the painting after restoration.

In general, global techniques such as the colour histogram are not effective for sub-image matching in their basic form. Some attempts have been made to retrieve sub-images using colour histograms by dividing the target images into pyramids of patches and recording the colour histogram for each. However, in our approach called the Multi-scale Colour Coherence Vector

(M-CCV) method, we use the colour coherence vector (CCV) rather than the basic colour histogram as the representation of the individual image patches as it carries some useful local spatial information. The colour coherence vector records the numbers of coherent and incoherent pixels of each colour in the patch where a coherent pixel is one which belongs to a patch of similar coloured pixels whose relative size is above a certain threshold. We pre-compute the colour coherence vectors for overlapping patches for each database image, repeating the process at reduced resolution until the reduced image corresponds to a single patch. The query sub-image is also sub-divided into patches and the best match is obtained by combining the CCV match scores. The CCVs are coded for rapid matching and can be compared at an average rate of 265,000 CCVs per second.

Results

In figure 1 we show a query image which is a fragment of the Moise présenté à Pharaon by Orsel Victor captured before restoration work at the C2RMF.

The best match is shown in figure 2, and it can be seen that the best match is with the correct parent image but after restoration. (The before restoration image was not in the database). Note that the position of the match is highlighted, a necessary facility when dealing with ultra high resolution images. This particular parent image is 6328 by 4712 pixels. It should be stressed that the parent image was scanned at 1.47 pixels per mm of the original painting whereas the sub-image query was captured at a quite different resolution, 0.92 pixels per mm. There were over 1000 images in the database varying in size from 440,000 pixels to 30,000,000 pixels and the retrieval process took about 46 seconds on a Pentium III 600Mhz PC. No multidimensional indexing has been used so far although it is intended that the process will be accelerated by its introduction.

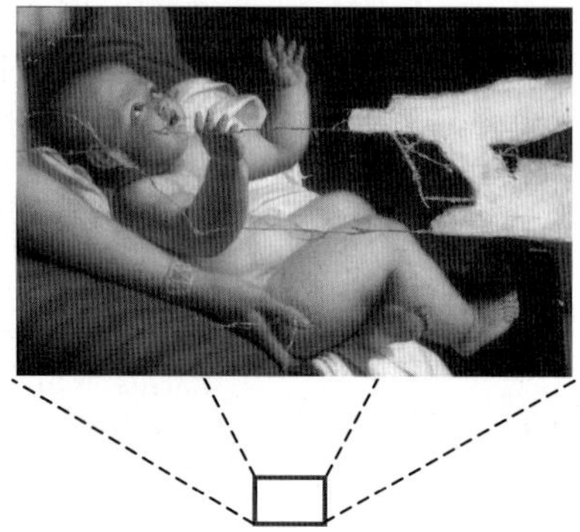

Query

Figure 1: The image of the baby represents a query. The small box below the image represents the scale in relation to its corresponding parent image in the database

Retrieved result

Figure 2: The result of sub-image matching is shown as a white boundary in the Moise présenté à Pharaon image (Orsel, Museum of fine arts, Lyon). The rectangular box beneath the query represents the relative size of the query image to the retrieved image

In figure 3 we show a further example of the multi scale CCV retrieval in action. The query image is a selection from a fabric from the V&A collection and the best retrieval shows the parent tapestry.

Query Result 1

Figure 3: A query image of a flower pattern. The first result of the retrieval locates the correct image and sub-region

In figure 4 we show the next two best matches for the same query. The retrieved results show tapestries with a similar motif to the sub-image query pattern. The retrieval process took 14 seconds to complete for the same database.

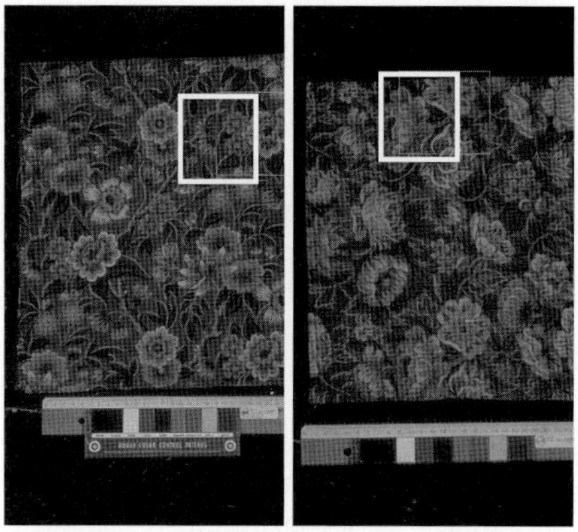

Result 2 Result 3

Figure 4: The next best retrieved results

CONCLUSIONS AND FUTURE WORK

In this short paper we have shown that the multi-scale colour coherence vector (M-CCV) technique can provide effective sub-image retrieval and is also applicable when the sub-mage and target are captured at different resolutions. The technique will form part of a battery of content-based retrieval and navigation tools which, together with integrated metadata based retrieval and navigation, will constitute the Artiste art image retrieval and analysis system.

ACKNOWLEDGMENTS

The authors are grateful to the European Commission for their support through grant IST-1999-11978 and to their collaborators (NCR (Dk), Giunti Interactive Labs (I), Uffizi Gallery (I), IT Innovation Centre (UK), The National Gallery (UK)) on the ARTISTE project for image data and useful conversations.

REFERENCES

1. Bonnie Holt and Ken Weiss. The QBIC Project in the Department of Art and Art History at UC Davis. In The 60th ASIS Annual Meeting 1997, Digital Collections: Implications for users, Funders, Developers and Maintainers, volume 34, pages 189-195. ASIS, 1997.

2. J. Matas, D. Koubaroulis, and J. Kittler. Colour Image Retrieval and Object Recognition Using Multimodal Neighbourhood Signature. In D. Vernon, editor, Proceedings of the European Conference on Computer Vision, LNCS volume 1842, pages 48-64, Berlin, German, June 2000. Springer.

3. Greg Pass, Ramin Zabih, and Justin Miller. Comparing Images Using Color Coherence Vectors. In MultiMedia, pages 65-73. ACM, 1996.

4. Arnold W. M. Smeulders, Marcel Worring, Simone Santini, Amarnath Gupta, and Ramesh Jain. Content-Based Image Retrieval at the End of the Early Years. In Transactions On Pattern Analysis And Machine Intelligence, volume 22 of 12, pages 1349-1380. IEEE, December 2000.

5. M. J. Swain and D. H. Ballard. Color Indexing. International Journal of Computer Vision, 7(1):11-32, 1991.

6. W. Vaughan. "Automated Picture referencing; Another look at Morelli" in Computers and the History of Art, Vol. 2 (1991), no. 2, pp.7 -18. [Harwood Academic Publishers ISSN 1048 - 6798]

ABOUT THE AUTHORS

Stephen Chan is a member of the Intelligence, Agents, Multimedia Research Group in the Department of Electronics and Computer Science in the University of Southampton. He is currently reading for a PhD in the area of video and image analysis.
E-mail: *scyc@ecs.soton.ac.uk*

Kirk Martinez gained a BSc in Physics from the University of Reading and a PhD in Image Processing at the University of Essex. He has been involved in several art imaging European projects: VASARI, MARC and Viseum. His research interests include content-based retrieval, imaging, parallel image processing, Web Hypermedia.
E-mail: *km@ecs.soton.ac.uk*

Paul Lewis is a member of the Intelligence, Agents, Multimedia Research Group in the Department of Electronics and Computer Science in the University of Southampton. His research interests are in image and video analysis and their application to multimedia information handling. He has been an investigator on numerous EPSRC grants most recently working on the development of content and concept based retrieval and navigation tools in multimedia environments. E-mail: *phl@ecs.soton.ac.uk*

Christian Lahanier. With a PhD in Physics, he managed the department of Physics in charge of X-ray technologies at LRMF for 15 years. Head of the laboratory in 1984, he set-up a particle accelerator for non-destructive material analysis; he then developed a multilingual documentary information system and a high resolution imaging system with the support of RTD EU programme (VASARI DGIII, and

NARCISSE Project in Telematics). Head of the Information and digital image department of the C2RMF in 1994, he installed an Intranet for access to the scientific documentation and high-resolution images of the laboratory and conservation-restoration department. His department participated to EU collaborative projects under the 4th Framework programme: VISEUM (ACTS - DGXIII), MENHIR (ESPRIT - DGIII), ACOHIR (ESPRIT - DGIII) and CRISTAL (Rafael DGX).
E-mail: *christian.lahanier@culture.fr*

James Stevenson is the Photographic Manager at the Victoria and Albert Museum in London. He has responsibility for all of the image creation in the V&A both analogue and digital, the management of those images and their exploitation though the V&A Picture Library. His interest lie in making all of the V&A's large archive of images available to the widest possible audience.
E-mail: *j.stevenson@vam.ac.uk*

Logical framework for information retrieval and conservation project. Geo-referential querying in integrated catalogues and notified cultural heritages

Luca Marescotti (*) *and Maria Mascione* (#)

(*) Politecnico di Milano, Dipartimento di Disegno Industriale e di Tecnologia dell'Architettura, Milano, Italy
E-mail: luca.marescotti@polimi.it

(#) Politecnico di Milano, Dipartimento di Disegno Industriale e di Tecnologia dell'Architettura, Milano, Italy
E-mail: m.mascione@archiworld.it

ABSTRACT

The design of an information system based on cataloguing actvities in a conservation project is istrumental for the operating aspects and procedures. Its definition is an example of some of the fundamental parts of a general project for the modernisation of Public Administration concerning cultural heritage. The main objective is not only to put together the different catalogues, but especially to enlarge then through access to heterogeneus documents, also trough access to multimedia application, and to all these economic informations that will be useful for the management of the patrimony. The cultural heritage conservation requires both planning and programming measures and operational activities. At the same time it requires the support of information system that can gather and distribute the information of 'values' and motivation for recovery, conservation and restoration are based, so that the different actors and the different administrations delegated to the task can operate as much as possible along to the same strategic lines. The definition of the logical criteria and conceptual design for the informatic storage of the cultural heritage inventory is based on the description of the operators and on the procedural needs of each in order to optimise the use of the software tools. That means to analyze the theme of information technology and Gis for cultural heritage through integrations of three aspects: the information system; the study of competences and rules of procedure about cultural heritage; the project and implementation of prototypal information system based on georeference.

KEYWORDS: catalogue; cultural heritage; gis; information technology; urban and regional planning

INTRODUCTION

Although the construction of an information system suitable for different users groups is based on a series of machines and programmes capable of managing large amounts of heterogeneous data, it also requires careful analysis of data and their functional aspects with a view to permitting not only geo-referential querying but also intelligent and friendly management. Overall, the logical framework for information retrieval is neither marginal nor banal, requiring the development of research capable of combining both theoretical approaches to the conservation project and database design capability.

The national programme entitled *Conservation Project: Methodological Guidelines For Preliminary Analyses, Action and Feasibility Control,* co-financed by the Ministero dell'Università e della Ricerca scientifica e tecnologica (Ministry for the University and Scientific and Technological Research) between 1998-1999 [1], represented an opportunity not only to retrieve results and solve problems resulting from previous activities connected with cultural heritage assessments, but also to combine different interdisciplinary points of view in order to consolidate these aspects by evaluating the knowledge acquired from a point of view of an information system project.

CONCEPTUAL PLANNING OF THE GEO-REFERENTIAL INFORMATION SYSTEM FOR ARCHITECTURAL AND ENVIRONMENTAL HERITAGES

The construction of an information system for territorial and environmental management and planning must fall within the Public Administration's general information system project, meaning that these should include different elements ranging from cultural heritages to real estate systems (buildings and land), from healthcare to traffic and transport, from use of urban to territorial or environmental land use.

This assumption involves not just interpreting the system as an expression of a particular point of view which employs a multi-user common database, but also depicting it, first and foremost, as a "public" information system, i.e. designed for information exchange between a number of users dealing with the same objects, although perhaps from different viewpoints.

Incidentally, three assumptions may be made:

- By definition the "public" information system is in direct opposition to the "private" information system, which is designed and used by a single user or by a homogeneous group of users.
- Although the information system is used for the management of standard procedural activities (files and procedures) and, at times, supports decision-making processes, it is, however, necessary to remember that this is based on knowledge that does not necessarily derive entirely from said procedures.
- Unlike corporate systems, the Public Administration's information system, must be open on two fronts: one for other Public Administrations and another for citizens and businesses.

These assumptions demand that the project not only responds to the logic of providing certified data, utilising clear, coded formats, managing standardized exchange protocols, utilizing standard updating criteria or adopting database integrity tests (not to mention security), but also that it may be structured on the basis of three analyses: the first on Public Administration work processes, divided according to specific skills; the second on identification of alphanumerical databases that have been distributed for use and management purposes; and the third, on identification of the detail levels necessary for geo-referential querying of objects and territorial transformation processes.

The *first analysis* represents the keystone of the project, in that it provides access to existing archives allowing not only for standardization of various information requirements but also of various transactions between work groups.
The *second analysis* highlights how the same object may be viewed by many work groups, but, in each group, with only part of the data: the project must ensure *database* consistency, avoid redundancy and permit the amalgamation of data from different groups.
The *third analysis* deals with the question regarding other aspects linked to cartography and geo-referential querying of objects: with regard to the above, it is necessary to dwell on a few fundamental questions, such as the need for "overlapping" between the following archive levels: regional cartography, provincial cartography, municipal cartography, land office cartography (ecographic code), ISTAT cartography (Registered sections and 2001 Census), where, in reality "overlapping" actually implies "geo-reference".

Benefits of the Information System

The success of the creation of an information system project depends on how useful it actually is. Although principles pertaining to data safety, integrity and ability to update are important in the decision-making process, actual implementation of the decision strongly depends on its utility both for the entire administration as well as the individual departments or individual officers.
System utility will depend on its compliance with office procedures, simplifying data access and transactions with other departments (for example, file transmission) and using efficient, intuitive, effective interfaces, commonly referred to as "user-friendly". However, interfaces and architecture respond to something that is secondary to the substantial question concerning system content and archive quality. Information systems related to real estate (buildings and land) must be seen as the result of a critical revision of databases normally used for management of same and, consequently, on the basis of the interaction between procedures and information sources.
Since territorial government bodies actually represent a unitary organization having specific tasks and skills, just as the territory is in fact a *unicum,* it would be reasonable to hypothesize that data should be able to be aggregated or split-up according to the possibility of territorial interpretation differentiated by the various points of view of the governing territorial bodies (State, Region, Province, Municipality, Park, Basin Authorities, Mountain Communities and similar bodies) with the aim of clearly reconstructing various territorial usage.

Feasibility Project Factors

The first step in the construction of the geo-referenced information system feasibility project is a "state of the art" study, the aim being to collect and assess a few examples of public administration information systems, with particular reference to the above mentioned problems, targeted at identifying success factors.
A second step is the strong identification of objectives and aims of the information system with a view to specific user groups, so that each group of users may define the procedures, while skills contribute to the definition of user groups. This phase considers existing distributed databases, ones that can either be retrieved or constructed.

The third step is the definition of the gradual implementation phases, which although conceived for individual user groups nevertheless allow for the expansion and sharing of information criteria and the data structure.
In any case, the information system feasibility study for building heritage management includes two criteria: the first is based on identifying new procedures or revising the concept of existing procedures; and the second on adopting compatible and reciprocally compliant development strategies for information sectors that are connected or undergoing transformation.
As elements of development, the geographical information sector can include land office reforms and the ISTAT Census, flanked by the results that are likely to emerge from the State-Regions Agreement currently being negotiated also with regard to cartography.
When designing the information system, the feasibility study must take into account the conceptual phase in order to subsequently deal with the logical and physical phase of the project, with the definition of IT aspects (hardware, software, data structuring and designing of system architecture) employing and maintaining a typical public works approach right up until tendering, construction and implementation. The *corpus* of information be specific i.e. related to the analysis of institutional tasks of the various public bodies, responsible, in some way, for architectural and environmental heritages; the study of the regulations in force is the first step along this path.
The second step consists of assessing the contents and layouts of a few samples of sheet drawn up by the regional, provincial and municipal bodies, the aim being to verify their utility as a source of information in proceedings concerning territorial organization and management. When the existing archives are vast, retrieval of same could prove to be so onerous that it could act as a deterrent to current management activities.

APPLICATION CASES [2]
Objectives and prototype implementation

The aim of the project is to build prototype work-stations based on GIS (*Geographical Information System)* and multimedia environments. These prototypes will serve to verify congruency between the definition of the general project framework and the effectiveness of the solutions, also with regard to the quantity of memory occupied, simplicity and speed in transactions and enquiries and the quality of the hardware and software architecture. The operation phases can be summarized as follows:

1. Analysis of competencies and administrative procedures within the provincial and municipal spheres;
2. Identification of internal and external databases: structure and interconnection analysis;
3. Conceptual database organization;
4. Logical planning of the information system according to the different levels of geo-referential querying;
5. Identification and standardization of queries;
6. Specific definition of geo-referencing;
7. Data access levels;
8. User interfaces.

Interpretation of the project has been applied to two cases in Lombardia, the Municipality of Monza and the Province of Lodi, through the construction of two GIS prototypes developed using Windows Microsoft with ArcView 3.1 (Esri) GIS software.

These are two studies which share a few common elements and objectives: both in fact deal with the geographic based consultation of cultural heritage as well as verification of the utility of this type of tool, the catalogue, within the context of Public Administration activities related to the management of historical towns and buildings.
The first objective pursued concerns the analysis of the possible integration of different databases on cultural heritages. The study starts with the analysis of contents and data structures and ends up by defining the relationships between different archives.
The purpose of accessing heterogeneous data banks through geographical querying and visualisation derives from the concrete possibility of integration with Public Administration activities: numerous diagrams with different layouts and different dates are used to assess the same building and the same P.A.'s, during the study and editing phases of town-planning tools which, in turn, promote specific surveys on building and environmental heritages. The phase related to prototype construction is targeted at their testing in order to assess their operating elements and updating possibilities.
The basic cartography for the two prototypes is the CTR (*Carta Tecnica Regionale* Regional Technical Map made by Regione Lombardia) 1:10,000, used as a reference to identify the heritages. This support is now considered a standard and is adopted both on a regional as well as provincial scale also for specific projects on cultural heritages.
Although the digitalized subjects relate to different topics, they are, nevertheless of specific use for inventory purposes, such as, for example, buildings assessed by the A(i) sheet, buildings present in regional databases, such as buildings where intervention measures are financed by special projects (e.g. those of the Regione Lombardia Frisl), boundaries of the Iccd territorial and town-planning forms, and summary elements of the town-planning scheme or roads. The catalogued subjects are combined with the census sheets, as well as other types of information, such as, for example, a photographic image or the reproduction of different documents relating to the building, or to a portion of territory.
Furthermore, the data concerning movable objects can be combined with the building housing them, the aim being to permit management of such heritages by also considering the territorial aspect of their conservation, useful also for reconstructing the entire collections scattered across the territory and for controlling their management. The diverse nature of the specific objectives falls within this context/general premise and is defined in relation to the specific competencies of the Municipal body in one case, and the Provincial body in the other.

Local level: a municipality (Comune di Monza)

Construction of a prototype for the Comune di Monza is underway for the management of files relating to the building heritage based on the analysis of relations between the main Public Administration offices involved: *Ufficio nuovo piano regolatore, Ufficio gestione urbanistica, Ufficio Edilizia privata* (New Town-Planning Scheme Office, Town-Planning Management Office and Private Buildings Office) [3].

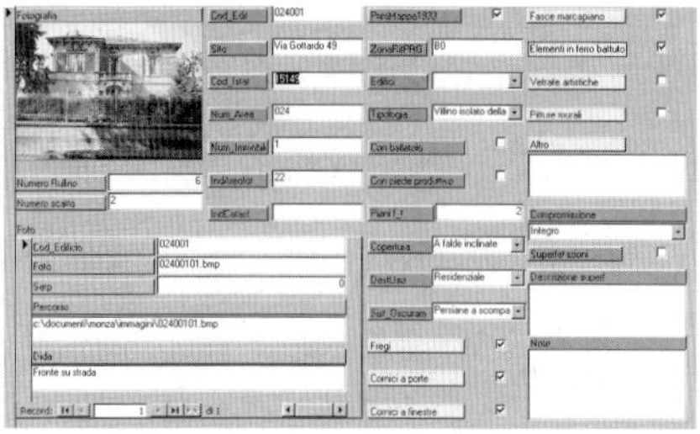

Figure 1: Data-entry layout (Access'97) of the "Valori storico-testimoniali" cardcatalogue, designed for the Monza master plan

The alphanumerical and geographical databases contained in the prototype are partly drawn up by various bodies (Comune di Monza, Provincia di Milano), partly newly-compiled or revised via re-elaboration of existing data, within the research framework.
In particular, cardcatalogue on "historical-cultural heritages" promoted by the *Ufficio Nuovo PRG* (New General Town-Planning Scheme Office) on the occasion of the drawing up of the Town-Planning Scheme in 1997 with the intention of including those government offices under restraint by the Scheme variation still for ratification, have been careful analysed as regards their conceptual structure, which has brought about the creation of a *database* using *Access'97 (Microsoft) software* that will serve to assess the possibilities of integration and use of the Gis functionality databases with regard to the activities that are in course at the Municipality's *Ufficio Nuovo Prg*. In this case, the cartography for geo-referential querying on the buildings has been obtained from municipal aerophotogrammetrical records (vectorial). Other information levels aiding in defining specific querying or creating themes refer to the cadastral cartography records (Land Office) and those of the General Town-Planning Scheme, both in numerical format. Furthermore, integration with contents of the specific cultural heritages databases [4] has enabled checking of overlapping of information on the same building deriving from different projects.

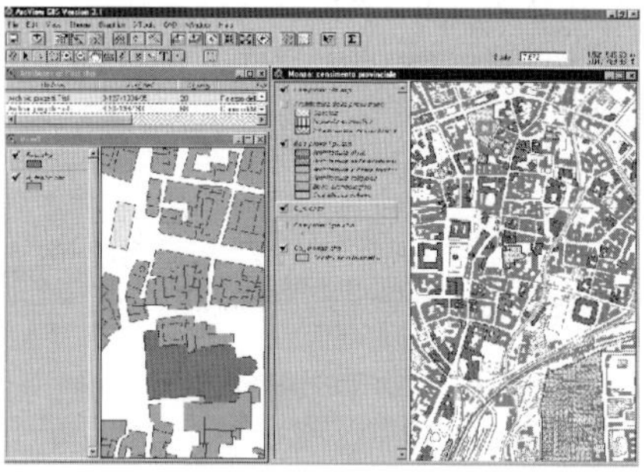

Figure 2: Elaboration of data relating to the census of the Provincia di Milano according to the typology of the catalogued properties.

The intermediate level: a Province (Provincia di Lodi)

The study's objective as regards the Provincia di Lodi consisted in analysing the Provincial body's specific competencies in the sphere of knowledge and management of cultural heritages and activities connected with co-ordinating the cataloguing activities in its territory, especially in view of the latest regional regulatory measures.
The project, drawn up and involving the Town-Planning and Cultural Activities Sectors, has an essential premise forming the basis of the definition of the prototype specifications in the study of the government body's specific competencies. Here too, analysis of projects and specific archives on cultural heritages relating to the examined territory proved fundamental, particularly exemplified by the Comune di Maleo [5]. Analysis of the databases existing for the territory under

examination [6], some in cardcatalogue, others in computers, represent the basis of the territorial information system.

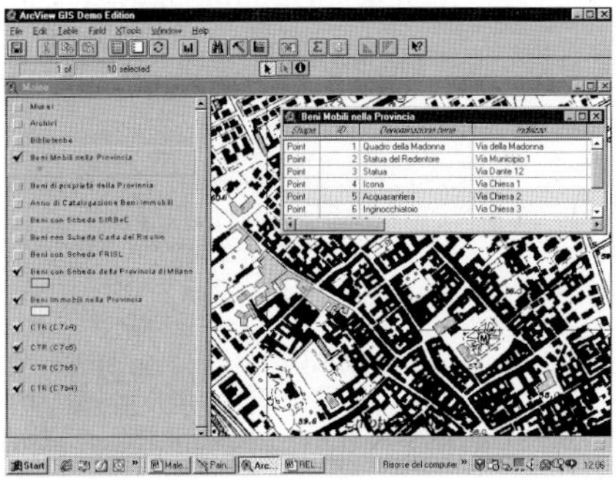

Figure 3: Maleo, architectonic heritages in the Provincia di Lodi inventory [7]

The prototype's logical project includes connecting ArcView and the various databanks, mostly set up using Access'97 [8]. The main cartographic queries and themes indicate how many times a property has been registered and with which kind of sheet, the year of the latest sheet and a visual photographic image. Also, the prototype experiments the connection with the SIRBeC application, the programme managing the data banks of the catalogued cultural heritage in the Regione Lombardia.

ORIGINAL RESULTS OBTAINED FROM SOFTWARE APPLICATIONS AND POSSIBLE DEVELOPMENTS

Theoretical in-depth studies and experiments have given rise to some reflections on the "state of the art" in the field of Gis software applications, which actually greatly involve the town-planning area, the choices made and strategies of intervention measures in the territory.

Situations which call for wide margins of improvement are pointed out as follows:

- The existence of numerous projects on knowledge of architectonic and environmental heritages, in some cases indirectly connected with specific utility objectives.
- The lack of knowledge of such surveys on the part of local government bodies and in sectors involved in town-planning.
- Increased interest in the potential of increasingly powerful, user-friendly and economical Gis software, which does not move at the same rate as clearly development of needs , requisites, performance of the system which has to deal with configuration, construction and implementation.

The problem of connection and integration of the different archives is not so much technical as qualitative, namely on the contents side, data validity, data comparability when deriving from different sources and data updating.

As regards this latter aspect, it is also noted that the same nature of some census operations, promoted by public administration bodies with a chiefly cultural type of mandate and not targeted at the management and control of the transformations, or of a fiscal nature, can in time become lacking insofar as updating is concerned. Definition of the method of querying must always bear in mind the heterogeneous nature at the start, trying to make as explicit as possible data search paths within the same "unit", but within the different databases. With regard to data structure a certain migration on the part of various lombard government bodies has been noted in favour of the Iccd standard sheets, at times integrated and improved as regards layout, in relation to specific information needed by the same government bodies.

Added to all this is the geographical

element, considered fundamental in the querying approach, which some archives do not combine nor hold in consideration.
The unifying factor for the various census operations is the geographical location of the registered property. The importance of geo-referential querying on the property is emphasized further if examined in relation to the scale of detail; immediateness in data searching, consultation, geographical identification of the property, represents the unifying factor and connection for combining heterogeneous data relating to the same object.

Constructing an information system: prospects

The experiments carried out using the two prototypes have largely confirmed the technical feasibility of the construction of a geographically-based information system project for cultural heritages.
Those requisites, which are still outstanding and important in order to really accomplish an integrated system that is distributed throughout the territory and for the use of different groups of users, concern strictly primary conceptual aspects:

Possibility of updating the system. Particular attention to the need to update the system, which is valid only if shared and continuously updated, if it is really used continuously for economical appraisal in the phases of planning and intervention implementation.

Integration with geographically distributed databases. Possibility of integration with external data banks (ISTAT, Land Office, Town-Planning Offices, Superintendence departments) in a hypothetical sharing of data, which requires specific methods of access to data and recognition of data ownership.
The archives are different and belong to government bodies that have collaborated in creating them; archives that are managed, administrated and updated by the government bodies/owners, but are accessible to public and private users. To this end, the study of an easy and highly reliable user interface acquires primary importance.

Possibility of geo-referential or multimedia querying. The system must be accessible online and available to a wide range of users.
The possibility of making specific information accessible to an ever-increasing public is one of the priority objectives in an area that privileges P.A.'s improved functioning both internally, as well as in its relations with citizens, through the supply of increasingly efficient services.
On-line consultation via Internet, already implemented in some regional and local offices, has been appraised in the prototype definition verifying the opportunity of producing theme cartography with mininum interactive play which, apart from visualisation of the cartographic data also enables consultation of combined alphanumerical data, as well as images.

Standards and data certification. The last aspect, yet quite a central theme in the data structuring project and among those aspects still to be resolved and common to the archive complex examined, is a problematical aspect, which consists in explicitly identifying standards or rulings adopted, the registration year and eventual updating, as well as criteria for compilation and selection of the realties registered. Prior to considering this aspect a "data-processing standard", it must be intended conceptually as a data characteristic and that it can be certified.
In conclusion, logical structuring of

information together with the enormous potential of new technology has brought to the fore that the essence is, above all, a theoretical issue oriented to design an effective information system to gain maximum efficiency.

1. The research project (co-ordinated by prof. Amedeo Bellini, Politecnico di Milano) comprises ten work groups engaged in research applied to different disciplinary fields. Further information is available on the website allocated to the research programme:

http://file-server.cilea.it/beni-culturali

2. All the government bodies are thanked for having made available and/or permitted consultation of part material examined and processed in the course of the research.

3. This study has been the subject of a stage within the framework of an agreement between the Comune di Monza and the Politecnico di Milano (stager: Anna Pietrocarlo, an undergraduate of the facoltà di Architettura). The workshop, of a month's duration, reviewed the activities of the Town-Planning and Programming Sector of the Municipal Administration.

4. Regione Lombardia-Politecnico di Milano, *Inventario dei beni culturali immobili e delle loro pertinenze, prosecuzione (schede A/I e PG/I).* The archive contains about 1,300 A sheets (architectonic works) and PG sheets (historical parks and gardens) relating to properties situated in Monza. (onsite investigation year: 1996). Provincia di Milano-Isal, about 400 sheets are historical buildings and environmental heritage registered in Monza.

5. Prototype study and construction is the subject of a degree thesis: *Un sistema informative territoriale quale strumento per l'espletamento delle competenze provinciali in materia di beni culturali – Un prototipo per la Provincia di Lodi,* authoresses: Nesta Ruth Sangermano, Marta Moraschi, (tutors: prof. Luca Marescotti, arch. Maria Mascione), Facoltà di Architettura – Campus Leonardo, Politecnico di Milano, 1999-2000.

6. The following catalogues have been appraised for use in the prototype: SIRBeC, Carta del Rischio and Frisl elaborated within the Regione Lombardia's projects, catalogues of the Provincia di Milano and of Provincia di Lodi, this last elaborated with the SIRBeC layout.

7. The image is part of the degree thesis *Un sistema informativo territoriale ...(op. cit.).* The authoresses are thanked for their collaboration in this work.

8. Software used in SIRBeC project, Carta del Rischio and Frisl.

About the authors
Luca Marescotti is Professor of Urban and Regional Planning Theory (Teoria dell'Urbanistica) at Facoltà di Architettura of the Politecnico di Milano) since 1982.
E-mail: *luca.marescotti@polimi.it*

Maria Mascione is Researcher at the Politecnico di Milano, DI.Tec Dipartimento di Disegno industriale e di Tecnologia dell'Architettura.
E-mail: *m.mascione@archiworld.it*

Quicklink: Managing Similarity Links in Cultural Heritage Archives on the Web

Isabella Gagliardi (*), *Bruna Zonta* (*)

(*) ITIM - CNR, Via Ampère 56, 20131 Milano, Italy
fax +39 02 70643292
e-mail: {isabella, bruna}@itim.mi.cnr.it
http://www.itim.mi.cnr.it

ABSTRACT

The retrieval of visual information from large archives of images is a need shared by many application domains. In some environments, such as that of cultural heritage, the pertinence and quality of the information are elements of primary importance. Retrieval based on so-called low-level features does not guarantee these properties: the image alone of a cultural artifact cannot convey all the necessary information; it must be completed by a description, and by the historical and environmental collocation, that is, by an accompanying text. However, systems that retrieve images through text in a completely unsupervised manner have been seen to produce results that are not only unsatisfactory, but difficult to understand for even experts in the sector to understand.

We present here a web extension of Quicklink, a system which: retrieves images similar to a query image in large web archives of artworks by dynamically matching their textual descriptions, adapts its own behavior to user requests and presents the retrieval results in HTML pages, where the images are ordered according to their similarity degree.

KEYWORDS: Quicklink, image retrieval, information retrieval, perceptual/ conceptual similarity, similarity link, web art archives

OUR APPROACH

The images are retrieved by assigning them a value of decreasing similarity to a given query image, calculated on the basis of their descriptive texts, generally brief and relatively homogeneous in form and content, as the cards of an art catalogue usually are. When this similarity is found, the texts are connected by a link of the associative type.

The similarity model used is the conceptual "contrast" type, which considers similarity an intersecting of features, in this case, of terms. The basic idea is that the presence of terms common to two different cards indicates that these can be considered similar to each other. The possible links thus identified are a function of the number of terms present in the two cards, and have a "strength" of (0,1). Precisely, the similarity between two texts is defined by the number of terms in common in proportion to the total number of terms on the two cards. This model clearly suggested the use of Salton's well known formula [7, 8], to which weights could be added:

$$sim_{i,j} = \frac{2(w_i term_i \cap w_j term_j)}{w_i term_i \cup w_j term_j}$$

wi is the weight associated with the termi throughout the catalogue, as we shall explain below.

The results depend, obviously, upon the terms chosen for comparison. This choice can be made in two contrasting manners: automatically, using lists of stopwords, or manually, by having experts in the domain indicate the more significant terms according to certain criteria.

To this purpose, three dictionaries were created, the first automatic, the second controlled, and the third weighted and controlled [5, 6]. These dictionaries differ both in the way they are built and in their semantic content. Their performance was tested by expert observers, who were asked to evaluate the perceptual similarity of the images retrieved with respect to the query image. The Weighted Controlled Dictionary was found to produce the best results; therefore it was the one employed in the subsequent experiments. A taxonomy was also created, grouping the terms in categories, and these in turn in higher level categories, for a total of three levels.

The weight can be assigned automatically, on the basis of the number of times the term occurs in the entire collection, following well-established procedures of Information Retrieval (IR), or manually, considering the importance of the term in the domain, or in the collection, regardless of its frequency. The former procedure was used here: the greater the number of times a term appeared in the whole archive, the smaller the weight assigned it by default in the dictionary.

The value of the weights may vary from 0 to 1, with a granularity to be agreed upon on the basis of the archive dimension and variety.

A further procedure was designed to allow the user to assign a greater/lesser weight to some terms (or categories of terms) at query time, in order to express his specific interest.

RELEVANCE FEEDBACK

After a query, when the system presents the retrieved images, the user is allowed to progressively refine its response by applying the mechanism of relevance feedback [4] to indicate the images' pertinence or non-pertinence. The user selects one or more of the images retrieved by the system to serve as positive, or negative examples. Following the user's choices, the system temporarily increases or diminishes the weights of the involved terms, and then reapplies the similarity algorithm. To do so the system constructs an "ideal query", the terms of which are the union of the terms of the positive examples, with their weights modified as follows:

- the weights of terms common to the positive examples are increased by as many degrees as the times they occur, up to a maximum of 1;
- the weights of terms common to positive and negative examples are

decreased by as many degrees as the times they occur, down to a minimum weight of 0.

The weights of terms found only in one positive example remain unchanged. Moreover, a double change may leave the weights of the involved terms unvaried.

QUICKLINK ON THE WEB

The web site of Quicklink provides an intuitive and user-friendly mode of interaction.

The first screen shows 20 thumbnail images with their identification codes (Figure 1); the user may run through the entire archive in this manner. To see the card corresponding to an image, he simply clicks on it (Figure 2). He selects the image for the first query by clicking on the YES button. He can modify the weights of one or more categories by pressing WEIGHT increasing, or decreasing them by one, or more degrees; otherwise the default weights are kept. He then uses the SIMILAR button to have the system calculate the similarity of the textual card corresponding to the query with the cards of all the other images in the archive, showing the user the query and the 19 most similar images ranked in decreasing order and accompanied by their coefficient of similarity (Figure 3). Whenever he wants to, the user can click on the TERMS button to look at the terms and relative weights that have led to the current result (Figure 4). If the user is not completely satisfied with the results obtained, he can indicate examples he judges positive and those he considers negative (selecting YES or NO on the new display). With the FEEDBACK button, the system builds a new query, using all the terms present in the positive examples, increased in weight according to the frequency with which they occur in these examples and decreased in weight according to the number of times they appear in the negative examples. The new calculation of similarity produces a display of the 20 images that most resemble the choices made by the user (Figure 5).

The catalogues of the Faenza International Museum of Ceramics form the principal part of the test bed. The study is part of the Italian National Research Council (CNR) project "Beni culturali: metodi e strumenti per la creazione di archivi multimediali nel settore della ceramica" (Cultural Heritage: methods and tools for the creation of multimedia archives in the sector of ceramics) developed at ITIM-CNR of Milan, in collaboration with IRTEC-CNR of Faenza [1].

CONCLUSIONS

The system, first designed as a stand-alone prototype, is presented here in its web extension. Some preliminary experimental results have been shown and commented.

Comparing Quicklink with CBIR systems, which use visual features for the retrieval of similar images seems to make little sense, especially since the choice to have Quicklink operate on the subjects – and therefore on the semantics of images specifically - produces results altogether different from those that can be obtained using low-level visual features. It appears more reasonable, instead, to compare the feature category of color in the two approaches. This is what the authors intend to do experimentally in the near future.

What is in any case desirable is a complete integration of the two types of system. A first attempt in this direction is already in operation [2, 3]. It is a general purpose tool that uses automatic dictionaries, that reduce the semantic power of the terms, thus drawing the two methods closer together. The results to date are encouraging.

REFERENCES

1. Carrara P., L'Astorina A., Ghiselli C., Padula M. Ceramica d'arte in Italia, (in Italian) *http://jargo.itim.mi.cnr.it/Ceramica*

2. Ciocca G., Gagliardi I., Schettini R., Quicklook: a content-based image retrieval system with learning capabilities, IEEE Multimedia Systems 99, IEEE Computer Society, Vol. II, (1999) 1028-1029.

3. Ciocca G., Gagliardi I., Schettini R.Quicklook2: An Integrated Multimedia System With Learning Capabilities, Journal of Visual Languages, submitted
4. Ciocca G., Schettini, R. A relevance feedback mechanism for content-based image retrieval, Information Processing and Management, Vol. 35, pp. 605-632, 1999 (SCI 3308)

5. Della Ventura A., Gagliardi I., Zonta B. The role of dictionaries in the Automatic Generation of Hypertextual Links, 2nd International Congress on “Science and Technology for the Safeguard of Cultural Heritage in the Mediterranean Basin”, Paris, 5-9 July 1999

6. Gagliardi I., Zonta B. Comparing dictionaries in automatic construction of hypertextual links: a case study in art catalogues, Visual Information and Information Systems (D.P. Huijsmans, Arnold W.M. Smeulders eds.), pp. 358-366, 1999.

7. Salton G., A. Singhal, M. Mitra, C. Buckley, “Automatic text structuring and summarization”, Information Processing & Management, Vol. 33(2), pp. 193-207, 1997, Elsevier Science Ltd.

8. Salton G., McGill M. Introduction to Modern Information Retrieval. McGraw-Hill, 1983.

ABOUT THE AUTHORS

Isabella Gagliardi took her degree in Physics at the University of Studies of Milan in 1985. She has been working at the National Research Council (CNR) since 1986. She has oriented her research activity in the field of multimedia Information Retrieval, particularly attending to the study of new methodologies for image storing and management. Her research interests are in multimedia databases and

hypermedia systems, especially in the Cultural Heritage and textile fields.
E-mail: *isabella@itim.mi.cnr.it*

Bruna Zonta received her Laurea degree in Classic Philology from the University of Studies of Milan in 1959. Her main interest is in the areas of: semantics, representation of knowledge, construction of indexes, glossaries and thesauri; particularly in the field of cultural heritage applications. She is a researcher of the CNR since 1961. In 1994 she joined the ITIM. In 1997 she retired maintaining, however, a research relationship with ITIM as an external collaborator. Co-author with Silvio Ceccato of "Linguaggio, consapevolezza, pensiero" Feltrinelli Editore, Milano, 1980
E-mail: *bruna@itim.mi.cnr.it*

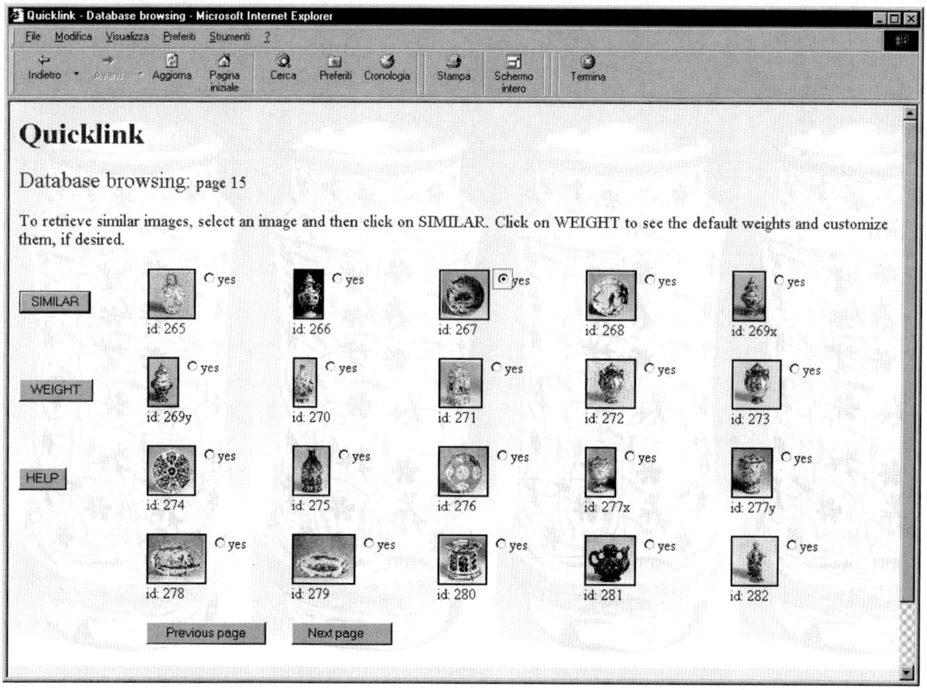

Figure 1: database browsing, with the initial query (id 267)

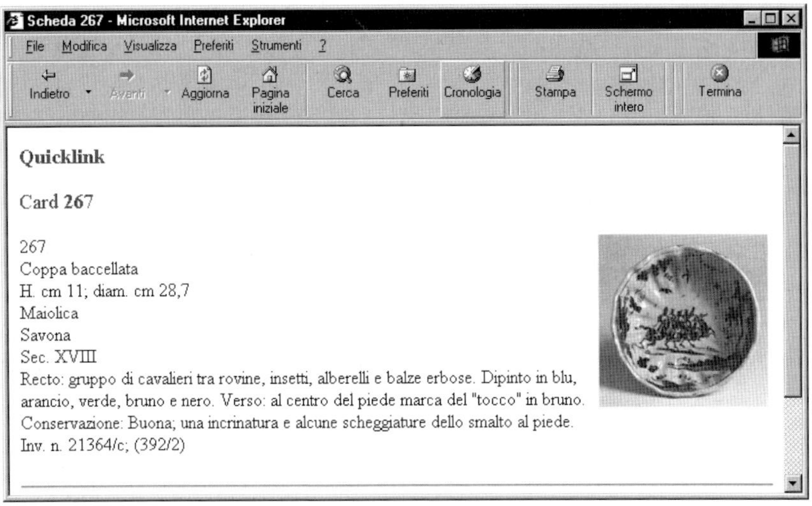

Figure 2: a card (card id 267) and its image

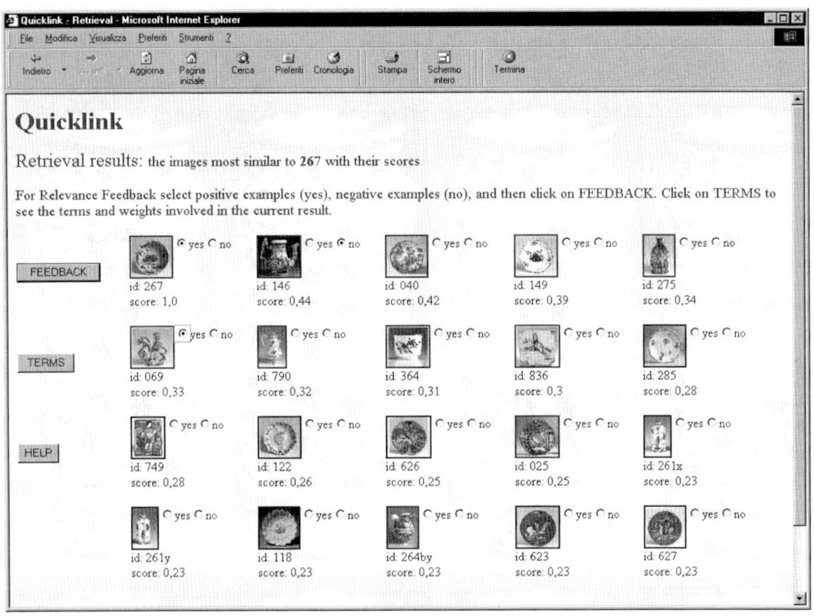

Figure 3: retrieved images as result of the query (id. 267), with positive and negative examples for feedback

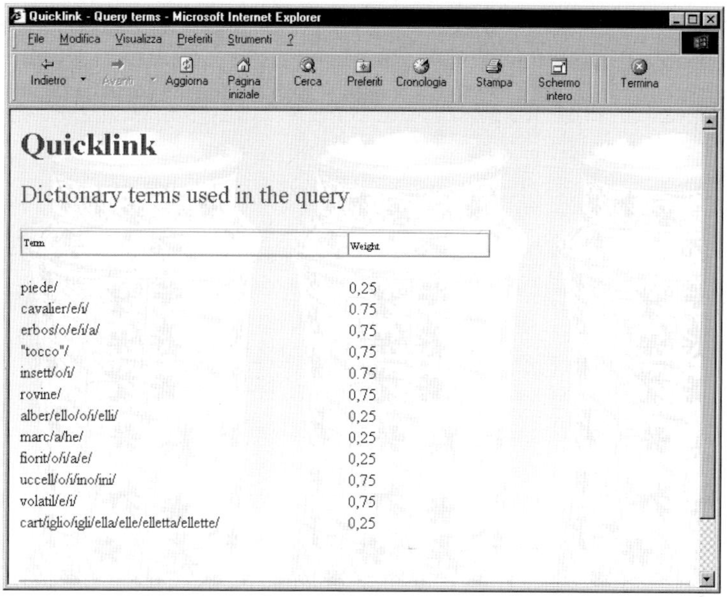

Figure 4: dictionary terms used in the query

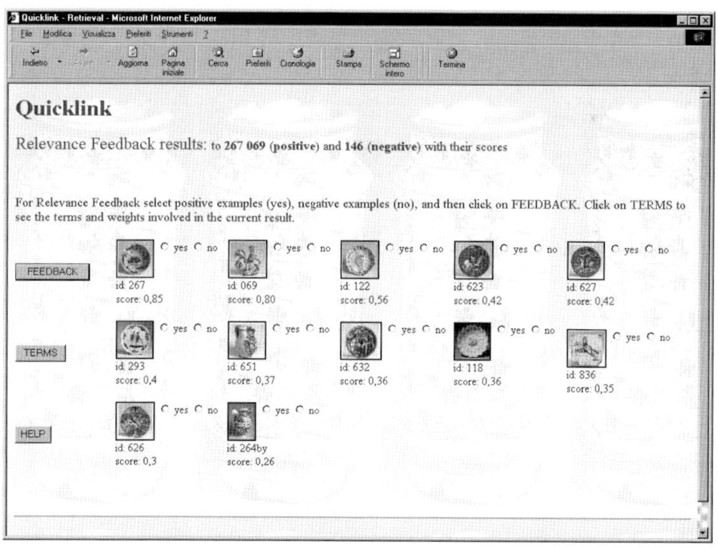

Figure 5: retrieved images as result of the feedback

Mobile Devices to Assist Cultural Visits

M. E. Bonfigli (*), *G. Cabri, L. Leonardi and F. Zambonelli* (#)

(*)Vis.I.T. Lab, CINECA Supercomputing Centre, Casalecchio (BO), Italy
E-mail: e.bonfigli@cineca.it

(#)Dipartimento di Scienze dell'Ingegneria, Università di Modena e Reggio Emilia, Modena, Italy
E-mail: {giacomo.cabri, letizia.leopardi, franco.zambonelli}@unimo.it

ABSTRACT
Traditional visits to environments interesting from a cultural point of view are based on technologies that do not permit flexibility and modification to the visit itself on the basis of the visitors' profile. Software agents running on personal mobile devices can overcome such limitations and lead to achieve personalised visits. The innovative technology based on mobile agents seems to well fit this scenario, where the agents on the mobile devices can interact with fixed HW/SW infrastructures. This paper proposes the reuse of several concepts from the mobile agent technology in the design of distributed applications based on mobile devices. In particular, coordination issues are addressed, and reactive tuple spaces are proposed as coordination means: thanks to their openness and flexibility, they well suit a scenario composed by mobile entities.

KEYWORDS: Mobile agents, Mobile devices, Assisted visit

INTRODUCTION

Cultural spaces are all those environments, outdoor or indoor, where some elements of cultural interest (*cultural goods* in the following) are present. A typical example of cultural space is a museum, but also a city with its monuments can be considered a cultural space. Currently, the visits to cultural spaces are based on technologies that, even if consolidated, do not allow the required flexibility to personalise a visit on the basis of different criteria, such as the personal tastes and knowledge, the already done paths, the time availability, and so on.

This work focuses on the possibilities of exploiting software agents running on mobile devices for the visit to cultural spaces. *Mobile devices* mean all those hardware devices that are autonomous and portable. They have a certain computational power that ranges from the lowest one of GSM mobile phones, which have a high degree of portability, to the highest one of the notebooks, which are less comfortable to be transported; in the middle we can find devices that have a medium computational power and a medium portability, such as the palmtops or the subnotebooks. Anyway, all of them have a graphical interface for the interaction with the user and an interface to exchange data with fixed infrastructures, for instance via an infrared port.

The *heterogeneity* due to the presence of several kinds of mobile devices calls for studying solutions that are so flexible as to adapt to the different situations. In particular, the same information must be shown by devices that can have different capability in terms of computation, portability and, above all, presentation.

This paper proposes the use of *agents* executing on mobile devices. Being on devices that dynamically change their environment, the agents can be considered *mobile* in their turn; so it is possible to reuse several concepts and results of the mobile agent technology. In particular, the paper will consider the coordination issue, since the interaction and the exchanging of information are fundamental activities in distributed applications based on mobile entities.

Starting from the existing HW and SW systems, we evaluate how to model the distributed application for the visit with mobile agents. In particular, we explore how much of the concepts and results of mobile agent technology can be reused in applications based on mobile devices. We claim that the coordination models based on tuple spaces grant the required adaptability to suit a general mobile scenario, no matter if the mobile entities are hardware or software. Further flexibility can be achieved by permitting the tuple spaces to react to the accesses in order to adapt their behaviour to the needs of applications or sites.

From the programming language point of view, the Java programming language is proposing as a fundamental tool for the development of applications in distributed and high heterogeneous environments, thanks in particular to the code portability on different platforms. Since the current implementations of the Java language require a significant amount of resources, its use in mobile devices must be carefully evaluated. A possible alternative that is gaining ground is the operating system WindowsCE, available for several palmtops. Obviously, it is at a different level with regard to Java, and it grants a more limited portability, but this operating system exhibits some interesting features that worth to be analysed.

The paper is organized as follows. Section 2 introduces the mobile agent technology and briefly discusses the coordination issue. Section 3 proposes the use of agents in mobile devices to assist the visit to cultural places. Section 4 concludes the paper and sketches future work.

MOBILE AGENTS

An innovative technology that is gaining interest in the distributed application area is the one based on *mobile agents* [5]. The mobile agent paradigm implies the definition of applications composed by proactive and network-aware entities (*agents*) capable of actively changing their *execution environment* by transferring themselves while executing (*mobility*). The current interest in distributed applications of this paradigm is widely justified by the advantages it provides over traditional approaches: *(i)* mobile agents can significantly *save bandwidth*, by moving locally to the resources they need, instead of requiring the transfer of possibly large amounts of data; *(ii)* mobile agents can carry with them the *code to manage remote resources* and they do not need the availability of specific servers on the remote node, leading to a more flexible application scenario; *(iii)* the execution of a mobile agent application does not require *continuous network connections*, because interacting agents can move to the same site when connections are available and interact without requiring further network connections. Consequently, *(iv)* the mobile agent paradigm intrinsically facilitates the integration of *mobile computing* systems in distributed applications: agents associated to a mobile computing device assume the role of mobile agents with regard to the fixed infrastructures.

In the distributed applications based on mobile agents, the flexibility is above all required with regard to *coordination* [7]. In fact, a fundamental activity in mobile agent applications is represented by coordination among agents and between agents and local execution environments. The distributed applications based on mobile agents are intrinsically collaborative and the definition of appropriate high-level coordination models to enable and facilitate interactions in the presence of mobility can have a strong impact on both application design and execution efficiency. The coordination models for the mobile agents that recently have been proposed with a high degree of success are those based on *tuple spaces* [1]. In these models, agent interactions occur via shared tuple spaces, residing on each hosting environment, which are used as information repositories where to locally store and retrieve messages. The most significant advantage of these coordination models is to achieve both spatial and temporal uncoupling in agent-to-agent interactions. When two agents need to exchange data, the sender agent can simply leave a message in the tuple space local to a given site, without needing to know neither where the corresponding receiver agent currently is, nor when it will read the message. When the receiver agent arrives at the given site, it can read the message left by the sender agent. This clearly suits a scenario where the position and the scheduling of the agents cannot be easily granted. In addition, being any agent-to-agent interaction forced to be performed via the tuple space, the hosting environment can easily monitor and control all interactions, thus leading to a more secure execution model. Furthermore, the accesses to the tuple space can be based on associative mechanisms, in a Linda-like fashion [2]: data can be retrieved from a tuple space on the basis of a pattern-matching scheme, without the need of explicitly naming the stored data but, instead, by accessing to it by its content. The well-known advantages of associative coordination models [4] particularly suit applications based on mobile entities. In fact, because of the dynamism and the openness of the environment, having a complete and updated knowledge of hosting execution environments may be difficult or even impossible, especially if it lacks a centralised control.

Further flexibility can be achieved by adding programmable reactivity to the tuple spaces: the behaviour of the tuple space can be dynamically programmed (by both administrators and agents) to be adapted to the peculiar needs of the sites or of the applications. The reactive tuple space models configure a technology suitable to facilitate the design and the development of distributed applications based on mobile agents. Moreover, they grant such flexibility as to result adequate to a scenario composed of mobile and heterogeneous entities.

ASSISTED VISITS TO CULTURAL SPACES

The adoption of a trendy technology, such as the mobile agent one, would permit to overcome the limitations imposed by traditional technologies. In particular, we propose the joining of the *physical* visit with a *virtual* visit, supported by innovative technologies. On the one hand, the *physical* visit permits the direct sight of a cultural good, which could be hardly substituted by the vision of the corresponding image, no matter if it is on paper or on video. On the other hand, the *virtual* visit permits to have more and updated information, which can be exploited adapting them to personal needs. The user that wants to visit a cultural space can therefore have a mobile device that guides him/her during the visit in a personalized, flexible and smart way.

The mobile device, put near the interesting cultural good, can exchange information with a fixed infrastructure. The software on the device (embedded or downloaded on demand), can explain the cultural good, selecting the pieces of information to be provided on the basis of the user personal profile; the profile could be defined by the user himself/herself or dynamically determined from previous visits. Moreover, keeping track of the done path, the software could, for instance, reference cultural goods already seen, and make some proposal about how to go on in the visit. The presence of more than one visitor must be taken into account, each of them with his/her own mobile device. In this case, they can "compete" for the visit to the same cultural good, or they can "collaborate" and decide to do the visit together, and in this case the software components loaded on the devices should agree on a path that can mediate the different needs and the different profiles.

We suppose that every cultural place is provided with a fixed infrastructure with HW/SW components to facilitate the local and remote visits. First of all, from the SW point of view, this infrastructure hosts a *database* that stores several kinds of information, above all about the cultural space, but it can also be considered as a repository for other useful information, such as the restaurant and hotels near to the cultural space. Then, there is a *Web server*, which is in charge of making the information in the database available for remote users, which, for example, want to know about the cultural space before performing the actual visit. From the HW point of view, the fixed infrastructure can provide a hardware component that permits the communication with the mobile devices near every cultural good (pictures, monuments, buildings, and so on); it can be considered as the local access point to the whole information system. From our point of view, the most interesting feature of the infrastructure is the capability of communicating with the mobile devices and exchange information, for instance via an infrared port.

There are two ways for exploiting the mobile agent technology in the applications based on mobile devices: the former one is to model the mobile devices as they were mobile agents; the latter one is to let mobile agents move from the mobile devices to fixed infrastructures and vice versa. The two alternatives are explained in the following subsections.

Agents Modeling Mobile Devices

The mobile agents can be used to model the presence of mobile devices in distributed applications [6]. In this case we can suppose that there is a fixed agent running on each mobile device. Since the device "moves", we consider that the fixed agent on a mobile device changes a part of its execution environment, so they have to face the peculiar issues of a mobile agent that changes its execution environment by moving from site to site. In this way, several concepts and results about mobile agent and mobile code [3] can be reused and exploited in taking into account mobile devices in distributed applications.

In the context of the assisted visit, a significant issue that must be taken into account is the exchanging of information between the mobile devices and a fixed infrastructure. Moreover, in the case that several users are attending the visit with their own mobile device, that fact that they can compete or collaborate implies that the exchanging of information can occur also among the mobile devices. Often, more than the simple exchanging of information is required, for example if synchronisation is needed, so we can talk about *coordination* of the involved entities, let them be fixed or mobile. In all cases, we propose the use of coordination models based tuple spaces. The fixed SW infrastructure provided with each cultural space can be considered as a site where the information useful to agents is stored in the form of tuples. Thanks to the associative mechanism, agents in mobile devices can retrieve such information in a flexible way, as explained in Section 0. In this case, the reactivity of tuple spaces can be used to recognise the user/device that has performed the request, and to prepare the information tuple to be returned on the basis of the user profile and the device capabilities. This leads to simplify the code of the agents and facilitate the programming task.

Agents moving from/to Mobile Devices

The latter way to exploit the mobile agent technology is to let personal agents move between the mobile devices and the fixed infrastructure provided with each cultural space. Mobile agents can perform their task (selecting information, for instance) also when there is no communication between the mobile device and the fixed infrastructure. For example, we can consider the users' home site, from which users can see a preview of the actual visit by means of a browser, and personalise it, for example by choosing the more appropriate path. In this way, users can also exploit several services, which can be directly related to the visit, for example the users can book visits, or even general services, such as the booking in a restaurant that is near the cultural space and is cheap enough for his/her funds. These services can be exploited *on-line* by a browser, but they can also exploited *off-line* by sending to the cultural space site an agent with the appropriate information, such as the preferred day for the visit, the range of prices for the restaurant, and so on. The same agent can negotiate the visit with the cultural space manager, and also with other agents to avoid too many people in the same place at the same time, allowing a more comfortable visit to all participants.

CONCLUSIONS AND FUTURE WORK

This work proposes the use of the mobile agent technology to assist the visits to cultural spaces. Users provided with mobile devices can be assisted by agents that not only explain the seen cultural goods, but also organise the entire visit on the basis of the users' profiles and provide the access to several related services. Sophisticated agents can perform more complex tasks on behalf of the users that sent them.

In context of virtual visits, the use of virtual reality technologies can improve the usability of the applications [8]. In fact, the users can access cultural goods that risk the damage, or which no longer exist, or of which we have only some fragments. Moreover, whole virtual environments can be built by putting together "objects" that are actually in different places and by simulating possible transformations in the time.

A future research direction can be a more strict integration with the Internet world, and in particular with the Web, so as to let users exploit different and complex services, also not directly related to the cultural space from remote.

Putting together the virtual reality and the Internet world, it is interesting to investigate the definition of a virtual-reality language for the description of cultural goods. Such language should be powerful to permit the description of complex picture or monuments, but at the same time it must be simple enough to fit both simpler mobile devices and slow Internet connections.

ACKNOWLEDGEMENTS

Work supported by Italian Ministry for University and Science and Technology Research (MURST) within the project framework of "MUSIQUE: Infrastructure for QoS in Web Multimedia Services with Heterogeneous Access" and by the University of Modena and Reggio Emilia with a fund for young researchers.

REFERENCES

1. G. Cabri, L. Leonardi, F. Zambonelli, "Mobile-Agent Coordination Models for Internet Applications", IEEE Computer Magazine, Vol. 33, No. 2, pp. 82-89, February 2000.

2. N. Carriero, D. Gelernter, "Linda in Context", Communication of the ACM, Vol. 32, No. 4, pp. 444-458, 1989.

3. A. Fuggetta, G. Picco, G. Vigna, "Understanding Code Mobility", IEEE Transactions on Software Engineering, Vol. 24, No. 5, pp. 352-361, May 1998.

4. D. Gelernter, N. Carriero, "Coordination Languages and Their Significance", Communications of the ACM, Vol. 35, No. 2, pp. 96-107, February 1992.

5. N. M. Karnik, A. R. Tripathi, "Design Issues in Mobile-Agent Programming Systems", IEEE Concurrency, Vol. 6, No. 3, pp. 52-61, July-September 1998.

6. A. L. Murphy, G. Roman, G. P. Picco, "Software Engineering for Mobility: A Roadmap", International Conference on Software Engineering, Limerik (IR), June 2000.

7. A. Omicini, F. Zambonelli, "Coordination for Internet Application Development", Journal of Autonomous Agents and Multi-Agent Systems, Vol. 2, No. 3, pp. 251-269, Sept. 1999.

8. K. H. Veltman, "Frontiers in Electronic Media", Interactions – New Visions of Human-Computer Interaction IV, 4, pp. 32-64, July 1997.

ABOUT THE AUTHORS

Maria Elena Bonfigli graduated with honors in Computer Science from the University of Bologna and received her Ph.D. in "History and Computing" from the University of Bologna. Since 1998 she has been collaborating with CINECA (InterUniversity Supercomputing Center), in a first time as external visitor and now as a research fellow. Her research interests include Virtual Reality and Human Computer Interaction applied to Cultural Heritage.
E-mail: *e.bonfigli@cineca.it*

Giacomo Cabri is a contract research associate in computer science at the University of Modena and Reggio Emilia. He received a Laurea degree in electronic engineering from the University of Bologna and a PhD in computer science from the University of Modena and Reggio Emilia. His research interests include tools and environments for parallel and distributed programming, wide-scale network applications, and object-oriented programming.
E-mail: *giacomo.cabri@unimo.it*

Letizia Leonardi is an associate professor at the University of Modena and Reggio Emilia, where she teaches basic and advanced computer science courses. She received a Laurea degree in electronic engineering and a PhD in computer science from the University of Bologna. Her research interests include object-oriented programming environments; parallelism and distribution issues, especially as they apply to object systems; and design and implementation of coordination infrastructures for agent systems.
E-mail: *letizia.leonardi@unimo.it*

Franco Zambonelli is an associate professor in computer science at the University of Modena and Reggio Emilia. He received a Laurea degree in electronic engineering and a PhD in computer science from the University of Bologna. He is a member of the IEEE and ACM. His research interests include parallel, distributed, and Internet programming; distributed algorithms for fault-tolerance; and debugging.
E-mail: *franco.zambonelli@unimo.it*

Data Models

A data description model for Cultural Heritage Hypermedia and Virtual Environments authenticated by archaeologists

Maria Elena Bonfigli (*), *Antonella Guidazzoli* (*), *Massimo Alessio Mauri* (*)
Antonella Coralini (#), *Daniela Scagliarini Corlàita* (#)

(*) CINECA –VIS.I.T Lab, Via Magnanelli 6/3, 40033, Casalecchio di Reno (BO), Italy
E-mail: e.bonfigli@cineca.it a.guidazzoli@cineca.it a.mauri@cineca.it
(#) Dip. di Archeologia - Università di Bologna, Pza S.Giovanni in Monte 2, 40124, Bologna, Italy
E-mail: ciwaco@tin.it scagliar@alma.unibo.it

ABSTRACT

In this paper we will show our data description model conceived for avoiding the risk of realizing misleading Virtual Heritage (VH) reconstructions and for making accessible in a standardized way the background research material that lies under a "VR cultural product". The basic idea is to face the problem of the lack of standards for documenting sources in order to realize not only nice fast models, but also historically accurate models, proposing a model conceived by an interdisciplinary team, that involves 3d modellers, computer scientists and archaeologists. Thanks to this kind of collaboration, it is possible to rely on a deep scientific knowledge of archaeological data both for their selection and representation as Hypermedia and Virtual Environments and for the authentication of interactive prototypes developed in order to access Cultural Heritage(CH) data through the Internet and the CINECA's Virtual Theatre. The proposed data model developed for our case study, the *Casa del Centenario* in Pompeii, is based on an extension of the Dublin Core (DC) metadata standard in order to describe a hierarchical structure conceived for describing archaeological objects and their relationships with iconographic sources - and digital reproductions.

KEYWORDS: Archaeology, Hypermedia, Virtual Environments, Dublin Core

INTRODUCTION

With this paper we intend to promote a methodology for avoiding the risk of realizing misleading 3D reconstructions and of transmitting erroneous information that could bring to a misinterpretation of the past. Despite the visual appeal of many virtual reconstructions, we think that new Virtual heritage is unfortunately at risk. It's often tilted up like a movie set on a weak foundation of battling digital standards, less than accurate data and it's doubtful many of these modern works will outlive the cultural heritage they seek to document. Dozen of virtual Pompeii's reconstructions now exist: there is a growing number of low resolution 3D and even semi-immersive virtual tours, but few really satisfy the preservationist's need for accurate documentation, the historian's need for archaeological interpretation or the public for engaging interaction and visual realism. Moreover, in our opinion the background research material that lies under a "VR cultural product" should be always easily accessible, and possibly, integrated with tools for analyzing the historical, archaeological and artistic sources. Consequently, the second aim of the proposed project is to develop an innovative interface system (implemented using a specific XML-based markup language) for accessing a corpus of Archaeological data from Pompeii, coded in tailored ways. From the point of view of applications the purpose is to allow different visitors (i.e. archeologists, students or museums) to access archaeological sources and explanations through the virtual reconstruction of buildings and objects of the past.

WORKING PHASES

In a period of great technological-digital evolution in all scientific fields, it is even more important to give particular attention to those interdisciplinary areas which will mark the *advant gard* of future research. In particular the great communicative impact that archaeology offers in itself is highly enhanced by using Hypermedia and Virtual Environments. In order to realize interdisciplinary projects it is fundamental to share different competences involving 3d modelers, computer scientists, and archaeologists. Thanks to this kind of collaboration it is possible to rely on a deep scientific knowledge of archaeological data both for their selection and representation as Hypermedia and Virtual Environment and for the validation of the prototyping activity.

Our work consists of realizing an authenticated virtual reconstruction of the *Casa del Centenario*, a very interesting double atrium house situated in Pompeii, as it was before the A.D. 79 eruption connecting this virtual reconstruction of the different rooms with the sources and the archaeological hypothesis on which they are based on. Moreover we consider equally important to reconstruct the house as it is nowadays in order to provide the users with a deeper understanding of what is the starting point of the archaeologists' studies (and also to give them a virtual access to an archaeological site in which and visitors are not admitted). The work is structured in three phases in which all the members of the interdisciplinary team are constantly involved and that are described in the following lines:

1. the first phase regards the creation of an appropriate data model and 3D modeling;
2. the second phase regards the definition of a markup language that describes archaeological objects, sources and 3D models for different user profiles and in accordance with different archaeological hypothesis and studies;
3. the last phase consists of designing and developing prototypes for visualizing virtual environments and related archaeological information through Internet and the Real Time semi-immersive display as CINECA's Virtual.

THE DATA DESCRIPTION MODEL

The *Casa del Centenario* is a large house with an impressive number of archaeological objects and sources. In order to set up a data model to describe archaeological objects -i.e. rooms, pavements, wall paintings, etc. - and their relationships with iconographic sources (i.e. photos, watercolour and drawings) that are on the basis of their virtual reconstructions. For example Figure 1.a shows a photo of the *lararium* taken in 1880 immediately afterwards the first excavations, while Figure 1.b shows the *lararium* as it is nowadays. It is worth noting that an area excavated over 120 years ago and not protected from raining and wind is now in ruins and so it appears very different and poorer of particulars. In the same way drawings and watercolors of details of wall-decoration painted in the 1880s can be very useful and interesting.

Figure 1: The lararium in 1880 and nowadays

The atrium has been selected as case study whether for his objective interest or his documentary situation, which offers stimulating possibilities of experimentation for 3D modelling and for data management. The atrium, one of the biggest in Pompeii, dates back to the original plan of the house, that probably dates back to the second half of the Third Century BC: it was a *tuscanicum atrium*, with *compluvium* and *impluvium*, surrounded on every side by rooms different in size and functions. In the following 300 years the atrium was rebuilt many times: we can recognize the traces of these reconstructions in the walls and, partially, in wall-painting and floor decoration. In the present situation, that corresponds to the situation shown by the excavation (1879-1880), the atrium shows the remains of the decoration realized nearly in the middle of the First Century A.D.: on the walls "Four Style" frescoes, including little pictures with tragedy and comedy scenes; on the floor, a simple black-ground mosaic, with parallel rows of *tesserae*. Almost nothing is kept from the original furniture: only traces (stains of rust on the mosaic) remain from the coffer placed in its north-eastern corner, on the left of the entering visitors. Certainly, in 79 A.D., the atrium was the main entrance of the house: the rooms gravitating toward it suffered instead a functional transformation, generally into store and service rooms. Once individuated the most important archaeological elements of the atrium we have organized them into a hierarchical structure (see Figure 2) conceived to be applicable to every room of the house. In general a room can be considered constituted by different archaeological objects as:

- founds retrieved during the excavations (e.g. pottery, statues, tools, candelabrums, jewels, lamps, etc.);
- single architectural elements (e.g. pavement, walls, roof, columns, etc.) that can be subdivided ulteriorly in subcomponents: i.e. in the pavement there can be fountains, thermal *suspensurae* or cisterns, a wall

can be considered as constituted by different wall-sectors and can include a door or a window;

- decorative elements relative both to founds and architectural elements.

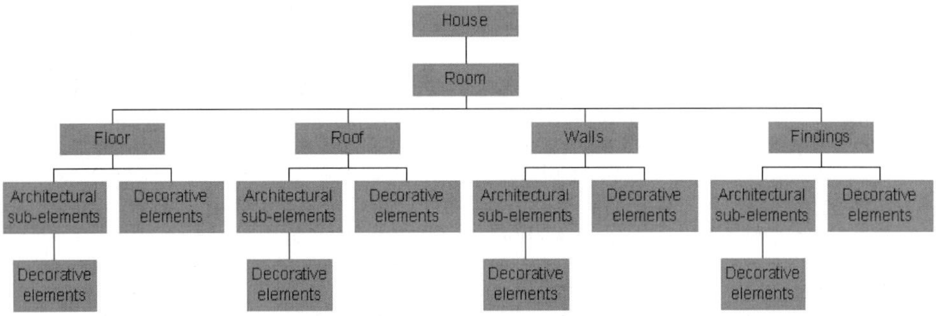

Figure 2: The hierarchical structure for a general room and for the case of the Main Atrium

Finally, each archaeological object of the hierarchical structure can be related with:

- **digital reconstructions**: **textures** reproducing disappeared or damaged wall paintings and decorations; **3D models** of archaeological objects reconstructing as they are at the present moment and as they were in AD 79.
- **iconographic sources**: **old and new photos**, **watercolors**, **drawings** or contact relieves;
  - **written sources** that describes archeological objects (for example that now doesn't exist).

Figure 3 shows some digital reconstructions and iconographic sources related to the Main Atrium West Wall.

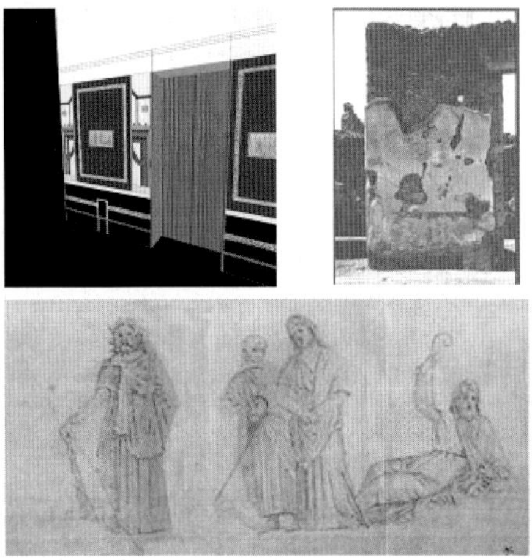

Figure 3: Digital reconstruction and iconographic sources related to the Main Atrium West Wall: a. the 79 AD 3D model, b. a recent Photo, c. an old Drawing

It is worth noting that the data model conceived for the Casa del Centenario can be extended to every kind of archaeological site virtually reconstructed. Basic steps of the resulting methodology are:
to individuate a hierarchical structure of archaeological object specific for the archaeological site itself,
to connect each "archaeological object" with iconographic or written sources and digital reconstructions (virtual textures and 3D models) following the DC rules explained in the next section.

3D MODELING

The 3D models that will be developed for the *Casa del Centenario* will regard the virtual reconstruction of the main *atrium*, *tablinum* (already finished) and of *peristilium*, *esedra* and *ninfeum* (work in progress) that constitute the most important area of the Roman house. The two resulting Virtual Environments (VEs) will show the central area of the house:

- as it was in AD79 at the time of the eruption that destroyed Pompeii in order to visualize archaeological hypothesis and theories based on the study of ruins, and findings and iconographic and written sources (e.g. the reconstruction of the roof that doesn't exist anymore, the reconstruction of decorative elements);

as it is nowadays on the basis of CAD maps and photos, in order to stimulate comparisons and to make possible the access in a area where tourists are not admitted. and as they were in AD 79.

Figure 5: VE of the main atrium as it was in 79 AD

Both the VEs are realized with Multigen Creator and then simplified in VRML lighter versions accessible through the Internet. The original high resolution Openflight models can greatly benefit from real life scale visualization through the CINECA's Virtual Theatre: a structure based on the Reality Center SGI Technology dedicated to the semi-immersive three-dimensional computerized visualization with a surround-sound system and a semicircular screen that allows the spectator to experience the illusion of a three-dimensional vision. The availability of these technologies opens up a new perspective and a new way of interpreting also already known aspects, realizing once again the more authentic sense of the word "theatre", that ancient Greeks intended as the space of observing in order to understand. Future environments for the fruition and study of Cultural Heritage could be conceived as a Virtual Portal, a virtual reference room linked to sources[6]. At the present moment we are investigating about different technologies to connect 2D data to 3D models in semi-immersive virtual environments developed for Cultural Heritage contexts.

THE DESCRIPTION LANGUAGE

The next work phase regards the implementation of a structured mark-up language describing archaeological objects, iconographic sources and digital reconstructions.

The starting point is the Dublin-Core (DC) standard [4], a metadata 15-elements set intended to facilitate the discovery of electronic resources. Actually it has also attracted the attention of formal resource description communities such as museums and libraries and its usability is tested by Consortium for Interchange of Museum Information (CIMI).

In order to describe in a coded way archaeological descriptions tailored on different languages (i.e. Italian and English) and on different users profiles, we have introduced some extensions: in particular we have considered to give different information to archaeologists and tourists and also to give the possibility to obtain "pure" DC records in order to assure the interoperability with other international Museum and historical archives and repositories.

Moreover we have introduced some extensions in order to include in the different archaeological descriptions hypothesis and comparisons with other finds or sources (in these cases the name of the archaeologist who made the hypothesis or the comparison is specified) and quoting (specifying author and references). The result is the definition of new tags and attributes

```
<description type="…" lang="…">
  <comp author ="…"> … </cfr>
  <hyp author="…"> … </hyp>
  <quot author="…"ref="…">…</quot>
</description>
```

The use of customized extensions of the DC standard grants the possibility to obtain in a very easy way "pure" DC records to assure the interoperability with other international Museum and historical archives and repositories. Moreover, using some "qualifiers" provided by DC itself to specify the type of relationship element[5], it is possible to describe either relationships among archeological objects in the hierarchical structure (using qualifiers *IsPartOf/HasPart* to describe the son-father relationships) or relationships among archaeological objects, iconographic and written sources and digital reconstructions (see the use of qualifiers *IsReferencedBy/References, IsBasisFor/IsBasedOn HasFormat/IsFormatOf* in Figure 6 and 7 below).

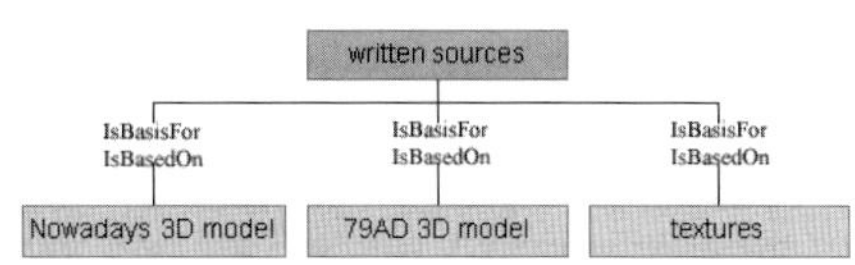

Figure 6: Relationships among written sources and digital reconstructions

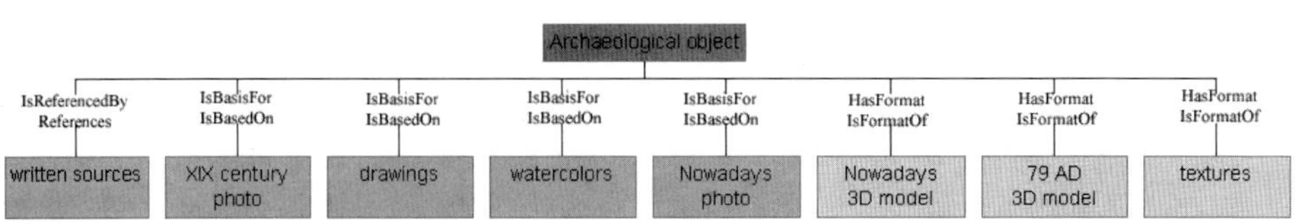

Figure 7: DC Relationships among Archeological Objects, written and iconographic sources, digital reconstructions

PROTOTYPING ACTIVITY

The last working phase regards the development of a web prototype in which the user, after having selected his profile and language, interacts with a VRML reconstruction of the main atrium. When the user selects an object in the VRML world a JavaScript function sends the corresponding XML file name to the server and the user's profile parameters to the server. On the server is installed Cocoon [5]: a Java publishing framework that returns to the JavaScript function a tailored versions of the XML file generated dynamically on the basis of the user's profile parameters. Finally the JavaScript function opens a window loading the resulting customized XML file and visualizing it as an HTML page using an XSL style-sheet.

At the present moment we have decided to visualize a pre-defined set of DC tags for each kind of users (archaeologists, tourists and DC users). Of course, using different XSL style-sheets it is possible to visualize the contents of a different set of tags. Figure 8 shows the information visualized for Italian archaeologists in the case of an archaeological object and of digital reconstructions (texture and 3D model): comparisons are visualized as lists, quoting in orange and hypothesis in green.

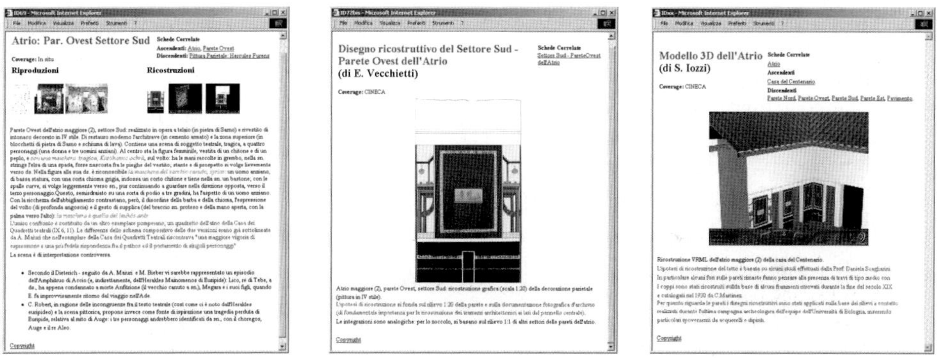

Figure 8: Some examples of information visualized for Italian archaeologists: the Atrium West Wall, a texture used for the same wall and the 3D model of the atrium

CONCLUSIONS

Summarizing, in this paper we have shown a methodology for avoiding the risk of realizing misleading Virtual Heritage reconstructions and for making accessible in a standardized way the background research material that lies under a "VR cultural product". In fact our prototype tries to face the problem of the lack of standards for documenting sources in order to realize not only nice fast 3D models but also archaeologically and historically accurate 3D models. Moreover the proposed work tests the efficacy of an extension of the Dublin Core metadata standard implemented with XML for a case study related to an archaeological site accessible through the Internet; but it is worth noting that we have worked in order to transfer the obtained methodology to other visualization platforms (i.e. the CINECA's Virtual Theatre) and also to other virtual reconstructions related to Cultural Heritage (i.e. the projects MUVI a Daily

Life virtual museum of the 20th century Bologna[8] and to NUME the historical reconstruction of a 4D city [1,2]). Currently this methodology and its XML implementation are used in the MUSE project[9], funded by MURST, whose aim is to realize a mobile multimedia system providing learning in museums and archaeological sites.

AKNOWLEDGMENTS
Authors wish to thank Tullio Salmon Cinotti (Bologna University), Fabio Sforza, Maurizio Malavasi (Boconsult) and all the people who collaborate to the MUSE project.

REFERENCES

1. M.E. Bonfigli, L. Calori, A. Guidazzoli, *Nu.M.E.: a WWW Virtual Historic Museum of the City of Bologna*, Proceedings of SAC 2000 - ACM Symposium on Applied Computing (J. Carroll, E. Damiani, H. Haddad, D. Oppenheim Eds.) Villa Olmo, Como, Italy, March 2000, Vol. 2, pp. 956-961.
2. M.E. Bonfigli A. Guidazzoli, ``A WWW Virtual Museum for improving the knowledge of the history of a City", in Virtual Reality in Archaeology, (J.A. Barcelo, M.Forte, D.H. Sanders Eds.),ArcheoPress, May 2000.
3. M.E. Bonfigli, L. Calori, A. Guidazzoli, M.A. Mauri & M. Melotti; "*Tailored virtual tours in Cultural Heritage worlds*", ACM SIGGRAPH2000 Conference Abstracts and Applications, New Orleans, July 2000, p.264.
4. CIMI *Guide to Best Practice: Dublin Core (DC 1.0 = RFC 2413), Version 1.1, 12 August 1999*, http://www.cimi.org/publications.html
5. COCOON, http://xml.apache.org/cocoon/index.html
6. B. Frischer, D. Abernathy, *The Use of Virtual Reality Technology for the Presentation of Cultural Heritage Sites,* Proc. of VAST 2000, Arezzo, November 2000.
7. VISualization Information Technology laboratory at CINECA: http://www.cineca.it/visit/
8. F. Bocchi, M.E. Bonfigli, L. Calori, A. Guidazzoli, M.C. Liguori, M.A. Mauri, M. Melotti & D.Vasetti; *Virtual Reality and Cultural Heritage: Some Applications*, Proc. of EVA 2000 , (V .Cappellini & J. Hemsley Eds.), Florence, March 2000, Pitagora Editrice Bologna, pp.162-165.
9. M.Malavasi, E.Romagnoli, T.Salmon Cinotti, F.Sforza, S.Summa, *MUSE: an Integrated System for Mobile Fruition and Site Management*, Proc. of ICHIM01; Milan, September 2001.

ABOUT THE AUTHORS

Maria Elena Bonfigli graduated with honors in Computer Science from the Bologna University in 1995 and received her Ph.D. in "History and Computing" from Bologna University in 2001. Since 1998 she has been collaborating with CINECA (InterUniversity Supercomputing Center), in a first time as external visitor and now as a research fellow. Her research interests include Virtual Reality and Human Computer Interaction applied to Cultural Heritage. E-mail: *e.bonfigli@cineca.it*

Antonella Guidazzoli graduated in Electronic Engineering from the Bologna University, Italy in 1988. She is the Project Manager of theVIS.I.T lab: the interdisplicinary laboratory at

CINECA (InterUniversity Supercomputing Center) dedicated to the visualization of massive or complex numerical data. Her main research interests include Computer Graphics and Image Processing Techniques, Innovative Implementations of Virtual Reality in non-traditional fields, such as artistic and cultural heritage and forensic applications.
E-mail: *a.guidazzoli@cineca.it*

Massimo Alessio Mauri graduated in Architecture from the University of Florence in 1995. He is a free lance architect and also writes for the architectural, design and graphics review "Ottagono". He is currently a fellow at the VIS.I.T Lab - CINECA as a 3D-world modeler. His activities include static and dynamic CAD systems and 2D-3D rendering for virtual theatres.
E-mail: *a.mauri@cineca.it*

Antonella Coralini graduated with honors in 1989 at the Bologna University, she received the specialization Archaeology in 1999 from Bologna University and the Ph.D. in Roman Archaeology at the Padova University in 1999. She had fellowships at Scuola Normale Superiore of Pisa, Italy, (1996) and the EPHE - Sorbonne – Paris, France, (1992-1995). Since 1998 she is the Scientific Secretary of the Bologna University "Pompei" Project.
E-mail:*ciwaco@tin.it*

VandA: a metadata model for supporting new usages of historical audio-visual archives material

Giuseppe Amato [(*)], *Donatella Castelli* [(*)], *Paola Venerosi* [(*)]
(*) IEI-CNR –Italy
E-mail: {G.Amato, D.Castelli, P.Venerosi}@iei.pi.cnr.it

ABSTRACT
The paper proposes a new metadata model (VandA) directed to support innovative exploitation of Historical AV Archives content. After an analysis of the needs of traditional and potential users, the model is illustrated through running examples to show at what extent the new usages can be supported.

KEYWORDS: metadata, digital libraries, video content description

INTRODUCTION

There is a growing demand for different and new ways of using the content of historical audio-visual (AV) archives. These range from different forms of TV programs, new on-line outlets in the Internet, CD-ROMs, etc., that illustrate and interpret events from the past, to research source of information exploited not only by historiographers, but also by teachers, that use it as interactive learning materials, and by program makers and film directors that are interested in observing how a particular actor/film director performed.

The aim of this paper is to highlight which are the cataloguing needs raised by the new supporting technology for historical AV archives and to illustrate how the use of an appropriate cataloguing metadata model can contribute to the satisfaction of these needs.

The paper begins with the analysis results of the new needs brought by the different classes of historical AV archives users (Section 2). Then, it proposes a new metadata model, called VandA, (Video and Audio) that combines descriptive features of traditional AV historical archives with features employed by advanced AV digital applications (Section 3). This model is illustrated through running examples that show to what extent it can support the new usages of the historical AV archives (Section 4). The paper ends by discussing how the model could be extended to overcome it limitations (Section 5).

USAGE AND REQUIREMENTS

The digitalization of historical audio visual archives allows the application of Internet advanced technologies, coming from interdisciplinary research, like the integration of historical archives enhancing new ways of content usage and distribution.

Three major classes of historical AV archives users, both usual and potential, were considered [6]. Here below we list these classes and for each of them we outline the Specific Functionality Required (SFR).

Educational institutions, teachers, specialists, students, researchers: they are characterized by an high level of domain expertise, but often a low level of video management systems knowledge.

- In the secondary education the audio visual material is employed both in lectures and in experimental activities. SFR: simple query interface (graphical appearance), reuse of content (browse-copy availability), multilingual capability.
- Historical researchers interests focus on exploiting historical archives of different countries with common access codes. SFR: finding aids, cross-search capability, personalization of presentation results, cross-language search.
- Cinema historians deal with film content and its expressive forms. SFR: information on framing typology, framing occurrence frequency, accelerated / decelerated camera movements, cutting effects, fades, etc.; feature based search; tools supporting automated analysis on film structure.
- Students experimenting film editing, the mixing of sounds, or the combination of both, learn how to handle the complex relations among media. SFR: highly qualified descriptors enabling "linking mechanism" for media synchronization.

Broadcasting companies, documentary makers, news editors, journalists, program researchers: they are the most involved in content management due to the need to produce new programs and reuse content in a competitive situation; they are also the most concerned with the "right issue".

- Journalist and documentarist activity leads often to the reconstruction of current or retrospective history about some subject, by composing video fragments. SFR: advanced finding aids, cross-document search for places, person, event, concept, etc.; annotation functionality; editor functionality.
- Programme researchers (broadcasting context) are involved to plan new manners of displaying archives, to organize the content in different views to satisfy the requests [16]. SFR: efficient repackage and scheduling methods of archive content; fast document/fragment identification; fast copyright issue investigation procedure [10].

Large Audio Visual Archives, archivists, librarians, marketing and sales managers: they are experts in video archive management systems and content, heavy users of existing systems and connoisseurs of articulated user needs.

- Archivists, librarians are engaged to modify their role from that of intermediary par excellence between documents and end-users to that of finding aids providers (inventories, registers, indexes, guides, collections), allowing people to search without the costly mediation of reference staff [2]. SFR: robust classification schemes providing to describe: all the aspects of an AV document in the whole, in its decomposition (audio, video, transcript), in its segmentation (sequence, shot, frame), in its physical structures (features); documents movement (migration) and preservation (refreshing); editor functionality.
- Large historical archives researchers are engaged to improve the archive

holding and to provide quality products.
SFR: granular description of content; finding aids (summaries, key-words, visual abstracts, story-board, thematic repackages of content, collections).

THE VandA MODEL

The new form of usages of historical AV archives reviewed above impose a re-thinking on the way of cataloguing documents. The VandA model presented in this section was designed to satisfy this need for a more complete cataloguing description.

VandA is an enlarged and revised version of the model designed within the European Chronicles On Line (ECHO) project[1] [19], [7], (*http://pc-erato2.iei.pi.cnr.it/echo/*).

The model is intended as a support for information discovery and as a descriptive surrogate of the AV resources disseminated to the users. It provides a necessary ground for supporting both the *traditional archive access services* - based on the descriptive fields about the content or the physical copy of the resource (e.g. access by name of the producer, by title of the series, by the tape collocation) - and a set of *advanced access services* - based on information automatically acquired through the manipulation of digital video (e.g. access by keyframe, by video abstracting, by words in the transcript, etc.).

The more authoritative reference metadata models for audio-visual resources [1], [5], [17] focus on specific aspects of the audio-visual resource description, and they do not provide a standardized harmonisation. VandA contributes to support the innovative AV historical archive organisation and usages.

VandA is an extension of a general metatada model proposed by the International Federation of Library Associations (IFLA) [14]. The extension allow to describe audio visual documents and to cope with the requirements discussed above. IFLA suggests to describe resources using four different entities: (1) *Work* entity to describe the abstract idea of a resource, (2) *Expression* entity to describe different versions of the same work, (3) *Manifestation* entity to describe different physical supports where an expression can be embodied, (4) *Item* entity to describe different copies of the same manifestation.

VandA, depicted graphically in Figure 1, extends IFLA with a set of sub entities and fields that are specifically tailored for describing historical audio visual documents. The entities are implicitly distributed on four description levels: each level corresponding to an IFLA base entity. The entities Work, Expression, Manifestation and Item are considered as the most general entity of the corresponding level. At the Work Level the entity *AVDocument* and its subentities *Newsreel*, *Documentary*, *Cut*, and *Reportage* are introduced for modelling an abstract AV historical work. AVDocument is a generic entity representing an intellectual or artistic

[1] ECHO - Project number 11994, Information Societies Technology (IST) Programme, funded by the European Commission. It aims to build a digital library of audio/visual historical documentary films. The first release contains selected materials from Institut National del'Audiovisuel (INA), Netherlands Audiovisual Archive (NAA), Istituto Luce (IL) and Memoriav.

work, while its subentities represent the different genres of historical audio-video documentary films.

The subentities Newsreel, Documentary, Cut, and Reportage model different kinds of works. These subentities turn out to be sufficient for our experimentation. However, other entities may be added at this level to represent other types of audio-visual resources (as fiction, entertainment, etc.) that require a specialised management. The important point here is that any extension of this type does not invalidate the descriptions made according to the previous model.

A newsreel may be composed of a sequence of reportages that correspond to various parts of the whole audio-video document itself. This situation is modelled by the relationship *ComposedOf*. This relation permits to describe independently the whole document and its individual parts. The elements in AVDocument may be related to the cuts from which are extrapolated through the relationship *CutFrom*.

Each document, belonging to the entities of the Work Level, may have one or more versions. For instance, an element of *Version* can be the audio realisation of a newsreel, another the audio-video realisation of the same newsreel. This is expressed by the relationship *ExpressedBy* between AVDocument and Expression. The entity Version is specialised by the subentities *Video*, *Audio*, *Audio/Video* and *Transcript*. These subentities represent, respectively, the following possible realisations of a work: audio only realisation, video only realisation, audio/video realisation, and transcription of spoken words. As the whole document and its parts share the same set of descriptive fields, the same entity is used for modelling both concepts. A relationship *PartOf* permits to associate a component of a document to the corresponding whole document; whereas the relationship *FollowedBy* express the temporal succession of the document parts and the type of transition between a part and the following one (cut, fading, etc.). The relationship *HasChannels*, defined on audio and audio-visual documents, models the different audio channels that an audio document may have. The relationships *HasAudio* and *HasVideo* create a link between an audio-visual documents and its audio and video components when these exist as separate resources. The relationship *HasTranscript* links the video document to the transcript of its audio (in the case that the audio document of that video document is missing) or links the audio document to its transcript.

At the Manifestation Level, the entity *Media*, with its specialisation *Analog* and *Digital*, represents different kinds of analog and digital supports that can be used to maintain audio visual document versions. Each version of the Expression Level can be stored on one or more Manifestation as described by the relationship *ManifestedBy*. The synchronisation among different elements of Media are modelled by the relationship *SynchronisedWith*. The entities Analog and Digital, that represent the entire document, are specialised in the entities *Analog-Seg* and *Digital-Seg*, that represent segments. The relationship *PartOf* permits to associate the part to its whole document.

At last level, the entity *Storage* describes general information about the individual copies of the documents.

Specific information about on-line and off-line document copies are modelled by the *On-Line* and *Off-Line* sub-entities. This partition permits, for instance, to represent a situation in which the same MPEG1 file may be stored on different servers, or in which several copies of the same CD-ROM or of the same VHS tape have different quality of preservation. The relationship *AvailableAs* links a manifestation to its items.

A set of specific descriptive fields are associated with each entity. They model the characteristic properties of each entity, i.e. they represent a description of the AV documents as perceived when looking at the AV document at a certain abstraction level. For example, a Version, which models a conceptual realisation of a work, is characterised by fields like: Title of the Version, Edition, Themes, Working Notes, Copyrights owners, etc. A Video, which is a particular Expression, is characterised by fields like: Subtitle Language, Video Abstract, Keyframe, Camera Movements, etc; whereas a Digital entity, which is physical embodiment of an expression, is described by fields like: Support Type, Format, BitRate, etc. The complete list of the descriptive fields can be found in [7].

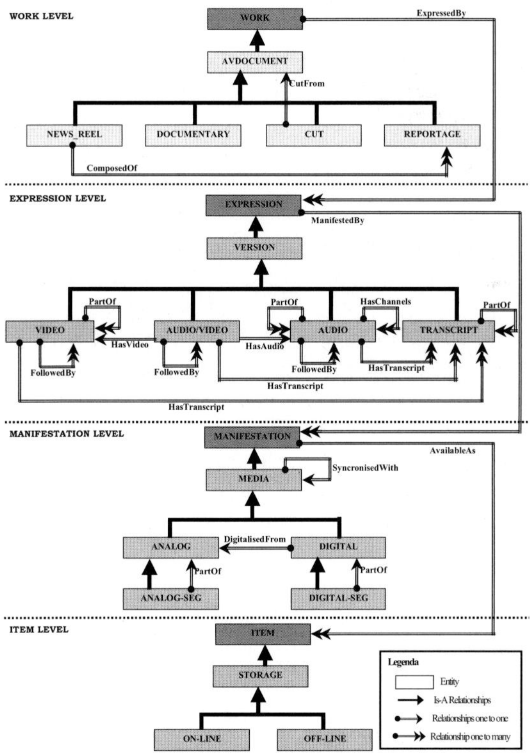

Figure 1: The VandA model

RUNNING EXAMPLE

The example discussed in the following is shown in Figure 2.

Work level

Let us suppose that we have a newsreel about the Landing of the allied forces in Normandy. This is represented by an instance of the Newsreel entity. The "Landing in Normandy" newsreel is composed of several reportages. For instance one where the US president announces the attack; one where the UK prime minister read an official declaration; one showing the soldier disembarking during the attack; one where aerial bombardments are shown. These are represented by instances of the Reportage entity.
We also suppose that we have another newsreel regarding the Second World War. This is again represented by an instance of the Newsreel entity. This newsreel is also composed of several reportages. One of these reportages, the one about bombardments, is the same used in the Landing in Normandy newsreel. In this case, the works corresponding to the landing in Normandy and the Second World War share the work corresponding to the bombardment. It is important to point out that this only means that the description (metadata) is shared. Indeed since the reportage is the same there is not reason to replicate this information. However, it could happen, and in most case it will really happen, that the files containing the two entire newsreels have the same reportage replicated. The result is that the data corresponding to the audio/video stream is replicated, while we have only one description of the reportage.
The existence of an explicit relationship between reportages and newsreels allows users to retrieve and have access to all contexts in which A/V documents or portions of them have been used. This feature, as previously stated, is one emerging type of usage of video material.

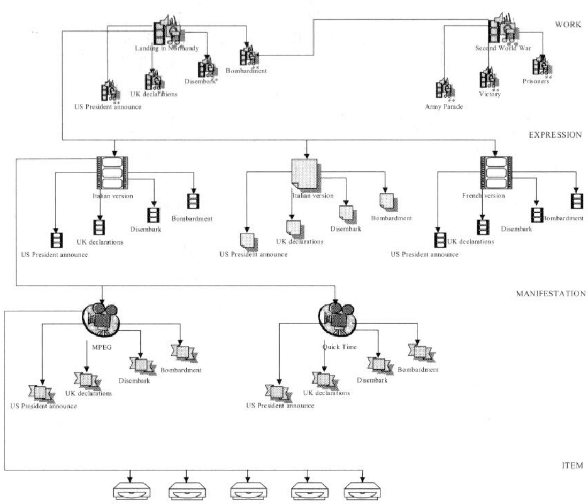

Figure 2: Running example

Expression level

Let us concentrate on the Landing in Normandy newsreel. Distinct version of a work can be modelled. These versions are implicitly related since they are conceptual realisations of the same work. Let us suppose that, in our example, it was produced in two different versions. One version was produced to be broadcasted in Italy, while the other to be broadcasted in French. The language is Italian in one version and French in the other. However, it is also possible that the Italian version and French version do not differ only for the language, but, for some reasons, some parts may have been cut or modified in both versions. This way, it is possible to inspect, not only how different phrases were translated in different language, but also the differences in which the same episode was reported in different countries. This type of usage is promising for video material accessed by teachers and students, when observing different versions of documents from a conceptual/analytic point of view.

In correspondence of the two versions, we also have two audio/video expressions and of course the corresponding audio only, video only and transcript expressions. In the figure, for simplicity, only the transcript expression corresponding to the Italian version was reported.

The description, contained at the expression level, characterises a particular version of a work. Among the others, here we can find attributes that allows us to perform similarity retrieval as for instance visual features (colour histograms, motion vectors, textures, etc.) of the video documents. In addition, attributes that give information on the style used for a particular version are included. For instance, in case of a video expression, information on the camera movement in a shot are specified along with information on the type of transitions between scenes (e.g. fade in, fade out, etc.). Using this information is possible to enhance the way in which documents can be retrieved, using the concept of similarity retrieval. As previously stated, specialists and researcher may need to analyse A/V material by using this kind of information.

Manifestation level

At the manifestation level, information on the physical support, on which expressions are available, is included. For instance, in our example, the Italian version of the Landing in Normandy newsreel is available in MPEG and QuickTime formats. This means that the "same" expression can be played using MPEG players and QuickTime players. The content is exactly the same, but is different the way in which it was encoded in the file. Different formats means also to favour different usages. If a document is requested for broadcasting maybe BETACAM is the right format.

Here also portions of the entire documents can be identified. For instance, the reportage corresponding to the bombardment can be precisely identified in the MPEG and QuikTime file.

The possibility to identify portions of A/V material also allows people (documentary makers, news editors journalists, etc.) to reuse these components by creating new documents that are created by extracting segments from other existing documents.

Item level

At the item level different copies of the same manifestation are described. In our example we supposed that the MPEG file, corresponding to the Italian version of the Landing in Normandy news reel, is mirrored on five different sites. That is, the file or its physical support (CD, tape, etc.) can be found in five different locations. Notice that these different copies may have, for instance, different access speed, preservation quality, therefore different access prices, and different authorisation information associated with them.

FINAL REMARKS

In designing this metadata model great care was taken to render it extensible. This is a key feature of any metadata model since it permits a successive expansion to cover new emerging needs.

Expansion may regard an addition of a new genre of *Work* form (in our case a new genre of film), or to be an expansion by specification when a certain type of usage claims for more articulated description level of a metadata (metadata qualification), or to simply represent the incorporation into the model of some further concept (metadata).

Now, we are involved to inspect the coverage level allowed by the model in respect to the stated usages and to provide the extensions when needed. In our model, for example, the metadata and attributes with which cinema historians are heavily concerned are only the essential ones.

Another aim in designing the model was the provision for interoperability. We are currently experimenting the mapping of this model into other metadata models to evaluate its interoperability level, starting with those belonging to the IFLA family.

About the reuse of AV material we will also remark that the current version of the model was designed taking into account the semantic approach rather the right approach of reuse. In fact the organization of the model permits to represent quite well the history of the resource content (programme, extracts) along the reuse phases and their context; meanwhile the right requirements for the reuse weren't kept in consideration in our inspection intents.

The reason is that right information often encompasses intellectual property rights, copyright, and other rights. Thus, if rights element is absent, no assumptions can be made about the status of these and other rights with respect to the resource.

To understand if a metadata model, directed to describe AV resources, should supply a rights specialization or an integration among models, applying each on a different aspect of the same domain (description of the resource and rights management), or furthermore assume the form of a e-commercial version [15], is a matter to inquire.

REFERENCES

1. ABC A Logical Model for Metadata Interoperability. *http://www.ilrt.bris.ac.uk/discovery/harmony/docs/abc/abc_draft.html*

2. American Heritage Virtual Archive Project. *http://sunsite.berkeley.edu/amher/proj.html*

3. A Metadata Model for Historical Documentary Films. Giuseppe Amato, Donatella Castelli, and Serena Pisani. ECDL 2000, LNCS 1923, 320-331.

4. ConText: Towards the Evolving Documentary. Glorianna Davenport, David Murtaugh. ACM Multimedia '95, 580-593.

5. Dublin Core Metadata Element Set, Version 1.1: Reference Description, 1999-07-02. *http://purl.org/DC/documents/rec-dces-19990702.htm*

6. ECHO User Requirement Report. Annemieke de Jong, Johan Oomen (NAA); Pasquale Savino, Paola Venerosi (IEI-CNR), Deliverable Number D1.2.1 of the ECHO Project, June 2000

7. Metadata Modelling Report. Giuseppe Amato, Donatella Castelli, Serena Pisani, Paola Venerosi (IEI-CNR); Philippe Poncin, Laurent Vinet (INA), Deliverable Number D3.1.1 of the ECHO Project, September 2000.

8. The FIAF cataloguing rules of Film Archives. *http://www.cinema.ucla.edu/fiaf/english/book.html*

9. Code of Conduct for Access to audiovisual Archives. Steve Bryant. FIAT/IFTA NewsLetter. *http://www.nb.no/fiat/2000-08/bryant.html*

10. How can Content and Asset Management systems solve the problematic issue of providing clarity on rights and copyright clearence for television archives? Catherine Lacken. FIAT/IFTA Vienna World Conference, 2000.

11. The Application of Metadata Standards to Video Indexing. Jane Hunter, Renato Iannella. Second European Conference on Research and Advanced Technology for Digital Libraries, Crete, Greece, September, 1998. *http://archive.dstc.edu.au/RDU/staff/jane-hunter/ECDL2/final.html*

12. A Comparison of Schemas for Video Metadata Representation. Jane Hunter, Liz Armstrong. WWW8, Toronto, May 10-14, 1999. *http://archive.dstc.edu.au/RDU/staff/jane-hunter/www8/paper.html*

13. The IASA cataloguing rules. *http://www.llgc.org.uk/iasa/icat/icat001.htm*

14. Functional Requirements for Bibliographic Records. Final Report. K. G. Saur München, 1998. *http://www.ifla.org/VII/s13/frbr/frbr.pdf*

15. Latest version of the model. *http://www.indecs.org/washington/offer.pdf*

16. The needs for and the means to exploit archives of television material. Sue Malden. DELOS Workshop on Audio Visual Digital Libraries, Santorini, Greece, 1998.

17. MPEG-7 Overview (version 3.0), Jose'M Martinez(UPM-GTI, ES) *http://www.darmstadt.gmd.de/mobile/MPEG7/Documents/W3445.htm*

18 The TV Programmes no one believed existed: Cooperation between Scholars and the Broadcasting Archives. Lasse Nilsson. 19th International Congress of Historical Sciences, Oslo, 2000

19 Building an Audio-visual Digital Library of Historical Documentaries: the ECHO Project, Pasquale Savino, D-Lib, November 2000

20 Trial Publication of RP 210A proposed SMPTE Recommended Practice-Metadata Dictionary. *http://www.smpte.org/stds/*

ABOUT THE AUTHORS

Giuseppe Amato graduated in Computer Science at the University of Pisa, Italy, in 1992. He is a member of the CNR-IEI in Pisa, working in the area of multimedia information systems. His main interests are model for content based retrieval of multimedia documents, access methods for similarity retrieval and approximate similarity search. He has published papers in many international conferences and journals in the areas of database information systems.
E-mail: *G.Amato@iei.pi.cnr.it*

Donatella Castelli graduated in Computer Science at the University of Pisa, Italy, in 1983. He is a member of the CNR-IEI in Pisa, working in the area of Digital libraries. Her current research interests include digital libraries systems and related interoperability problems, metadata models and formal methods. She has published papers in many international conferences and journals in the areas of information systems.
E-mail: *D.Castelli@iei.pi.cnr.it*

Paola Venerosi gratuated in Political Sciences in 1974 at the University of Pisa, Italy. She was member and now collaborator of the CNR-IEI in Pisa, working in the area of indexing, knowledge organization and library automation. She is interested in the following topics: document representation, e-learning. She published several papers and technical reports.
E-mail: *P.Venerosi@iei.pi.cnr.it*

From Parchment to the Network: Manuscripts Management and Cataloguing

Maria A. Alberti (*), *Giliola Barbero* (#) *and Cesare Pasini* (#)

(*) Dipartimento di Scienze dell'Informazione, Università degli Studi di Milano
E-mail: alberti@dsi.unimi.it
(#) Biblioteca Ambrosiana, Milano
E-mail: cpasini@ambrosiana.it, gbarbero@hotmail.it

ABSTRACT
Conservation libraries, which hold manuscripts and are willing to provide catalogues, face the problem of their diversity: each manuscript, document or archive has its own features, due to its history. Information technology helps librarians both to record ancient documents descriptions, dealing with highly non-standard data, and to allow their dissemination. Biblioteca Ambrosiana collaborates with DSI of the Università degli Studi di Milano to find solutions to a variety of problems related to management, catalogue and description of these documents. In this paper we present some results: a topographical inventory and a filter to interface and make compatible archives of manuscript documents in *Manus*, a relational DB distributed in Italy, and XML descriptions, adopting the DTD for manuscripts defined by the EU Project MASTER.

KEYWORDS: Digital library, Manuscript catalogues, XML

INTRODUCTION

Manuscripts collections preserve the entire knowledge of ancient western culture from Greek to Latin literature, philosophy and history. Also medieval vernacular culture and its products have survived, before the age of print, thanks to patient and often unknown copyists' work. We can access our cultural roots only by reading the manuscripts and the comments that have been written on them during the years, often in different temporal layers.

Codices, closed in libraries, archives and private collections have been waiting for hundreds of years their liberators, as if they were in prisons. The word *prison* has been used for the first time by an Italian humanist, Poggio Bracciolini in 1416, to refer to manuscripts' confinement, and in a way it sounds right also nowadays. After visiting St. Gall's monastery, where he found a complete copy of Quintiliano's *Institutio oratoria*, Poggio wrote: *Erant enim non in bibliotheca libri illi, ut eorum dignitas postulabat, sed in teterrimo quodam et obscuro carcere.*

Still at present many codices are preserved but left in a state of forgetfulness. Sometimes we don't know the exact contents of these treasures because they have not been studied; and even if they are, sometimes we have lost the information elaborated in the past by scholars who worked on them. Each manuscript copy of a work is indeed different and bears a unique witness of a particular stage of the tradition, due to the author himself, to the copyist or to some reader. Therefore philologists need to take all copies into account: they need to compare and collate every existing manuscript copy to edit a critical text of a given work.

This is the first reason why it is very important to catalogue manuscripts and their contents. If we want to come in touch with manuscripts before moving to the institutions where they are preserved, we must read all ancient and modern printed catalogues describing libraries' collections. Even these catalogues are not easily found but in few institutions. That's why information about manuscripts is often available only to scholars who can travel and move to libraries reading rooms. Information technology (IT) can help librarians and scholars to change this situation.

Biblioteca Ambrosiana, which preserves since the beginning of XVII century about 15.000 manuscripts and 10.000 documents on parchment [8], at the end of a complete restoration asked the Dipartimento di Scienze dell'Informazione (DSI) of the Università degli Studi di Milano to introduce IT in the library, particularly in relation to the management and cataloguing of ancient funds. Some efforts have been necessary to establish the communication between specialists and library's employers, not yet involved in IT. This phase was conceived as a necessary training, in order to manage all the human resources involved in the process. The collaboration that followed provided a fruitful ground for further developments.

Indeed, from the co-operation has descended a research activity concerning manuscripts cataloguing, with particular attention to XML, developed in connection with different national and European institutions.

MANAGEMENT OF ADMINISTRATIVE INFORMATION

Libraries already make use of IT to solve administrative tasks: such as loans management or library memberships. All the software products so far have a target of printed material more than manuscripts documents. These type of documents require different services. For instance, loans are not granted to individuals but to institutions for exhibitions; sometimes, for conservation concerns, digital or microfilmed copies are offered to the public in substitution of original documents; often, individuals are asked to justify their manuscripts requests and it may be useful to keep track of these users. These administrative tasks are simple in principle but traditionally require heavy duplication of procedures and back in 1997 no software system was known devoted to management of manuscript documents. Rarely, for instance, libraries have topographical lists of the ancient documents they own, which would allow to manage, by their shelf-mark, any movement that occurs to the single item: loan for exhibition, restoration, reproduction on microfilm or digitization. To keep track of all these procedures a curator, traditionally, at each step has to fill a form, write a card or several sheets with all the data required.

A topographical list for Ambrosiana

A topographical list of manuscripts has been implemented, with the aim to fulfill users and librarian's needs to save time and to ease library's management. It is the list of ancient documents preserved by Ambrosiana, where each manuscript is denominated by its shelf-mark, and shelf-marks are listed as they are *topographically* distributed in the library's rooms. After restoration all ancient funds have been checked, item-by-item. Cataloguers added information about custodial, restoration status and existing reproductions have been recorded. In the context of Italian conservation libraries such an electronic topographical list introduces an innovation.

The electronic topographical list is based on an object-oriented database, chosen because it allows producing an open archive, where many sorts of documents can be represented and related in different ways, by using the polymorphism and inheritance properties of object-oriented paradigm.

In this archive manuscripts, identified by their current shelf-mark, constitute objects holding data, such as: location in the library (room, floor, shelf number), reproduction (microfilm, CD-ROM), status of restoration (restored or not restored), custodial status (preserved in modern conservation box or not), availability (manuscripts are not available when they are send to restoration or exhibition).

The database is also a reference archive for library's users. When they join the library, they receive an identification number, which they will use for any later consultation. Users have no more to fill forms, which usually implies writing two or three copies of the same data like name, surname and address. Now people digit only their identification number and the shelf-mark they are asking for. The query output tells if the manuscript is available, suggests to use microfilm reproduction if the codex is too much damaged, pretends a special justification for studying a particular manuscript, if it is a very precious one.

The system provides some useful features: it recognizes the most typical mistakes in the shelf-mark digitations, as omissions of some parts or of punctuation marks. For example, if the right shelf-mark is *A 1 inf.*, the system accepts input also in different formats, such as *A. 1. inf.* or *a. 1 Inf.* This is most important because in the handwritten copies of catalogues, known to the public, the format may not have been used consistently. Moreover, if an inexistent shelf-mark is being given in input, the system suggests similar shelf-marks present in the database[1] [6].

MANUSCRIPT DESCRIPTION

A second important task of IT is to support catalogue management and publishing. There were many efforts to standardize manuscript description even before introduction of IT. Several scholars tried to codify the information necessary to give a complete description of each codex, but in practice cataloguers can find new type of data at any new manuscript description. This happens because manuscripts are documents produced by human hand, not by machine, and their typologies depend directly upon each producer. In printed catalogues data difficult to classify can be easily expressed in prose and may be found only in few cases: any record contains, i.e. the physical support or the number of folios, but other information can be found only in few records. Just to give an example:

> *Membr., ff. IV + 121 + IV'; mm.183x115 (124x80), 26 long lines.*

A line like the above holds typical information about physical description of the codex and almost any cataloguer places them in the same order in a record. More difficult is to give a codified description of a last page of a codex where, for example, a scholar has

[1] The system has been recently further developed by the company AISoftw@re based in Milan.

given a false date and a false attribution and some other scholars, deceived, apposed their final approval, like in the following excerpt:

> *At f.130v the text ends with "Januarii X M543": the last three numbers have been added over the original ones. Three notes follow, written in the XVIth century by three different hands "explicit opus lucubratum et compositum per me Franciscum Arethinum Theanensem", "Ego dominus Ioannes Antonius de Gallutio laudo", "Hoc opus bene recteque esse compositum et quidquid in eo continetur laudabile esse confirmo. Ego umanitatis magister Lucas Capuanus manu propria".*

The first attempts to produce electronic catalogues were made using database technologies. These attempts questioned the efficacy of database technology for cataloguing manuscripts, whose descriptions are typically very complex, diverse and highly unstructured, as we see in the above example.

Moreover, the systems implemented are so many and diverse that, while moving in the direction of manuscript wide accessibility, they open a new problem: the incompatibility between different systems that makes impossible to move data across different platforms. Finally, some systems may be run only on some hardware or require some particular software platform.

XML: A STANDARD FOR MANUSCRIPT DESCRIPTION

An alternative to database technology is the application of XML (eXtensible Markup Language), which is recently emerging as a standard also for manuscript description [1] [2]. The advantage of adopting the XML language consists in providing the possibility to tag technical and precisely defined information and to structure, in a less rigid scheme, also a prose description, frequently encountered in printed catalogues. In traditional database the verbose part of a handwritten catalogue either is lost or is inserted in generic *notes* fields, where the information lay unstructured and therefore un-searchable. Turning to the previous record, we exemplify how XML tags allow structuring precisely the information in the first line of the record:

```
<material>Membr.</material>,
<extent>ff. IV + 121 +
IV'</extent>;
<dimensions>mm.183x115
(124x80)</dimensions>,
<layout>26 long lines
</layout>
```

and how others tags, like the tag <name> can help to codify the story about the false attribution, as well, inside a generic prose tag <p>:

```
<p>
At f.130v the text ends with
"Januarii X M543": the last
three numbers have been added
over the original ones. Three
notes follow, written in the
XVI century by three different
hands "explicit opus
lucubratum et compositum per
me <name>Franciscum Arethinum
Theanense </name>", "Ego
dominus <name>Ioannes Antonius
de Gallutio</name> laudo",
"Hoc opus bene recteque esse
compositum et quidquid in eo
continetur laudabile esse
confirmo. Ego umanitatis
magister <name>Lucas Capuanus
</name> manu propria".
</p>
```

The definition of a standard allows making different archives compatible and therefore to exchange information, as well. Several institutions in USA and in Europe are strongly committed to elaborate a standard for electronic description of medieval manuscripts to allow communication between different catalogues. The main institutions are:

- Electronic Access to Medieval Manuscripts (EAMMS). The Andrew W. Mellon Foundation funds this project at the Hill Monastic Manuscript Library in Collegeville, Minnesota. It ends to determine the standards for on-line cataloguing of medieval and renaissance manuscripts and develops the tools for the two implementations of MARC and SGML/XML format.
  http://www.hmml.org/eamms/index.html
- Text Encoding Initiative (TEI). An international and interdisciplinary project, established in 1987, to develop guidelines for the encoding of textual material in electronic form for research purposes. TEI became a consortium in December 2000. Several projects use the TEI encoding scheme, also in archival and museum information field, but above all in publishing and managing electronic historical and literary texts. The University of Oxford, the Brown University, the University of Bergen and the University of Virginia host the TEI consortium. A sub-group is devoted to manuscript cataloguing.
  http://www.tei-c.org/
- Manuscript Access through Standards for Electronic Records (MASTER). A EU funded project, attempting to define and implement a general-purpose XML based standard for the description of manuscript material. Master is collaborating closely with TEI.
  http://www.cta.dmu.ac.uk/projects/master

EAMMS has two working groups: the first one focusing on a standard definition in MARC (Machine-Readable Cataloging) [9], the second one studying a SGML/XML DTD (Document Type Definition) for manuscripts description. As MARC is a standard created for the exchange of bibliographic records, the EAMMS group is trying to find proper solutions also for manuscripts' information.
The TEI Description Group is elaborating a DTD for manuscript description, while MASTER is using its own, similar but not identical to the American one. The two organizations are now working to come to a definition of a common DTD; the result of the collaboration is a revised MASTER DTD, compatible with TEI DTD [4].

Filter Manus/XML in Ambrosiana

We implemented a filter between the so-called *Manus* system and the MASTER DTD [2] [10]. *Manus* is a relational database, based on Microsoft Access, defined by the Italian Istituto Centrale per il Catalogo Unico (ICCU) and implemented by Finsiel/Telecom. Its design follows the traditional rules of manuscripts cataloguing, but being a database, as oppose to XML, it bridles cataloguers in a quite rigid scheme. For this reason we tackled the problem of providing compatibility between ICCU database and XML description of data. The implementation was carried over by student Stefania Smaldone as part of her thesis [10].
We analysed the correspondence between *Manus*' fields and the tags defined by the MASTER DTD and implemented an application to import into *Manus* archive descriptions

originally created in XML. This is done transferring the XML files in the Ambrosiana's object-oriented database, in the form of a syntactic tree: the tree root represents the document, the nodes are the XML tags and the leaves are the tags' content. The tree structure allows to search tags and to manage their contents, in order to transfer them into *Manus*' fields. We can therefore produce manuscripts description in XML, conserve them inside the Ambrosiana archive, and still be compatible with ICCU project as well.
Another procedure allows transferring the manuscript descriptions created in *Manus* into XML format, once more using the object-oriented database, as interface between the two different data description.
These procedures, written in Java, guaranty the compatibility of ICCU application, started as a national tool, with any other in the world adopting the XML standard. This approach represents a good solution for all those manuscripts cataloguing projects, which do not use *Manus* for data input, but still want to be compatible with this archive[2].

CATALOGUES ON-LINE

There are few catalogues on-line, but they are sufficient to demonstrate how much they can help scholars. One of the best examples of catalogue on-line is *Digital Scriptorium*, a project started in 1996 with a grant of the Andrew W. Mellon Foundation to the Bancroft Library of the University of California at Berkeley, and the Rare Book and Manuscript Library of Columbia University, in New York. Digital Scriptorium has been created with an Access database compatible with the TEI DTD. As Consuelo W. Dutschke, Curator of Medieval and Renaissance Manuscripts at Columbia, says: *Digital Scriptorium is a* web-based visual union *catalogue of medieval and renaissance manuscripts:* web-based, *in that it was intended from the start as an expandable and correctable tool, in the way that the static media of print or even CD-Rom cannot be;* union, *because it meshes into one database the holdings of a number of institutions; and* visual, *because it includes at least one image of every manuscript described, thus allowing the user to test and eventually correct the catalogued information by means of his own specialized knowledge*. At present Digital Scriptorium contains about 2200 entries, and 8000 images [3].
The aim of disseminating manuscripts' information through the Web is also the concern of the British Library's Manuscripts Catalogue, a big archive, started in 1994, by retrospective conversion of previously printed catalogues (the catalogue is, therefore, not a first-hand product). Rachel Stockdale, head of manuscripts cataloguing, explains the rationale: *"... [users] who now expect information to be available in their own homes and universities literally at the click of a button"*. To realize this project they scanned and parsed the data of about 60 volumes, and used them to populate a database, now accessible to the public [11].
References to other significant projects can be found in the EAMMS web page with links to major international web sites devoted to manuscripts and catalogues
(http://www.hmml.org/eamms/links.htm

---

2 The filter is in its prototypal phase. ICCU and Finsiel/Telecom are planning to extend Manus with the filter. Little steps like this one are most useful in the direction of standardization.

l). There are also some Italian projects [12] of electronic manuscript catalogues and we refer to some of them:

- *Progetto Galilei*, started by the Biblioteca Nazionale Centrale, Florence, in collaboration with the Max-Planck-Institut für Wissen-schaftsgeschichte, Berlin. The archive contains one of the most important documentation about history of science in XVII and XVIII centuries. http://www.bncf.firenze.sbn.it/
- *Dante on-line*, started by Società Dantesca Italiana. It offers manuscripts' descriptions and images of codices of works by Dante. http://www.danteonline.it/
- *Donne e cultura scritta nel Medioevo*. It begun at the end of 2000 by the Universities of Cassino and "La Sapienza" of Rome, in order to gather documents written by women till the end of XV century. http://edu.let.unicas.it/womediev ;
- the *Microfilms' Catalogue* of the Istituto di Storia del Diritto Italiano of the Università degli Studi of Milano is a catalogue of juridical manuscripts' reproductions owned by the institute http://www.historia.unimi.it/istituto/microstat.ht;
- the Biblioteca Universitaria Alessandrina, Rome has published an HTML version of the descriptions of manuscripts 236-450, with an historical introduction and some indexes http://www.alessandrina.librari.beniculturali.it/;
- the Biblioteca Pannizzi, Reggio Emilia has created an archive with some description of its medieval manuscripts. http://panizzi.comune.re.it.

All these experience pose the problem of distributed data-bases: it is obviously a strong point to be able to access an archive through the network, but even more would be to be able to send your same query to different archive and to be able to search different archives in a uniform way. To work in this direction standards are to be accepted by the community and an effort to open database and to provide filters has to be placed. The standard XML to describe data provides a solution to this and appears very promising but a consensus in the community is still needed.

MAINTENANCE IN TIME

Dissemination is the main advantage of IT applied to manuscripts. Standardization is a chief pre-requisite, that allows fulfilling this aim. Another condition is considering the persistence in time of electronic supports and the management and maintenance of systems.

Libraries should calculate the costs of replacing software and hardware in time, following the technological evolution, which at times is really so fast to force a breath taken technological run. Any digital project must be kept alive and it is not done once for ever. The software has to maintained, the platforms have to be updated, the hardware on which they run becomes obsolete and even the physical support of digital data may become obsolete. This experience differs from traditional experience of conservative libraries, where the supports of texts, like parchment and ancient paper, are still alive after hundreds and hundreds of years and we don't need to keep on changing them. Manuscripts can be read after centuries and sometimes they look stronger than our freshly printed bestsellers. What happened to the first electronic archives implemented in the recent years of the XX century, according to time scale of manuscripts, but in the prehistory according to the electronic evolution?

An archive rescue in Ambrosiana

Another result of the collaboration between DSI and Ambrosiana allowed saving the content of an old tape, the last witness of an early and far-sighted database realized in the 70's, under the direction of Angelo Paredi, Director of the library at the time. He had imported in Stairs the content of an old handwritten catalogue on 10.000 documents on parchments, dated from the 10th to the 19th century, and composed by Alessandro Bianchi [5]. The electronic archive for the time was a very innovative one and has been very much consulted. Some organizational changes and the end of funding left the archive totally un-supported, soon became unavailable and was forgotten[3]. Once the tape was found again, there was no machine to read it anymore and no one knew exactly what it was all about: only 20 years later. It required months of research, just to be able to find a machine still working to read the tape.

Finally, the data were exported in text format and patiently analyzed to find both their syntactical and layout features, which marked different kinds of information. For example: each description of a single document began with a number corresponding to a shelf-mark, while the second line represented the data of the document, and the third one the place where it was subscribed. The data were thereafter logically subdivided and were used to populate a new object-oriented database, compatible with the Ambrosiana archive [7].

CONCLUSION

Since the advent of the World Wide Web and the effort of initiative like TEI, important advances in digital catalogues of manuscripts have been made. The promise of a universal accessibility of manuscripts may still be a goal but it is in broadening in ways never envisaged before. The vision by Poggio of freedom of manuscripts by a too long secret custody can now count on technological development, like XML, which appears to be quite appropriate. There still is a gap between what is technically feasible and what is practically realized, due to a number of constraints that vary from economical limitations to librarian awareness. We suffer from a loss of communication between specialists of contents, and that's why we have undertaken dissemination initiatives. Recently at DSI, in collaboration with the MASTER project, the Università Cattolica di Milano and the Biblioteca Ambrosiana, we organized a workshop for manuscripts curator to introduce them to the XML DTD, developed in the MASTER project, and to experience the possibility of uploading manuscript description to a server on the web for accessibility. For most of the librarians this was a very new and intriguing experience.

Even though a universal catalogue may never be feasible, what is strongly desirable is that several national

---

[3] Most information about this archive have been found in a leaflet, originally printed for the use of library's students, which survived almost by chance, because found in the pages of the ms catalogue by Bianchi [5]

initiatives can meet other experiences and agree on a standard.

ACKNOWLEDGMENTS

The authors would like to acknowledge the initial impetus given by Gianfranco Prini of Università degli Studi di Milano to the collaboration between DSI and Biblioteca Ambrosiana; Mons. Pier Francesco Fumagalli, member of the Collegio dei Dottori of the library, for supporting the project and the MASTER initiative, and Peter Robinson, Director of the Centre for Technology and the Arts, De Montfort University UK, and leader of the MASTER project.

REFERENCES

1. Barbero G., Manoscritti, basi di dati e le riflessioni di Paul Canart, *Biblioteche Oggi*, 18/10, dicembre 2000, 32-35

2. Barbero G., Smaldone S., Il linguaggio SGML/XML e la descrizione di manoscritti, *Bollettino AIB*, 40/2, giugno 2000), 159-179

3. Barbero G., Digital Scriptorium, *Biblioteche Oggi*, 19/1, gennaio-febbraio 2001, 69-71

4. Burnard L. (ed), *Reference Manual for the MASTER Document Type Definition*, January 2001. *http://www.hcu.ox.ac.uk/TEI/Master/Reference/*

5. Bianchi A., *Inventario delle pergamene della Biblioteca Ambrosiana*, 12 ms volumes with shelf-mark K 201-212 suss

6. Fava E., *La Veneranda Biblioteca Ambrosiana: un sistema informatico per la gestione della movimentazione dei documenti e sua interazione con i cataloghi*, Tesi di laurea in Scienze dell'Informazione, Università degli Studi di Milano, AA 1996-1997

7. Ghezzi G., *Recupero di un regesto di pergamene e sua pubblicazione in Internet,* Tesi di laurea in Scienze dell'Informazione, Università degli Studi di Milano, AA 1998/1999

8. Pasini C., Catalogazione e conservazione nel fondo manoscritto dell'Ambrosiana, *Bollettino ABEI*, 8/1, 1999, 46-54

9. Pass G. A. (ed), *Descriptive cataloging of ancient, medieval, renaissance, and early-modern manuscripts.* Draft March 2000. *ftp://165.134.156.3/vatican*

10. Smaldone S., *Manus e Master: due differenti approcci alla descrizione dei manoscritti e conversione tra i rispettivi formati*, Tesi di laurea in Scienze dell'Informazione, Università degli Studi di Milano, AA 1998-1999

11. Stockdale R., Retrospective conversion of the British Library Manuscripts Catalogue: a description of the project, *Journal of the Society of Archivists*, 21, 2, 2000.

12. Zamponi S., Iniziative di catalogazione di manoscritti medievali, *Studi medievali*, s. III, 40, 1999, 369-393

ABOUT THE AUTHORS

Maria Alberta Alberti is professor at DSI, Università degli Studi di Milano. Her research field is multimedia systems and user interfaces. She is interested to the applicative field of cultural heritage.
DSI, Università degli Studi di Milano
Via Comelico 39, 20135 Milano
E-mail: *alberti@dsi.unimi.it*

Giliola Barbero, Ph.D. in Medieval and Humanistic Philology, since 1995 in Biblioteca Ambrosiana as manuscripts cataloguer. She collaborates with the MASTER Project and with DSI. She is editor of the column *Manoscritti & Computer* for the review *Biblioteche Oggi*.
E-mail: *gbarbero@hotmail.com*

Cesare Pasini is Vice-Prefetto of Biblioteca Ambrosiana, in charge of the Library. His research focuses on Greek paleography and patristic.
Bilioteca Ambrosiana
Piazza Pio XI, 2
20123 Milano
E-mail: *cpasini@ambrosiana.it*

The library: a cultural mine

Paola Carrara (*) *and Giuseppe Fresta* (#)

(*) ITIM-CNR, Milan, Italy
E-mail: paola.carrara@itim.mi.cnr.it
(#) CNUCE-CNR, Pisa, Italy
E-mail: g.fresta@cnuce.cnr.it

ABSTRACT
The paper proposes to consider virtual libraries on the Web as repositories of complex, digital documents to be not only viewed and searched, but also manipulated by users of different typologies. It presents a prototypal environment, the virtual desktop, in which content may be compared, manipulated, edited, etc. at the browser side. The solution is based on XML technologies such as RDF for designing purposes and SVG to render vector graphic.

KEYWORDS: virtual libraries, semantic Web, conceptual design, XML-RDF, XML-SVG

INTRODUCTION

Libraries, in their twofold function of collecting and safekeeping knowledge, are similar to Museum collections or, more often, to their stores; they are in charge of maintaining the integrity of their heritage as an evidence of the activities of producers of each single document, and as a memory support, of facilitating direct examination and research and also of improving popularization and education.

While the first two aspects are, from ancient times, matter of study and of huge efforts in order to suitably solve the general problem of cataloguing - just remember the Encylopédie of the Enlightenment movement -, the third is the most neglected, due to both the troubles linked to the direct vision and the lacking of reliable tools.

The aim of a virtual library is to put at users' disposition not only its catalogues - which are nevertheless indispensable -, but also its 'objects', exhibited on a virtual desktop instead of a precious case, unreachable to the hurried tourist urged by the following visitors or to students often badly involved in the communication project: all of them unable of critical meditation due to the shortness of visits.

This paper shows how the use of some technological standards recommended by the W3C (XML, RDF, SVG) may represent an adequate solution also to the needs of widespread fruition of the library as a part of the heritage, as long as tagging is guided by a deep analysis of the application and of its context performed by means of conceptual tools (like the E-R approach and heuristic rules from the IR field). Moreover XML-like technologies allow today to implement user Web interfaces comparable to those developed by proprietary tools, miming in a virtual space the most suitable metaphor, and offering a set of tools to manipulate objects directly on the browser.

DIALOG MODELS AND METAPHORS

The conversational (also called sequential or computational) and the asynchronous (also computational event-based) schema, are among the

many models of dialog suggested in the literature [1] and more often used in the human-computer interaction.
In the first case, the dialogue is of the kind query/answer and the user navigates in the hyperspace following one path among those suggested by the author. This type of Web applications exploits the fundamental characteristic of hypertext, that is navigation: the author is in charge of structuring contents so that the reader may explore them by following a personal path corresponding, in some way, to her or his needs, attitudes, interests, etc. The user maintains a passive role with respect to the content creation and manipulation, being only a 'reader', though a non-sequential one.
In the second case, the man-machine interaction is based on the contemporary availability of more tasks corresponding to different paths and metaphors. It is possible to strengthen user interactivity by, for example:

- manipulating pieces of content and reassembling them on the basis of needs and aims
- contemporary managing of documents under various format perspectives (texts, sounds, videos, graphs, images, etc.) to reach a level of expressiveness higher than in a separate, sequential display or to meet user requirements (i.e. temporary or permanent disabilities)
- following a link offering more destinations which the user can interactively select
- clipping and exporting some texts, images, etc. to compose a new document to be elaborated and integrated with other information
- allowing collaboration in the hypertext development which can be performed by different people, also working in remote sites, at different times, by means of network facilities, in order to exchange materials, discuss, and investigate.

The complexity is further increased if applications are conceived also as adaptive environments [4; 11], able to tailor content depending on the user's characteristics as a member of a group and as a person. Good solutions proposed in the past, however, were usually limited to local realizations on CD-Rom developed by means of proprietary software. Recent XML derived technologies allow to extend to the Web, in an open software environment, applications comparable with those obtained by Asymetrix Toolbook, Apple Hypercard, Macromedia Director, etc. on CD-Roms. It is now possible to follow the paradigm of the *virtual desktop* which provides for

- having all the wished "objects" at the same time on the desktop, each of them rendered by means of the best suitable metaphor (i.e., catalogue, map, space or time scales, etc.)
- building and keeping personal folders with documents, objects, annotations, sequences of selection, etc.
- managing "objects" on the desktop by a set of tools, offered by the browser; therefore, information is no more the final result of a procedure usually just performed elsewhere, but a semimanufactured product to be directly elaborated, arranged, commented, integrated, depending on users' needs, background, abilities, objectives, etc.
- modifying offered tools or building new ones (no designer nor

conceptual model can foresee all the future necessities)

This paradigm is particularly attractive in the field of popularization and fruition of the cultural heritage, with its constraints of geographic distribution, its access and conservation requirements, and, at the same time, its constant and intrinsic impulse to personal research, investigation, creativity. It requires, however, a reflection on the meaning and role of 'documents' in the digital world.

THE ELECTRONIC DOCUMENTS

The concept of document, which is fundamental for information professional, is no more easily defined in the electronic publishing perspective. Electronic documents are manipulable, linkable, searchable, and these characteristics are unusual in the traditional documents. A new definition must be found, such as: *document is a unit consisting of dynamic, flexible, non linear content, represented as a set of linked information items, stored in one or more physical media or networked sites; created and used by one or more individuals in the facilitation of some process or project* [10]. Within the above definition, by example, also bibliographic descriptions; paratextual characteristics (glosses, notes, etc.); metatextual characteristics; arrangement annotations, call numbers, etc. may be viewed as a unique document together with the information (books or other) to which they are related. Furthermore also display requirements and information structure are essential parts of the electronic document representation. The example in figure 1 illustrates the idea of a cultural object as composed of many pieces of knowledge, which must be suitably represented as far as digital format and representation metaphor.

The example considers material related to a Codex of the XIII Century, such as images, images of texts, their transcriptions in Ascii formats, and also videos, sounds, voices, etc. having one or more structure, one or more display representations, many links to other pieces of the same cultural object or to other objects.
Other related information is: context descriptors; searching constraints and requirements (in the content, in the structure, in some sections, etc.); instructions to aggregate and represent information in significant ways.

Illustrations (image format):

Original texts (image format):

Transcriptions Of The Original Texts, Such As:

MADONNA, IL FINO AMOR CHED EO VO PORTO
mi dona si' gran gioia ed allegranza
(ch'aver mi par d'amore)
che d'ogni parte m'aduce conforto,
quando mi membra di voi la 'ntendanza,
a farmi di valore,
a cio' che la natura mia me mina
ad esser di voi, fina,
....

Bibliographic Information (with structure):

Document Id: Canzoniere Vaticano Latino 3793
Reference: Letteratura italiana, diretta da A. Asor Rosa
Le opere. Vol I. Dalle Origini al Cinquecento
Publisher: Torino, Einaudi, 1992

Figure 1: An example of document as a cultural object

XML AND THE CONCEPTUAL MODEL

XML (eXtended Markup Language), the W3C recommendation for structured data on the Web, is a suitable solution to consistently build a virtual desktop for above documents, and separates display concerns from the content, which is mandatory to reuse data in different navigation environments without replicate them [12].

However, XML calls for a precise, a priori definition of the markup items (tags) for each document (or set of documents) to be created: it is the Document Type Definition (DTD) which states the valid elements and attributes and their constraints [8]. The creation of a DTD is usually naif, mainly based on the personal experience and background of the developer, whereas it must be considered, in our opinion, a complex process.

The following piece of text illustrates some examples of the troubles usually met in tagging a document in the cultural heritage field.

> Il Vaticano Latino 3793 (V) e' il più ricco dei canzonieri duecenteschi (sec. XIII), contenendo 999 componimenti, a differenza di L (Laurenziano Rediano 9), che ne conta circa 430, e di P (Banco Rari 217) con 180 testi. [........] Della straordinaria importanza di V si accorse l'erudito marchigiano Angelo Colocci (1474-1549), che lo commentò, realizzandone una copia integrale, ...

The range of dates 1474-1549, for example, may be associated to three different events: the life of Angelo Colocci; when the Codex was commented; when it was copied.

A possible tagging choice like <life_time>1474-1549</life_time> is therefore a simplification as tags should take into account multiple layers of significance. As far as the time in which the Codex was assembled, it can be identified in the first sentence by the terms “sec. XIII”, but also by the adjective “duecenteschi”: the choice of tagged terms is not, in this case, trivial as it may have a deep influence on searching the document.

In our opinion, the complexity of

modelling such documents, identifying all objects involved in the applications, calls for trusted design methodologies, such as the Entity-Relationship approach (E-R), to describe the 'universe' of the application [5]. Furthermore, once the objects of the application have been recognized, they have to be described as far as content (and sometimes their structure), in a way suitable to be matched with users' queries, thus producing the so called surrogates or auxiliary data [2]. Experience in surrogates creation may help in defining tags, to give them the role of metainformation able to represent in a less ambiguous way "knowledge" and "meaning" of the document content: methods derived from the Information Retrieval (IR) practice could be therefore suggested to increase the significance of search results [9].

If the system is designed so that XML documents have to be manipulated by specialized personnel, tag definition may be oriented to user understanding. In the following new example, the tagging model is consistent with the interests and skill of experts in the field:

> *... the basilica is mentioned for the first time already in the <dating>year 375</dating> and its importance ... The existing building almost entirely dates back to the <construction>Xth century</construction>* ...

On the other hand, markup items may be machine-oriented, if their main aim is to be automatically managed by some programs (agents, page builders, etc.), as in the following example, where tags partially lose their meaning with respect to human readers, but present richer structures suitable for automatic filtering:

> *... the basilica is mentioned for the first time already in the <item type="date" role="citation">year 375</item> and its importance ... The existing building almost entirely dates back to the <item type="date" role="foundation" >Xth century</item> ...*

From the above two examples it may be argued that there may be multiple XML documents which are referred to the same pieces of information, and IR experience is not sufficient to guide to a unique representation. This creates problems, for example, in querying, as a user query might be independent of the way documents are defined.

In the perspective of the semantic Web design, the Resource Description Framework (RDF) gives a standard way to identify, beyond all possible document structures, the 'objects' involved in the applications and their relationships [3]. In fact, the first tagged group of the two statements above results in the same RDF assertion:

> *triple (date_of_citation, religious_building, year 375)*

This type of annotation points out available resources like a subject (*religious_building*), a direct object (*year 375*), and their predicate - another resource - (*has date_of_citation*), where all labels represent Uniform Resource Identifiers (URIs) [12].

A PROTOTYAL DESKTOP ENVIRONMENT

Following above considerations, a prototypal virtual desktop has been developed to allow access to hypermedia documents regarding an ancient Codex. The aim is to test weather this environment presents advantages with respect to a traditional HTML hypermedia application (from the user point of view), in which the generic interaction process traditionally comprises the following steps (not necessary in consecutive order):

1. accessing to an information node;
2. reading or viewing or listening;
3. activating an available tool offered by the host;
4. accessing to new nodes by selecting available associations;
5. performing again one or more of the actions from 1. to 4.

However, the interaction process is influenced by the particular purpose of the application: the interface must enable the user to interact in a suitable way (with respect to her mental schema), by suggesting various associations and adequate metaphors (with respect to the user requirements and to the digital format of the data). In the particular case of the relation user-library a possible process (once a user profile has been identified) could be:

1. to identify a subject and a period: by example the Italian literature of the Middle Age;
2. to obtain the lists of available objects regarding the selected choices;
3. to select and display an object of interest: by example the text on G.Guinizzelli;
4. to inspect available associations as far as type and metaphor;
5. to follow one or more associations;
6. to search for objects related to a particular subject;
7. *to add annotations;*
8. *to establish new associations;*
9. *to modify some object by elaborating its forms and contents by means of offered tools;*
10. *to compare objects;*
11. *to arrange objects on the desktop;*
12. *to build user-tailored paths;*
13. *to save new material;*
14. to perform again, if necessary, some or all the previous processes.

The steps 1.-6. correspond to available actions in a traditional Web application; the 7.-13. are feasible (by a user and in the Web) only with browsers allowing modifications by users.

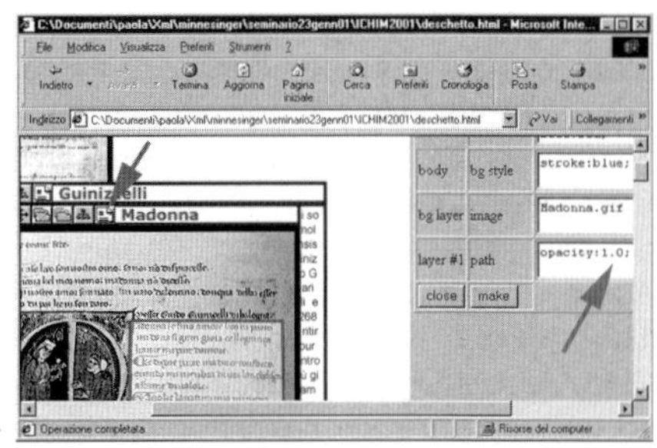

Figure 2: In the desktop space more windows are open; the arrow on the left points to the buttons and in particular to the editor; by the editor the user may change the opacity value (see the arrow on the right) of the image in the red square to increase readability

The prototype presents to the users a rectangular space, the virtual desktop, and some menus, allowing to select available tools and pieces of information. They appear on the desktop as windows (see figure 2) that the user may close, reduce to an icon, move in the desktop space, put forward-backward, duplicate, by using buttons on top of the window itself. The button next to the window title activates an editor: depending on contained data and user's permissions, the editor allows to

modify the content of the forward window or its display form. All these tools correspond to the points 9. and 11. of the second list above; also operations like 7. and 8. become possible.
More than one window may be contemporary present on the desktop, in accordance to point 10. A "tools" menu contains, for some users, a "save" facility, in order to save on the local/remote archive data elaborated by the user at the browser side (see point 13.).
Within the text, hotwords are in blue (in figure 3, "i modelli poetici siciliani"). They indicate source anchors that, once clicked, may show to the user a set of possible choiches such as, in the figure 3, "biography" (a text), "iconography" (an image), and "history" (an URL to another site). The destination content may be different, like in the example, or may present the same content under various metaphors or reproduction media.
An example of content modification suitable to the user purposes is illustrated in figure 2. "Madonna", i.e. an image, is accessed: by clicking on the editor icon (see the green arrow on the top left), the user may change the opacity value (see the purple arrow on the right) of the image within the space enhanced by the red rectangle so that characters in its background may be clearly read.
The prototype shows the feasibility of some of the facilities listed above for new hypermedia applications in the cultural heritage field. It is important to remember that the effects of the user dialog with the application are, in most cases, limited to the browser side, having no effects on the content and organization of the materials stored on the server (unless the user is an authorized one).

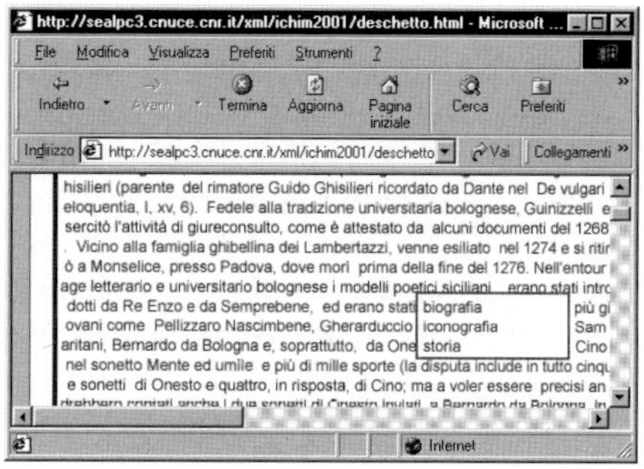

Figure 3: A single source anchor corresponds to more choices to different destinations

In this application development the SVG standard has been adopted: proposed at the end of '99 (and now at the stage of recommended candidate). It represents vector graphic on the Web, and produces many advantages with respect to similar proposals and approaches: it agrees with the 'open source' principle of Internet; it is XML; it is highly and carefully documented; it allows to identify and to group objects helping modification, transformation, animation, and accessibility; it is fully compliant with the DOM (Document Object Model) [12] and proposes a wide set of events, methods and properties to improve interactivity and programming; it is an ASCII, not a binary format, thus allowing to associate meta-information directly to objects or object parts.
The overall metaphor presented by the prototype interface is windows-like (as far as both display and behaviour of active elements, of windows, icons, buttons, etc.) which is extremely diffused and avoid interpretation and training problems by the users: however, SVG [12] is apt to design interfaces following more expressive metaphors.
As regards our application SVG allows:
a) to build metaphors for the retrieval of

textual or non-textual formats in a homogeneous way [6] (for example, parts of an image can be accurately addressed and their properties modified (see the example in figure 2);
b) to render the same XML fragment of information under different perspectives (for example, a poetry to be read - text - or listened - played -);
c) to let user modify display properties, content, position of the retrieved information;
d) to design and develop objects to be programmed also at the client side;
f) to execute all above operations, and many others, not in a closed application developed by proprietary binary software, but on the Web by means of standard browsers and open source software.

CONCLUSIONS

Whereas in traditional hypermedia applications on the Web the user usually maintains a static role with respect to the content creation and manipulation, in the virtual desktop the man-machine interaction is based on the contemporary availability of more tasks corresponding to different paths and/or metaphors: the user may choose and follow one of them, then suspend to perform a new task, resume the first one and so on, in an asynchronous fashion; in this way it is possible to strengthen user interactivity. In this perspective the user may be 'creative': the complexity of such an application is so increased because the reader is able to self-tailor content.
Recent XML technologies allow to extend to the Web, in an open software environment, the design and development of interfaces following the paradigm of the *virtual desktop*.
The risks of the well known hypertext drawbacks of "cognitive overhead" and "to be lost in the hyperspace" are however increased, as they are associated to the degree of freedom allowed to the user. Therefore design issues become particularly important for the development of robust, reusable, reliable structured hypermedia applications, that is applications having a coherent information structure both at logical and presentation levels, resulting in a regular, consistent, and predictable navigation environment. Also system controls on users' actions are needed to keep the integrity of the application. It is wise, for example, to maintain separation of pieces of information from links to easy check and avoid the user generation of dangling, inconsistent or duplicated links [7]. Also navigation tools must be emphasized in terms of context recognition, backtracking, etc.
The aim of this contribution, however, is to give prominence to users' role in interaction with hypermedia applications on the Web.

REFERENCES

1. ACM Hypertext 89, *Proceedings* (Pittsburg) 1989, Special Issue of the SIGCHI bulletin

2. Agosti M., Gradenigo G., Marchetti P.G. A hypertext environment for interacting with large textual databases. In *Information Processing & Management*, 28(3), 1992, 371-387

3. Berners-Lee T. Semantic Web Road map, [Online] 1998, *<http://www.w3.org/DesignIssues/Semantic.html>*

4. Brusilowsky P. Methods and techniques of adaptive hypermedia. In *User Modeling and User Adapted Interaction*, 6(2-3), 1996, 87-129

5. Elmansri R. and Navathe S.B.

Fundamentals of database systems, The Benjamin/Cummings Pub. Co. Inc., 1989

6. Grosky W. Multimedia Information System. In *IEEE Multimedia*, 1(1), 1994, 12-24

7. Leggett J.J. and Schnase J.L. Viewing Dexter with open eye. In *Comm. ACM*, 37(2), 1994, 77-86

8. Maler E. and El Andaloussi J. *Developing SGML DTDs: From Text to Model to Markup*, Englewood Cliffs, NJ: PTR Prentice Hall, 1996

9. Salton G. and McGill M. *Introduction to modern Information Retrieval*, McGraw-Hill, 1983

10. Schamber L. What is a document? Rethinking the concept in uneasy times. In *Journal of the ASIS*, 47(9), 1996, 669-671

11. Signore O., Bartoli R., Fresta G. Tailoring web pages to users' needs. In *Proc. of the Workshop on Adaptive systems and user modeling on the WWW - UM97* - Chia Laguna (CA), 1997, 85-90

12. W3C, W3C Consortium, [Online] 2001, *<http://www.w3.org>*

ABOUT THE AUTHORS

Paola Carrara has been working at the Italian National Research Council (CNR) since 1986. Her researches include Information Retrieval Systems, design of hypermedia applications, geographic and spatial information systems. She is also concerned with interoperable technologies on the Web and in particular with open software proposals for vector graphic.
E-mail: *paola.carrara@itim.mi.cnr.it*

Giuseppe Fresta has been associated with the Italian CNR since 1973, when he was concerned with methods and systems for the elaborations of graphic data, and information systems to manage environmental data. He is now engaged in the design and development of user interfaces, hypertext/hypermedia, and Web publishing; in particular he is interested in the evolutions of the XML standard with its associated technologies (DOM, XSLT, SVG-XML).
E-mail: *g.fresta@cnuce.cnr.it*

MauroTeX - A Language for Electronic Critical Editions

Paolo Mascellani(*) *and Pier Daniele Napolitani*(#)

(*) Università degli Studi di Siena, Dipartimento di Matematica, Siena, Italy
E-mail: p.mascellani@dm.unipi.it

(#) Università degli Studi di Pisa, Dipartimento di Matematica, Pisa, Italy
E-mail: napolita@dm.unipi.it

ABSTRACT

MauroTeX, an extension of the well-known LaTeX typesetting system, is a language designed in order to completely describe philological critical editions of ancient mathematical and scientific works. It provides a set of *macros* that the editor can use starting from the early stages of the edition (the transcription and the collation of the different "witnesses" of the edited work), and ending with the building of several "critical apparata".

The outcome of the process, is a *marked* text, suitable both for (semi-) automated analysis, such as the extraction of the transcription of the text of a "witness" (as long as it critical apparata), or the analysis of concordant errors, and for the automatic production of typographical quality (postscript and PDF at now) and hypertext (HTML) versions of the edition.

MauroTeX is thus a major step in transforming the traditional techniques of the ecdotics into more modern techniques, which allow editors to work better, obtaining better results, without substantially changing their acquired work skills.

KEYWORDS: critical editions, electronic publishing, history of science.

INTRODUCTION

Some years ago, our research group faced the problem of building the electronic critical edition of the mathematical works of Francesco Maurolico (1494-1575), an important Italian scientist of the Renaissance [7] This group included several Italian researchers as long as people form France, Japan and Switzerland; moreover, the mathematical works of Francesco Maurolico amounts at more than 5,000 pages; hence, one of the first problem we encountered was to use a unified work methodology. After some discussion, we decided to try to define a language everyone of us should have followed in its work, based on the well-known LaTeX typesetting system [5], [6], [10]; this language has now been pompously named MauroTeX (or MTeX, for short). This language and its applications have been developed by the authors and by Paolo d'Alessandro (University of Chieti - Italy), Ken Saito (Prefectural University of Osaka - Japan), Jean-Pierre Sutto (Revel - France), and Roberta Tassora (University of Bari - Italy). A complete manual of MauroTeX [3] is available at the site of the Maurolico Project [7].

In its early versions, MTeX was only a small (but growing) set of LaTeX macros, used mainly as useful shorhands of their respective expansions, but soon we observed that their use allowed us to obtain a *mark-up* of the edited works, that, in our opinion, is now the most important of its features. Marking-up the critical edition, is a way for including in a standard ASCII file (LaTeX and MTeX use only characters codes from 0 to 127) all the information about the considered witnesses, the sources of the work's author, the editor's notes, ... and so on, without forcing any *a priori* typographical convention, but leaving to the final stages of the work the possibility of deciding them. In turn, this allows everyone to build analysis tools that exploit this information, such as programs that find all the concordances in error, useful in order to build the "stemma codicum" (the dependency tree of witnesses).

One of the principles that we have always tried to fulfil in the phase of the definition of the syntax of the macros, has been the ability to obtain back the original text of every witness from the marked text: this guarantees us that no information have been lost in the mark-up process.

BUILDING A CRITICAL EDITION

MTeX is not only a description language, but it follows the editor at every stage of its work. Let us give a (simplified) schema of this job.

Transcriptio. Transcription of the first basic text; normally the editor chooses the witness, or a preceding edition, that he/she thinks (or guess) to be the closest to the critical text; however, this cannot be decided at this stage, hence it is better to make no assumptions in this sense.

Collatio. Collation of every other witnesses against the collation of the witnesses already collated.

Eliminatio Codicum Descriptorum. The witnesses that clearly depend from other available witnesses, are discarded.

Stemma Codicum. Classification of the variants and derivation of the dependency tree between the witnesses.

Constitutio Textus. Editorial interventions such as: choice of the critical text, corrections, conjectures, integrations, expunctions, "cruces", paragraphs delimitation.

At the end of this work, it should become clear that MTeX could be useful in all of these phases.

During the "transcriptio" and "collatio" phases, the editor have to refrain itself to make decisions, such as conjectures that are not justified in those phases, but that can only be made in the third phase. However, the language have to support him in this: for instance, in presence of a variants between some witnesses, the editor should wait for the collation of the last witness, before formulating any hypothesis on the critical text; moreover, it should be always possible (and easy) to change the choice of critical text.

In principle, every phase following the transcription of the first witness, should consist simply of some (possibly many) addition to the marked text, in order to take into account the differences between it and the new witness (in the second phase) or the critical choices of the editor (in the third one). Of course, the language has to be specifically designed in order to obtain this.

Example
We have three witnesses (***A***, ***B***, and ***C***), that report the following parallel phrases.

A: *Sit data gratia, sit datus cubus.*

B: *Sit data latio, sit datus cubus.*

C: *Sit data ratio, sit datus cubus.*

After the transcription of ***A***, the corresponding edited text should be:

```
Sit data gratia, sit
datus cubus.
```

During the collation of ***B***, the editor realises that the phrase is interested in a variant and produces the following edited text:

```
Sit data
\VV{{A:gratia}{B:latio
}}, sit datus cubus.
```

After the collation of ***C***, the edited text becomes:

```
Sit data
\VV{{A:gratia}{B:latio
}{C:ratio}}, sit datus
cubus.
```

It is only at this point that the editor can make its choice and promote "ratio" as critical text, shifting it at the first place, thus producing:

```
Sit data
\VV{{C:ratio}{A:gratia
}{B:latio}}, sit datus
cubus.
```

Transcriptio and Collatio
Let us see in more details the transcription and collation phases. There are a number of situations that the editor can fall in and that MTeX handles naturally.

Comments. The editor can find a nonsense phrase, or an evident error, or any other situation that requires to be signalled and/or further investigate. In order to produce a faithful transcription, but also, to be sure to return on situations that need further studies, at least during the editorial phase, the editor can signal them using a LaTeX comment or, better, using an MTeX macro, which produces also and index of such comments.

Text Description. If the text presents some significant characteristics, the editor can also write notes that are intended to appear in the critical apparatus. This can be done, adding such comment between the "siglum" of the witnesses interested by the comment and the corresponding "lectio". Other situations that can be described are marginal additions, vacancies of text, changes of page, and many others. Finally, if the editor doesn't find any predefined macro that fit into its case, he/she can add any comment he/she likes.

Dates. Hidden in phrase or explicitly given, some date can appear in a work. It is normally of great interest to take note of such date, in order, for instance, to establish when the work had been written.

Citations. A similar case is that of the citation of another work. Once again it is important to record these situations, in order to study, for instance, the chronology of the works. However, if we wish simply say that something is the title of a work, we can signal it without any further specification.

Mathematics, Astronomical and Other Special Symbols. Of course, we can write mathematics. Here, of course, the power of LaTeX is used; unfortunately, the way we write mathematics today is quite different from that used by ancient writers and we cannot use a uniform notation for all this world: this implies that this problem have to be solved "ad hoc" for each situation. Nonetheless, MTeX provides a number of special symbols that can be useful:

- astronomical symbols for the sun, the moon, the planets and the zodiacal signs (the graphical definition are that of [9]);
- geometrical abbreviations for triangle, square, and some other figure;
- single and double bars, that some author uses in order to divide his reasoning;
- some other special symbols, like uppercase and lowercase "radix".

Theorems-like environments. Almost every mathematical work, is divided into propositions which have a title and a description; this can be coded in MteX.

Long Variants and Transpositions. All we have seen so far, works well if the variants between witnesses are local, i.e. if only one or a few words differ; on the contrary, if there are differences between witnesses which interest several rows or whole pages, some special care have to be taken. For instance, there must be a way to signal to the reader the beginning and the end of the variant, in the critical text, without repeating all the text. One can argue that this is an output issue, rather than a mark-up problem; however, it implies that some additional data have to be coded in the critical edition. Another special case is that of transpositions, i.e. parts of text that appear transposed in different positions. In principle, they are simply special cases of (long) variants; however, it is of great interest to distinguish them from other kinds of variants. MTeX has a special macro, for marking transpositions.

Eliminatio and Stemma

During these two phases, the editor has to carefully analyse the outcome of the previous ones. Here, the advantage of using MTeX is essentially the possibility of having a powerful set of analysis tools.

Classification of Variants. During the transcription and collation phases, the editor has to be very precise in signalling the variations between the witnesses; however, in the following phases, the editor can be a little smarter, in order, for instance, to not flood the critical apparatus with meaningless notes. On the other hand, the precision of the editor is not meaningless itself and must not be wasted: for instance, someone may wish to study the orthographic variations between the witnesses or between a work and another. In such situation, the editor simply has to change the name of the macro and the corresponding variants will not appear in the apparatus. If this is not enough, and the editor wishes to classify the variants in a more precise way (distinguishing, for instance, between formal, genetic, adiaphorous, etc. variants), in the next version of MTeX it will be possible to classify variants in 26 different classes.

Constitutio Textus
This is the phase during which the editor adds his/her comments, notes, warnings, and so on.

Choosing the Critical Text. We have already seen this operation in the Example; hence, it is not worth to spend many words here. It is important to note that the editor can, during this activity, completely revise his/her preceding assumptions and, if there are new evidences, even change the basic text. This is very important, because it is only now that the editor has all the needed information. In cases where a new witness is found after that some work or, even the whole edition, has be done, MTeX allows the editor to adapt its edition to the new situation with the minimum effort possible (which, of course, can be still not little).

Conjectures and Corrections. We have seen that the editor can choose the critical text, between that of the witnesses, simply shifting the corresponding variant at the first place; however, there are situations in which we cannot find in the witnesses any satisfying "lectio". In these situations, the editor has to change something, in order to obtain a readable edition, but, of course, he/she has to signal his/her interventions clearly and possibly to give reason of them. A variation of this situation is when there is a mathematical mistake in the witnesses.

Integrations, Expunctions and Cruces. In same cases, the editor can choose to integrate a phrase (some words may have been missed by the copyist), or to delete it (the phrase has been erroneously copied there by the copyist), or simply have to admit that he/she is unable to find a reasonable amendment for some phrase.

Propositions and Paragraphs. One of the decisions the editor has to do is about the "structure" of the work he/she is editing. Normally, a mathematical work can be divided into "propositions", which not necessarily correspond to the propositions of the author (maybe, the editor wish to group some propositions altogether, or to disaggregate some other). This, in order to provide some search facility, can be integrated with keywords or arguments. Moreover, in order to allow anyone to cite with no ambiguity a pass of the edited work, the editor can also choose a finer division into paragraphs or units. MTeX provides macros for all this needs.

ANALYSIS
MTeX allow us to think at a number of tools we can provide to the researchers in order to analyse critical editions, both for internal such as verification, and for external, such as special studies, uses.

We can make an (unfinished) list:

1. syntax check: it is well known that TeX and LaTeX, despite all their good characteristics, are not very friendly with their users when they take same mistake; this can be overcome with a program which performs a syntactical analysis of the source and which emits warning messages concerning more frequently mistakes, such as brackets mismatch or misspelled macros; this tool is also the basis for many of the others;

2. lists of variants: all the variants concerning a group of witnesses can be listed, in order to study

their dependencies and to formulate (or check) a candidate “stemma codicum”

3. witness extraction: the text of a single witness, as long as its critical apparatus, can be extracted for specific studies;

4. collation composition: if we have different witnesses collated separately against the same basic text, it is possible to obtain a unique collation of all the witnesses against the basic text (this is a real case: it happens, for instance, when we have different people working in parallel to the same edition);

5. subject search: all the propositions concerning a group of subjects can be retrieved;

6. text retrieval: all the propositions or units that match a query can be retrieved;

7. cited works retrieval: the works or the authors cited in an work, or group of works, or group of propositions, can be retrieved;

8. all the textual units where there are concordances, either in error or not, between two witnesses can be retrieved, in order to study the dependencies between them and establish the “stemma codicum”, i.e. the dependencies graph;

9. exportation: the whole work or some part of it can be exported in another formalism, such as TEI [10], or DBT [8], or XML [13];

Not all of this tools are yet developed, however, MTeX allow us, or, why not, someone other, to create them. At the current stage of the MTeX development, we have built the syntax checker, which is based on a LALR grammar (a subset of context free grammars) and implemented using the “flex” lexical analyser generator and the “bison” syntactical analyser generator [1]. However, the syntax checker is a milestone in the development process, since it provides, as a by-product, the syntax tree of the document, which, in turn, is essential for all the other analysis programs, as well as for the HTML converter mentioned in the following section.

PUBLISHING

MTeX is not only a description language; its descent from LaTeX allow us to obtain a real, printed, clearly readable, version of our critical edition, or, at least, a postscript or PDF version of it. Moreover, using a conversion program, it is possible to produce automatically an HTML version of the critical edition, ready to be browsed from a cd-rom or directly from the Internet.

Print Quality Publishing

In MTeX, the print capability is provided by a LaTeX package (mauro.sty) that includes the definitions of all the new macros of the language. Considering that LaTeX implementations exist for almost every computer platform (Windows, MacOS, and Unices at least) [11], we can say that MTeX too can be used on all these platforms.

It is noticeable that LaTeX and, by consequence, MTeX can, not only produce a printed version of the document, but also produce electronic standard format versions, like Device

Independent (DVI), Postscript (PS) and Portable Document Format (PDF). Moreover, it is not difficult to think at a program that produces a Rich Text Format (RTF) version of the edition (it can be built using the syntactical analyser, in a way similar to the HTML converter).

If we give even a short look at some critical editions, we can see that their styles can be very different: someone uses different "apparata" for editorial notes and for citations, someone not; someone uses notes marked by line numbers, someone marked by paragraph numbers, someone marked by note numbers; ... and so on. Of course, we have chosen our preferred style; however, we can say that our set of macros implements *a* transformation of an MTeX source, which is very general, into a printed quality representation of the critical edition: if someone doesn't agree with our representation choices, he/she can always write his/her set of macros, or, better, make his/her modifications to our package (MTeX is available at source level under the GNU General Public License [4]), and obtain a different, and hopefully nicer, representation of his/her edition.

Moreover, we plan to provide in the next release of MTeX a set of *style* macros, which will make easier to customize the output. A little subset of these macros is already provided by the package, mainly for transcription checking purposes: we call it the "draft style".

HTML Publishing

Using the full power of electronic publishing and ending only with a printed critical edition is not very satisfactory. Moreover, having the postscript or PDF version of the critical edition is not very much more (provided we don't have bookmarks nor hyperlinks in the PDF version). The real big step of electronic publishing is having an hypertext critical edition, browsable and linked both internally to the edition itself and externally to external related sources. MTeX can help also for this.

Although also PDF has hypertext capabilities, the world "de facto" standard for hypertexts is HTML; hence, in order to make things easier for the editor, as well as for the (hyper-) reader, we have built a program, which transform an MTeX source into a set of linked HTML documents. Once again, there are many different ways or styles, one can use in an hypertext: we have chosen our preferred style, but anyone can build a different translator, directly from the published grammar, or using the syntax tree provided by our syntax checker, or simply, modifying our translation program (Another way, can be that of modifying the MTeX package, in order to generate bookmarks and/or hyperlinks, when used with PDFLaTeX [11]).

The program M2H is written in C and is composed by a parser, which builds the syntactic tree of the MTeX document, and by an HTML formatter, which transforms the syntactic tree into a set of linked HTML documents. Every proposition, is translated into a main HTML document, containing, essentially, the critical text, and some others, containing the different apparata. Each variant, for instance, have its own numerical mark, and, clicking on it, the corresponding note appears into another window (or frame, depending on the browser capabilities); conversely, clicking on the note's mark, we can go back to the corresponding critical text. Different fonts, sizes, faces and colours, allow an easy reading of the critical text and of the apparata.

The program M2H compiles under almost every Unix machine, using the GCC compiler and under DOS/Windows machines, using the DJGPP porting of GCC. No great effort should be done in order to compile it under MacOS.

Some more tools allow to make indexes of propositions and of works, to pack the critical text window with the apparata windows into a unique frameset, and, in general, to manage the HTML documents into a site, which can reside into a cd-rom or a web site. For instance, the Maurolico site [7] has been developed using MTeX.

CONCLUSIONS

Electronic publishing allow us to do things that have never been possible before; however, we have to be very careful in order to exploit these benefits, and MTeX tries to help to do that to anyone who wish it.

A critical edition is a tool that researchers can use in order to investigate different aspects of the edited work, as well as of the author, of the period or of something other. It is important that useful information can be extracted easily from it and, in turn, that the transcription doesn't hide them. This is why the editor has to be very precise and this is why MTeX allow him/her to be so; these two things allow us to have an increasing set of powerful analysis tools for the edited works.

Moreover, we cannot think that our work is final: some other witness can be discovered, which can obsolete our work, or, after all, we too can take a mistake. Of course, in such cases, we don't wish to restart from scratch, but, instead, to reuse as much work as possible. MTeX has been designed with this always in mind.

Someone argued, and we somehow agree, that this language is complicate and that philologists, which in general are not too acquainted with computers, should have many difficulties in using it. Our answers are:

1. this is true only for situations that are inherently complicated, which, in general, are not the majority of the situations an editor can face during the edition of an work; in these cases, we suggest the editor to begin with the easier ones an to let those situations for a second time, when he/she will have greater confidence with the language;

2. we are working to integrate MTeX with the Philological Workstation [2], designed by Andrea Bozzi, that has a nice WYSIWYG user interface, in order to make transcriptions; maybe in the future, also the work of the editor will be done using this interface;

3. having a single portable document, written in a publicly available documented language (we plan to distribute the BNF syntax of MTeX, as well as the source of the basic analysis tools, under the GNU General Public License [4]), is so important that it is worth to

make some effort in order to have it.

Much work has still to be done; especially in the field of analysis of editions; however, we think that MTeX can be a major step in the path that leads to a "computational philology" (we have to thank Andrea Bozzi, for this term), i.e. a philology carried out using the full power of information technology.

REFERENCES

1 Alfred Aho, Ravi Sethi, Jeffrey Ullman. *Compilers: Principles, Techniques and Tools*. Addison-Wesley, 1986.

2 Andrea Bozzi. *Philological Workstation web site*. Istituto di Linguistica Computazionale, CNR Pisa - Italy. *http://www.ilc.pi.cnr.it/philwork*.

3 Paolo d'Alessandro, Paolo Mascellani, Pier Daniele Napolitani, Ken Saito, Jean-Pierre Sutto, Roberta Tassora. *Elementi di MauroTeX: un linguaggio per le edizioni critiche*. Dipartimento di Matematica – Università degli Studi di Pisa - Italy. 2001.

4 *GNU General Public License*. Free Software Foundation, 1991. *http://www.gnu.org*.

5 Donald E. Knuth. *The TeXBook*. Addison-Wesley, 1986.

6 Leslie Lamport. *LaTeX: a Document Preparation System*. Addison-Wesley, 1994.

7 Pier Daniele Napolitani (editor). *Francesco Maurolico web site*. Università degli Studi di Pisa - Italy. *http://www.maurolico.unipi.it*.

8 Eugenio Picchi. PiSystem web site. Istituto di Linguistica Computazionale, CNR Pisa - Italy. *http://www.ilc.pi.cnr.it/pisystem*.

9 Peter Schmitt. *AstroSym* - Version 1.00. Institute of Mathematics - University of Vienna, Austria. 1992.

10 Text Encoding Initiative. *http://www.tei-c/org*.

11 TeX Users Group. *http://www.tug.org*.

12 Martin L. West. *Textual Criticism and Editorial Technique: Applicable to Greek and Latin Texts*. Stuttgart, Teubner, 1973.

13 WWW Consortium. *http://www.w3c.org*.

ABOUT THE AUTHORS

Paolo Mascellani is an independent software engineer and a Ph.D. student at the University of Siena, Italy - Department of Mathematics. His scientific interests are in the fields of software development, computational logic, natural languages processing, data mining and history of mathematics.

Pier Daniele Napolitani is professor of history of mathematics at the University of Pisa, Italy – Department of Mathematics. His scientific interests are mainly in Renaissance and XVII century mathematics. He is the director of the "Maurolico" Italian interuniversity project for the critical edition of the mathematical works of Francesco Maurolico..

The "XML/Repetti" Project: Encoding and Manipulation of Temporal Information in Historical Text Sources

Fabio Grandi (*) *and Federica Mandreoli* (#)

(*) C.S.I.TE.-C.N.R. and D.E.I.S., University of Bologna, Italy
E-mail: fgrandi@deis.unibo.it

(#) D.S.I., University of Modena and Reggio Emilia, Italy
E-mail: fmandreoli@dsi.unimo.it

ABSTRACT

The paper deals with the deployment of XML-related technologies in Cultural Heritage applications concerning the encoding of temporal semantics in the digital version of historical documents. Since written sources have often the same importance as material evidence in medieval archaeology, our approach can be applied to the development of tools for the support of archaeological research. In previous work, we developed an XML/XSL infrastructure called "The Valid Web" for the definition and management of historical information within Web documents. In this paper we describe the application and extension of such an approach to the realization of the electronic version of Repetti's historical-geographical dictionary of Tuscany. The extension concerns the uniform management of temporal indeterminacy, the use of multiple calendars and granularities and the proposed solutions have been inspired by similar research done for temporal query languages. From the user viewpoint, the proposed XML extensions allow the addition of historical metainformation to the encoded text sources and their "intelligent" temporal navigation via standard Web browsers. The project also involves the definition of optimized search algorithms, storage and temporal indexing of XML-encoded Repetti's Dictionary items, implementation of a prototype. As a byproduct, also a tool for computer-aided temporal XML-encoding of text sources will be developed to be used by Cultural Heritage operators (e.g. archaeology researchers).

KEYWORDS: XML, semantic Web, temporal data management, digital libraries.

INTRODUCTION

In the information processing field, XML [21] is becoming a new standard for data management and exchange over the Internet [1]. In particular, a great deal of interest concerns its adoption for the representation and integration of structured and unstructured data. Moreover, an outstanding (and very appealing for Cultural Heritage applications) feature is the capability of easily encoding semantic metainformation in XML documents, to be automatically used by advanced computer tools, like "intelligent" search engines (towards a Semantic Web [23]). In this context, our research interests focused in recent years on the introduction of temporal aspects into the Web, by adapting and extending concepts and techniques deriving from a more than decennial experience in temporal database research [12,14]. To this end, we developed an XML/XSL infrastructure, named "The Valid Web", for the management of temporal documents and data on the Web [6,9]. The proposed techniques enable the explicit encoding of distinguished temporal/historical information within XML (or legacy HTML) documents, whose contents can then be selectively accessed according to their temporal validity with any XML-compliant Web browser (like MS Ie5 [22]). In order to show its potentialities, the infrastructure has been implemented on a demo prototype [7] (also available on-line [8]) showing, as application example, the functionalities of a temporal fine-arts Web museum [19], that is a virtual environment in which it is possible to cut personalized visit routes for a specific epoch of interest. In this paper, we introduce an interesting application and extension of such an approach for the management of historical text sources in digital form. The general framework of this research is a collaboration with the Computer and History group at the University of Florence [18], which has been involved for a couple of years in the study of a new edition, in electronic form, of Repetti's Dictionary. The "*Historical-geographical dictionary of Tuscany*" by Emanuele Repetti is an encyclopedic collection of information about Tuscany published in the XIX century concerning notable places (from large towns to small villages, providing historical, archaeological and artistic information), physical land attributes (viz. mountains, rivers, lakes, wetlands, etc.) and special items (like statistical tables). It has been anastatically reprinted many times and its digital edition under study should incorporate hypertextual links and additional multimedia data and should be made available on the Web to be worldwide accessible through the Internet. In our case, this aspect has a noteworthy relevance from a Cultural Heritage and also scientific point of view: it frequently happens in medieval archaeology that written sources have the same importance as material evidence. It has already been pointed out how the role of Internet in archaeological investigation is continuously increasing, as wide, fast and easy sharing of information on the Web has a substantial impact on the archaeological methodology [3,13], which could be ever boosted by the deployment of XML-related technologies (as also evidenced in [2,11,17]).

THE "XML/REPETTI" PROJECT

As a starting point, the straightforward application of "The Valid Web" approach to the "XML/Repetti" project allows the uniform classification and encoding of the temporal information contained in the dictionary and the availability of techniques for temporal search support. However, advanced functionality and efficiency specifications (Repetti's Dictionary is a quite large collection of text) require an improvement of the basic approach. Such an improvement actually provides for three goals:

1. The extension of the XML/XSL infrastructure to deal with the specificity and semantic richness of the temporal information stored in Repetti's Dictionary and similar textual sources;

2. The redesign of the overall system architecture, including an efficient organization of the temporal search engine (with optimized search algorithms) and the XML document repository (with temporal indexing

facilities);

3. The design and implementation of a user-friendly tool for computer-aided temporal encoding and document mark-up, which could save history or archaeology researchers from "manual" editing as much as possible.

In particular, the infrastructure extension requires an enhancement of the markup scheme and search mechanism in order to be be able to capture the semantics of temporal expressions (widespread in Repetti's Dictionary indeed) involving **indeterminacy** (as in: "around 1456"), **multiple calendars** (e.g. use of the Julian calendar) and **different granularities** (e.g. months versus years). Special attention has been devoted to the indeterminacy problem, which has interesting theoretical implications and requires the most consistent infrastructure extensions.

Advanced XML Encoding of Temporal Information

An analysis of Repetti's Dictionary contents evidences the large use of imprecise temporal expressions that could only be represented with the original "The Valid Web" infrastructure with a unacceptable oversimplification. For instance, the following examples can all be found in the "A letter" section of Repetti's Dictionary (the expression are evidenced in boldface; an English translation is added in square brackets):

1. Un terzo giro fu tracciato con ampio pomerio, profondi fossi e più regolari vie, **circa il 1276** [about in 1276], per ordine del vesc. Guglielmino degli Ubertini,...

2. Quindi, **intorno al 1530** [around 1530], il francese Marcilla dipinse a vetri colorati le belle finestre...

3. Fu Comunità sino **alla fine del secolo XVIII** [at the end of the eighteenth century], compresavi la popolazione della piccola borgata di Petrognola.

4. ...Cosimo III fondò un convento ai frati Minori della riforma di Spagna, mantenuti a spese del R. erario, soppressi **sulla fine del secolo XVIII** [near the end of the eighteenth century].

5. Con simile vocabolo fu **designata nei secoli intorno al mille** [in the centuries around the first century A.D.; the sentence talks about a name given to a church] la pieve di S. Cresci a Maccioli

6. La grandiosa e ricca cappella della Madonna fu fondata **sulla fine del sec. XVIII** [near the end of the eighteenth century] nella parete sett. del Tempio...

7. ...con una porzione di quella Maremma stata già donata **verso l'anno 830** [towards the year 830] dall'imp. Lodovico Pio.

8. ...nell'ultima ricostruzione delle mura aretine ordinata da Cosimo I, che di nuovi baluardi e cortine **fra il 1549 e il 1568** [between 1549 and 1568; the sentence talks about building of city walls] le fortificò.

Similar expressions correspond to the denotation, affected by some imprecision and/or indeterminacy, of the occurrence time of a fact that can be

$$\Pr[t_1 < t_2] = \sum_{i<j} P_1(i) P_2(j) \qquad (1)$$

instantaneous (in this case we properly call it an event), as in Ex. 5, or a period with non-null duration, which can be described by a temporal interval, as in Ex. 8. Obviously, the concept of instantaneousness of a fact depends (in addition to the interpretation of the text, which could also be ambiguous) on the choice of the reference temporal granularity (scale unit on the time axis), which in the case of Repetti's Dictionary ranges from the day to the century. For example, the adoption of the day as reference granularity implies that any historical fact whose validity can be located within a given date has to be interpreted as an event (i.e. ideally *instantaneous*, being the reference unit indivisible). On the contrary, the validity of historical facts whose duration can be expressed via granularities coarser than the day can also be expressed via an interval of days in which we surely know the event happened, though we do not know exactly when. The solution we adopt for the representation of indeterminate dates in the "XML/Repetti" project is based on the *probabilistic approach* introduced for the temporal query language TSQL2 [20] and further developed in [5]. In this approach, every indeterminate date t is represented by a tuple (t_L , t_U , P), where t_L and t_U (resp. lower and upper support) are the boundaries of an interval in which t may have occurred and P is the probability distribution for the occurrence of t in such an interval. For query evaluation, two dates can be compared by evaluating their precedence probability: where $P_k(x)$ is the occurrence probability of t_k at time x.

The encoding scheme. In particular, we developed an enhanced encoding scheme for indeterminate dates, based on piecewise-constant probability distributions, which is described in detail [10]. It is based on the classification of textual expressions into four main categories associated to predefined probability distributions according to the correspondence shown in the Table which follows:

Ca	Prototvpe	Shape	Distrib
C_1	*about in...*	Flat	`DURIN`
C_2	*at the*	Decrea	`EARLY`
C_3	*at the end*	Increas	`LATE`
C_4	*around...*	Bell-	`AROUN`

The (very frequent) C_1 case corresponds to a so-called *granularity mismatch* [5], where a determinate expression with higher granularity is used to denote an indeterminate expression with lower granularity. The probability distributions used are *piecewise-constant* functions over a small number of equal *base intervals* between the lower and upper support. It is shown in [10] how this choice, which is fairly correct from a semantic point of view, yields extremely efficient comparison algorithms between dates without any storage space overhead, which, on the contrary, would make unfeasible the application to Repetti's Dictionary of the basic approach in [5]. For the distributions but `DURING`, we also consider variants implying a different number of base intervals (i.e. `VERY_EARLY`, `VERY_LATE`, `STRICTLY_AROUND` and `WIDELY_AROUND`). In particular, indeterminate dates are represented in our encoding scheme by means of a (I, P) pair where I is the *principal interval* and P is one of the available distributions. The principal interval is the base interval where P is maximum and exactly corresponds to the temporal expression originally written in the text (and, thus, is a more intuitive parameter to identify than the lower and upper supports).

XML encoding. In order to implement temporal search facilities for Repetti's Dictionary, the dictionary items are used as target units for the search engine. To this end, the textual contents of every XML-encoded item will be enclosed in a tag pair `<ITEM>` ... `</ITEM>`, which can be selected on the basis of the encoded temporal references they contain. For example, if we are interested in a particular time period, we have to look for every item containing at least one temporal expression which overlaps such period. The temporal expressions of interest include single dates representing the validity of an historical event and time periods, which can be specified through their beginning and end dates. To this purpose, we introduced a "basic type" DATE, to be used alone or in pairs to represent events or intervals, respectively. By means of the DATE type, we are able to define the <EVENT> and <INTERVAL> tags. The <EVENT> tag contains the <AT> XML element with DATE type, while the <INTERVAL> tag will contains the <FROM> and <TO> elements, both with DATE type. In this way, events can be represented via structures like:

```
<EVENT> <AT *DATE* />
 text of the temporal
expression (event)
</EVENT>
```

whereas the interval markup will be like:

```
<INTERVAL> <FROM *DATE* /> <TO
*DATE* />
 text of the temporal
expression (interval)
</INTERVAL>
```

The base type DATE has several attributes, some of which are specific for the indeterminacy support:

- GRANULARITY, which allows to specify the granularity used to express the date value as "DAY" (default), "MONTH", "YEAR" or "CENTURY";

- VALUE, which allows to specify the date expression (in a way consistent with the assigned granularity);

- INDETETERMINATE, with values "YES" or "NO" (default), which allows to specify whether the date is expressed in indeterminate format or not; in case the attribute value is YES, do the further attributes have meaning:

- DISTRIBUTION, whose value can be one of the seven supported probability distributions (with their variants);

- DURATION, which expresses (with default "1"), as granularity multiples, the amplitude of the principal interval;

- CALENDAR, which allows to reference a specific calendar, as explained in the next section.

The principal interval is expressed in *implicit* way, by means of an interval having as lower boundary the first day of the VALUE temporal expression (e.g. 1456/1/1 for VALUE="1456") and an amplitude which can be evaluated as the specified GRANULARITY, value converted to days, multiplied by the DURATION value. In this way, a sample expression *around year 1622* can be simply encoded as:

```
<EVENT>
  <AT VALUE="1622"
INDETERMINATE="YES"

DISTRIBUTION="STRICTLY_AROUND"
/>
  around year 1622
</EVENT>
```

The choice of an implicit encoding makes it a bit more "transparent" and user-friendly, so that the user (i.e. the history researcher) can best concentrate on the choice of an intuitive "form factor" among a few available alternatives rather that on mathematical details of distributions like the support computation or the variance. Notice also how a uniform encoding choice, in which value and granularity exactly correspond to the textual expression used in the document (which we call **rigorous encoding rule**), represents itself metainformation, on the original form of the text contents, to be used for advanced searches.

Use of different granularities and calendars. The possibility to express dates at different granularity levels [15] mirrors the richness of forms in which dates are actually recorded in written historical sources, although it could make more difficult their management by temporal reasoning tools. The rigorous encoding rule we introduced allows to maintain such a richness also in the XML format for the benefit of the researcher. For instance, the temporal expressions which follow (evidenced in boldface) can all be found in the A-letter items:

[day] La memoria più remota di Anghiari sino a noi pervenuta spetta a un istrumento del **13 nov. 1083**, in forza del quale Bernardo, soprachiamato Sidonia...

[month] ...sino da quando il C. Guido figlio del C. Alberto, stando nella canonica del pievano di Stia, **nell'aprile del 1054**, donò alla vicina chiesa di Sprugnano...

[year] Il cast. di S. Angelo fu diroccato **nel 1282** dai Senesi, che lo diedero in feudo ai Salimbeni.

[century] Fu feudo de' Nobili di Pugliano, i quali vi possedevano torre e palazzo anche **nel secolo XV**.

It should be noted how there often exist nuances of meaning, for which the context is determinant, between the presence of indeterminacy and the real use of different granularities. As a matter of fact, in the second example which concerns a deed of gift, it is quite likely that it happened on a certain date located in April 1054, rather than it actually went on as an activity lasting for the whole month. In this case, we are in the presence of *granularity mismatch*, which corresponds to a C_1-type indeterminacy and requires the use of an indeterminate date with distribution `DURING`. The fourth example is indeed different, as it concerns the possession of estates by a noble family, that we could guess it was kept for the whole fifteenth century. In this case, the possession validity is more correctly expressed via a determinate interval. In any case, in order to introduce the most appropriate expression, it necessary a very careful choice (a support tool could possibly help by automatically prompting for the seemingly more likely alternatives). In our approach, the management of different calendars does not present particular difficulties, as all the calendars in use have the day as base granularity and seemingly use the same lattice (the only granularities of interest are in practice always: day, month, year and century). Hence, in the context of the `DATE` type, we can simply use the attribute `CALENDAR` to declare the name of the reference calendar for the date specification (e.g. `GREGORIAN`, which is the default value, `JULIAN`, `ROMAN`). The additional use of special calendar styles [4] can be encoded via variants of the calendar (e.g. `GREGORIAN_FLORENCE_STYLE` for the "Florentine" style of the Gregorian calendar). In order to be processed, all the dates found in a temporal XML document, possibly specified via different calendars, are automatically converted to a common reference calendar (e.g. the Gregorian) before successive operations like comparisons.

DESIGN OF A DEVELOPMENT TOOL

An important part of the "XML/Repetti" project is also the design and implementation of a development tool for the computer-aided encoding of temporal XML documents. Starting form an electronic version of Repetti's

Dictionary (in HTML format), such a tool tries to find out, using regular expressions, as many temporal determinations as possible. After this pre-processing phase, the tool assists the user in producing the XML markup, by proposing solutions strictly dependent on the temporal expressions found. Such a semiautomatic process leaves anyway to the user the freedom of choosing whether to accept the encoding prompted by the tool, or to change it on the basis of his/her own interpretation of the text. Moreover, the user will always be able to select pieces of text to encode as he/she thinks they contain historical information, although the tool did not succeed in pointing them out. The system will be made available on the Internet and usable through a standard Web. A prototype version of the tool is under advanced development. Its user interface is divided in two parts: a main area (upper window) where the document under processing can be seen with the temporal expressions evidenced by the pre-processing and a lower window which allows the user to specify structure and attributes of the XML tags to be inserted via friendly input forms. If the user, when browsing the document in the main window, finds a temporal expression which has not been automatically spotted by the tool, he/she can proceed with a manual selection of the corresponding text and its markup. The successive design phases of the tool functionalities and its graphic interface will be effected in strict collaboration with the history researches who are working to the "XML/Repetti" project, as they represent the final users of the tool.

IMPLEMENTATION OF A SEARCH ENGINE

The technological infrastructure developed for the "XML/Repetti" project must include an advanced temporal search engine. The design of the search engine has been guided by three main requirements:

efficiency - the engine must have an optimized implementation to cope with the Repetti's Dictionary dimension and the complexity of the retrieval procedures; this will impact on the overall system architecture, on the dictionary storage organization and on the temporal query processing algorithms;

powerful semantics - the engine must exploit all the temporal semantic richness supported by the adopted XML-markup scheme, including indeterminacy, multiple calendars and granularities, full TSQL2-like temporal query expressiveness; furthermore, also the use of "second-order" metainformation (e.g. based on the rigorous encoding rule) should be supported;

availability - the engine must be accessible on the Internet with a standard Web browser, it must have easy-to-understand functionalities and a friendly interface for user interaction. Special care was devoted to the fulfillment of the first requirement, as it plays a primary role for the success of the "XML/Repetti" project.

The design solution we propose is based on the use of a temporal index structure and optimized search algorithms on a server-side architecture. In particular, the solution is based on the adoption of a MAP21 temporal index [16], which relies on standard "off-the-shelf" technology. In the overall system architecture, the search engine runs on the server side: in our first prototype implementation, it is implemented by means of PERL scripts, which are activated by the client through a standard CGI mechanism. The retrieval work is effected by compiled C^{++} programs that access the only MAP21 index leaves that may contain qualifying dates to fetch the exact expressions of the (indeterminate) dates which are passed to the function doing probabilistic comparison. For all the dates passing the test at the assigned plausibility level, the pointers found in the index leaves are used to access the dictionary items on disk. The search results to be returned to the Web client are assembled by the PERL script, according to some user's option, either as a single XML/HTML file explicitly containing all the retrieved items, or as a "digest" containing a list of links to the selected items. It should also be noted that the temporal indexing via the MAP21 is a secondary indexing, which does not prevent other indexes to be built (e.g. on names, places, etc.). The client-side functionalities only require the execution of controls for the management of the query formulation, including the full specification of the temporal period of interest, possibly including all indeterminacy, granularity and calendar parameters. In particular, the user-interaction will be based on a friendly interface, so that it could be easily used by non computer-experts, which can be similar to the one designed for "The Valid Web" approach (mainly based on a Java2 applet and Javascript functions) [6,9].

CONCLUSIONS

In this paper we presented the evolution of "The Valid Web" approach for the representation and management of temporal information in XML documents and its application to the "XML/Repetti" project. Such project involves the computer-aided encoding of a large collection of historical text sources, for which a support tool is under development, and the efficient implementation of an "intelligent" temporal search engine available on the Web.

REFERENCES

1. S. Abiteboul, P. Buneman, D. Suciu, *Data on the Web: From Relations to Semistructured Data and XML*, Morgan Kaufmann Publishers, San Francisco, CA, 1999.
2. A. Benvenuti, F. Niccolucci, S. Baragli, C. Carpini, "Advances in XML Treatment of Historical Documents", in *La Historia en una Nueva Frontera*, AHC, Toledo, Spain, 2000.
3. I. Bogdanovic, O. Vicente, J.A. Barcelo, "A Theory of Archaeological Knowledge Building by using Internet: the DIASPORA Project", Proc. of Intl' CAA'99 Conf., Dublin, Ireland, 1999.
4. A. Cappelli, *Chronology, Chronography and Perpetual Calendar*, Hoepli, Milan, 1998 (*in Italian*).
5. C.E. Dyreson, R.T. Snodgrass, "Supporting Valid-time Indeterminacy", *ACM Trans. on Database Systems*, 23(1), 1998.
6. F. Grandi, F. Mandreoli, "The Valid Web: it's Time to Go...", TIMECENTER Technical Report TR-46, 1999, *www.cs.auc.dk/research/DP/tdb/TimeCenter/*
7. F. Grandi, F. Mandreoli, "The Valid Web©", Proc. of Software Demonstrations Track at the EDBT'2000 Intl' Conf., Konstanz, Germany, 2000.

8. F. Grandi, F. Mandreoli, "The Valid Web©" Home Page, *degas.deis.unibo.it/ValidWeb/*

9. F. Grandi, F. Mandreoli, "The Valid Web: an XML/XSL Infrastructure for Temporal Management of Web Documents", Proc. of Intl' ADVIS'2000 Conf., Izmir, Turkey, 2000.

10. F. Grandi, F. Mandreoli, "Effective Representation and Efficient Management of Indeterminate Dates", Proc. of Intl' TIME'01 Symposium, Cividale del Friuli, Italy, 2001.

11. F. Grandi, F. Niccolucci, "XML Technologies for the Representation and Management of Spatiotemporal Information in Archaeology", Proc. of Intl' CODATA Conf., Baveno, Italy, 2000.

12. F. Grandi, M. R. Scalas, "Extending Temporal Database Concepts to the World Wide Web", Proc. of Natl' SEBD'98 Conf., Ancona, Italy, 1998.

13. S. Hermon, F. Niccolucci, "The Impact of Web-shared Knowledge on Archaeological Scientific Research", Proc. of Intl' CRIS 2000 Conf., Helsinki, Finland, 2000.

14. C.S. Jensen, J. Clifford, R. Elmasri, S.K. Gadia, P. Hayes, S. Jajodia (eds.) et al., "A Consensus Glossary of Temporal Database Concepts - February 1998 Version," in *Temporal Databases – Research and Practice*, LNCS 1399, Springer-Verlag, Berlin, 1998.

15. A. Montanari, E. Maim, E Ciapessoni, E. Ratto, "Dealing with Time Granularity in the Event Calculus", Proc. of Intl' Conf. on Fifth Generation Computer Systems, Tokyo, Japan, 1992.

16. M. Nascimento, M.H. Durham, "Indexing Valid Time Databases via B$^+$Trees", *IEEE Trans. on Knowldege and Data Engineering*, 11(8), 1999.

17. F. Niccolucci (ed.), *From the Sources to the Network*, Proc. of Workshop on XML Applications to Historical Source Encoding, Florence, Italy, 2000.

18. F. Niccolucci, A. Zorzi, M. Baldi, F. Carminati, P. Salvatori, T. Zoppi, "Historical Text Encoding: an Experiment with XML on Repetti's Historical Dictionary", Proc. of Intl' AHC-UK'99 Conf., London, UK, 1999.

19. N. Pioch, "The Web Museum", *www.cnam.fr/wm/*

20. R.T. Snodgrass (ed.) et al., *The TSQL2 Temporal Query Language*, Kluwer Academic Publishers, Boston, MA, 1995.

21. The eXtensible Markup Language (XML) Home Page, W3C Consortium, *www.w3.org/XML/*

22. The Microsoft Internet Explorer Home Page, *microsoft.com/windows/Ie/*

23. The Semantic Web Agreement Group Home Page, *swag.semanticweb.org*

ABOUT THE AUTHORS

Fabio Grandi is Associate Professor of Information Systems at the University of Bologna. His main scientific interests include temporal databases, Web extensions and schema versioning.
E-mail: *fgrandi@deis.unibo.it*

Federica Mandreoli is Research Assistant at the University of Modena and Reggio Emilia. Her interests include information retrieval, Web extensions and schema versioning.
E-mail: *fmandreoli@dsi.unimo.it*

Restoration and Preservation

Automatic Feature Recognition Technique on Stone Monuments Using Visible and IR Photography

José Luis Lerma (*)

(*)Depto. Ing. Cartográfica, Geodesia y Fotogrametría.
Universidad Politécnica de Valencia, SPAIN
E-mail: jllerma@cgf.upv.es

ABSTRACT

Visible photography has long been used in the field of architecture for documentation and restoration tasks. However, the automatic extraction and analysis of facade features, i. e. materials and alterations, is unusual in most of multidisciplinary teams involving documentation, conservation and restoration projects. This paper shows the profitability of applying digital image analysis on stone monuments, in order to provide exhaustive, global and accurate information from visible and near-infrared photographs. Output data are rectified and automatically classified images from which it is possible to obtain, among others, detailed positions of materials and degradations on facades, real measures of distances, areas of architectural features, as well as statistical documents of the accuracy achieved in the whole process of automatic recognition.

KEYWORDS: architectural photogrammetry, digital image processing, feature recognition, multispectral classification

INTRODUCTION

Before any conservation and restoration work, it is advisable to compile all kind of information regarding the stone monument. The monuments are in complete interaction with inner and outer factors that contribute to its maintenance and, inevitably, to its deterioration. Whatever the actual state of the monument, it is necessary to carry out multidisciplinary studies in order to chose the right present/future action. Furthermore, the higher the quality and the completeness of the analysis the better diagnosis of the Cultural Heritage.

The recognition of constituent materials and the evolution of their damages over the time is an important task in the documentation work [6]. Besides, it is not a quick process; it takes long to get all the necessary sampling data, mainly by means of direct methods. Data capture depends either on the accessibility of the monument, its emplacement and dimensions, and on the way of inspection and analysis.

The objective of this paper is to present a way of analysis –objective, global and non-destructive– based on the automation of topographic surveys by means of digital photogrammetry and digital image processing. The output data are just metric rectified images (orthoimages) and classified images showing the different kind of materials, features and alterations of stone facades.

This study applies spectral classifications of visible and near-infrared images on stone monuments as pointed out in [1] and [2], although the methodology follows the guidelines that appear on the latter paper. Additionally, it was unnecessary the application of both spectral and textural analysis [3][4] because there were not many different materials on the facades; limestone was the predominant material.

STUDY SITE AND FEATURES

The study site is the *Saint Agustin's* Church located on the old part of the city of Valencia. The constituent material mainly consist of limestone. Two facades are studied, the main one and a piece of the lateral facade. After a direct observation of the main facade, several masonry wall materials and damages (features) were identified: (1) *washed limestone*, (2) *polluted limestone*, (3) *very polluted limestone*, (4) *dark areas* (lampposts, indoors and shadows).

Regarding the lateral facade, similar and additional features were found: (1) *washed limestone*, (2) *polluted limestone*, (3) *white limestone*, (4) *humid limestone,* (5) *spot* covering the limestone, (6) *painting* on the limestone, and (7) *plants* in front of the facade.

METHODOLOGY

The main facade was photographed with a conventional colour digital camera requiring three shots to acquire the whole facade, meanwhile the lateral facade was photographed with two cameras, the digital one and a standard analogue 35mm camera. This second camera was used for the reflected near-infrared radiation.

Within the digital preprocessing stage some tasks were performed: image enhancement (histogram modification of individual images), multispectral image rectification (Fig. 1) and image extraction of the whole set of visible and near-infrared images and, finally, the construction of a digital mosaic from the set of overlapping rectified-images.

Besides, image rectification also allowed the creation of the mosaic image for the main facade (Fig. 2). The applied method of rectification always was the two-dimensional transformation. The mosaic image was a metric image (orthoimage) from which real data could be obtained afterwards (distances, areas of interest, etc.).

(a) (b)

Figure 1: Visible (a) and near-infrared (b) images of the lateral facade after rectification

Supervised classifications were performed on both facades. Spectral classes were assigned to each of the four features from the main facade and seven features from the lateral facade. Training sets were collected from well-distributed and separately placed polygons over the study sites. The classifier used was the maximum-likelihood with a 95% confidence level threshold [5]; the non-classified pixels were assigned as Null class. Fig. 3a displays the multispectral classified image of the lateral facade; the main facade appears spectrally classified in Fig. 3b.

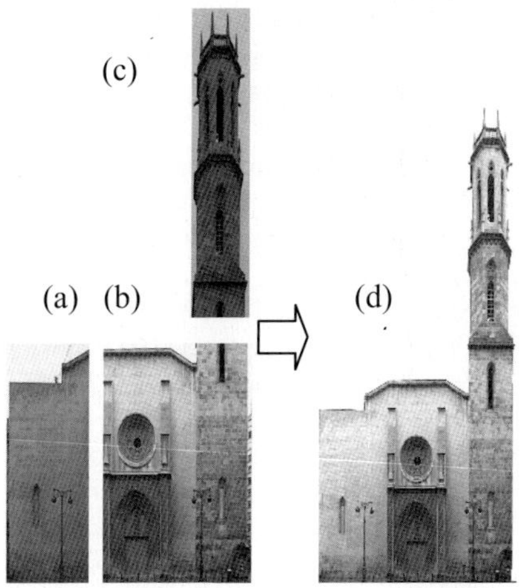

Figure 2: Mosaic image (d) of the main facade constructed from rectified images (a,b,c)

Figure 3: Classified images of lateral (a) and main (b) facades

RESULTS

The accuracy assessment among the actual classes (inspected areas) and the predicted classes (classified data) was evaluated by means of an error matrix for the lateral image. Null classes were not taken into account for the accuracy assessment; only the assigned classes after their classifications were evaluated. The overall accuracy achieved is around 96 percent. Regarding the main facade, identical statistical analysis was carried out; the overall accuracy reached is approximately 84 percent.

As regards the user´s accuracy, most of the classes achieved an accuracy around 95 percent. Only the user´s accuracy of *Washed limestone* class is below 90 percent although it is not worrying because of the spectral closeness with its partner *Spot* class. Furthermore, the producer´s accuracy of *Washed limestone* class is approximately 80 percent because the excess errors coming from *Spot* and *White limestone* classes.

The overall accuracy of the main facade is lower than the accuracy achieved on the lateral facade mainly because the near-infrared band was left (not available). Omission errors within *Washed limestone* class influenced negatively the classification accuracy; however, its user´s accuracy is high. Other important fact is the similar spectral response between *dark areas* and *very polluted limestone* classes. It might help us to focus on the areas needing inevitably restoration.

CONCLUSIONS

This study shows that the recognition and characterization of features by means of digital processing tools on stone monuments not only are possible but also offer a high degree of completeness and accuracy. Furthermore, when photogrammetric techniques are used to create the metric photographs, output data obtained from them are ready for measurements and fit to exact positions.

The near-infrared image is essential to distinguish properly the different features of the lateral facade. Results from the main facade could have been even better using more spectral images (not only visible).

REFERENCES

1. Grunicke, J. M., Sacher, Strackenbrock, B., Image processing for mapping damages to buildings. In *Proceedings of CIPA International Symposium*, Cracow, 1990. 257-263.

2. Lerma, J.L., Herráez, J. Reconocimiento y Cartografiado Automático de Monumentos Arquitectónicos. In *Proceedings of the XI International Congress of Graphical Engineering*. Logroño–Pamplona, Spain, 1999. Vol. II, 732-740.

3. Lerma, J.L., Ruiz, L.A. Análisis de Fachadas de Edificios Históricos mediante Métodos Espectrales y de Texturas. In *Revista Electrónica de Visión por Computador*, April 2000. Vol. 2, *http://www.cvc.uab.es/revc*

4. Lerma, J. L., Ruiz, L. A., Buchón, F., Application of spectral and textural classifications to recognize materials and damages on historic building facades. In *International Archives of Photogrammetry and Remote Sensing*, July 2000. Vol. XXXIII, Part B5A, 480-483.

5. Niblack, W., *An Introduction to Digital Image Processing*. Prentice/Hall International, 1986.

6. Rossi-Doria, P., Mortars for restoration: basic requirements and quality control. In *Materials and Structures*, 1996, 19 (114), 1-8.

ABOUT THE AUTHORS

José Luis Lerma received his Tech. Ing. degree in Topography in 1992, a Ing. degree in Geodesy and Cartography in 1996 and a Ph.D. in Photogrammetry in 1999 all of them from the Polytechnic University of Valencia (Spain). He is the author of four books regarding photogrammetric (terrestrial, aerial; digital, analítical). His research activities include digital architectural and archaeological photogrammetry. URL address:
http://www.top.upv.es/asignatu/fotogram/lerma.htm;
E-mail: *jllerma@cgf.upv.es*

A Digital Image Processing and Database System for Watermarks in Medieval Manuscripts

Emanuel Wenger (*), Victor N. Karnaukhov (#), Alois Haidinger (§), and Maria Stieglecker (§)

(*) Austrian Academy of Sciences, Commission of Scientific Visualization, Vienna, Austria
E-mail: emanuel.wenger@oeaw.ac.at
(#) Russian Academy of Sciences, Institute for Information Transmission Problems, Moscow, Russia
E-mail: victor.karnaukhov@iitp.ru
(§) Austrian Academy of Sciences, Commission of Paleography and Codicology of Medieval Manuscripts, Vienna, Austria
E-mail: alois.haidinger@oeaw.ac.at or maria.stieglecker@oeaw.ac.at

ABSTRACT
Watermarks are the most important tool for dating old, not explicitly dated paper documents. Hence, catalogues and databases of watermarks play an eminent role for the work of medievalist and paper historians. This article presents an integrated software system developed for storage, retrieval, manipulation, digital processing, and identification of watermarks in old manuscripts. The whole processing pipeline beginning with the scanning of watermarks up to their identification is described. Most of these watermarks were captured from manuscripts kept in the Klosterneuburg monastery (Austria).

KEYWORDS: watermark, database, dating, medieval manuscripts, paper, digital image processing

INTRODUCTION

Medieval manuscripts, and incunabula represent an important part of our cultural heritage. Investigation, cataloging, and restoration of these books are necessary in order to preserve our heritage for the future. Many of the old books (documents) are not dated explicitly, although the knowledge of the date of their production would be essential for historical research. The comparison of dated watermarks with undated such is the major method for dating undated medieval handwritten paper documents. Several standard catalogues exist containing thousands of hand-drawn sketches of watermarks [2,6]. The identity of a watermark with one in the standard catalogues is a good indicator for the age of the watermark and document in question.

There are some essential hindrances to the precise dating of the watermarks in a document using the standard catalogues containing hand-drawn sketches of the watermarks. In many cases the watermarks of a document are covered by the written text such that it is impossible to produce good hand-drawn sketches. But even if it is possible to make a perfect sketch of the watermark, it is still time-consuming and tedious to find an identical watermark among the

hundreds of similar ones in the catalogues. Furthermore, all catalogues are incomplete and there is no guarantee that the search for an identical watermark will be successful. Additionally, it appears that the proof of the watermark identity is not sufficient for a reliable dating result. Some additional data must be supplied and taken into account. A way for overcoming most of these drawbacks is the use of computers, which are ideal tools for cataloguing, comparing, and retrieving of huge amounts of data. An image and text database [4,7,8] combined with special software tools for tracing, image processing, and comparison can facilitate significantly the identification task.

ACQUISITION AND DIGITAL PROCESSING OF WATERMARK IMAGES

Watermarks are small deviations in the thickness of paper. Prior to inputting the watermark into the computer, a hardcopy has to be produced. Beta radiography and electron radiography are considered the best methods for clear recording of watermarks because of their accuracy and sensitivity. Minute deviations in the density of the paper are recorded sufficiently well by them. However, even these methods produce hardcopies with low contrast, except some very bright or dark spots and areas caused by holes in the paper or special color ink.

The watermark processing starts with the input of the prerecorded watermark hardcopies. A sensitive flatbed scanner with a transparency extension is used for their scanning. In the preprocessing stage the scanned images are enhanced in order to increase the contrasts, to compensate for scanning artifacts and to reduce the noise. *Figure 1* shows a typical watermark image scanned from a beta-radiographic hardcopy. The objects which are used for the watermark identification are: watermark (a), chain lines (b), and laid lines (c).

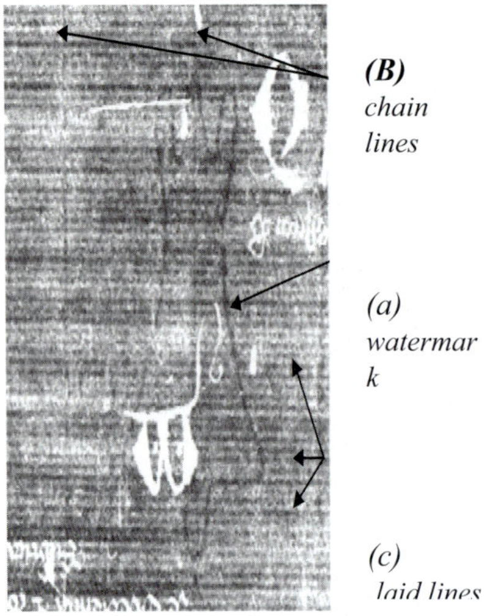

Figure 1: watermark image: watermark (a), chain lines (b), and laid lines (c)

COMPUTER-AIDED EXTRACTION OF WATERMARK CONTOURS

Watermark contours cannot be extracted automatically due to the many artifacts in the images. A semi-automatic procedure was chosen in order to achieve good and fast results. The semi-automatic procedure is designed to minimize user interaction and to optimize the quality of contour tracing.
The procedure for watermark contour extraction is implemented as an interactive one. At first, the watermark motif, which defines the overall watermark shape, is determined by user interaction. The watermark motif (e.g. bell, scales, bow and arrow, ox head

etc.) with subtypes down to four levels is selected from a hierarchically-structured list and predetermines the number and positions of control points, which have to be finally positioned manually. Their number is kept minimal as far as possible (typically between ten and twenty). Furthermore, the motif also sets geometric restrictions for the variability of the contour.

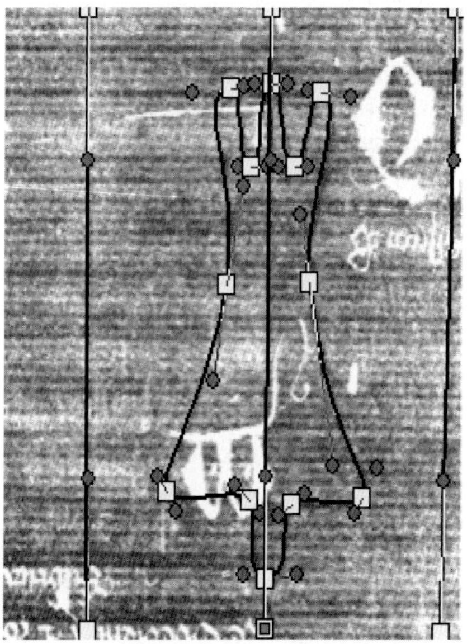

Figure 2: Watermark image with drawn contours and control points

Then, a semi-automatic procedure based on parametric cubic Bézier curves [1] is used for the approximation of the watermark contours [9]. The control points as chosen by the user divide the watermark contours into sets of adjacent contour segments. The resultant Bézier curve interpolates the two control points of each contour segment and approximates the other two. These additional control points are calculated under the criterion to fit the watermark contour as good as possible. The two main steps of this procedure are demonstrated in *Figure 2* and *Figure 3*. By dragging the control points, the user has the opportunity for fine-tuning of contour segments. The final set of control points is stored in a relational watermark database.

The Bézier curve is affine invariant. This means that any linear

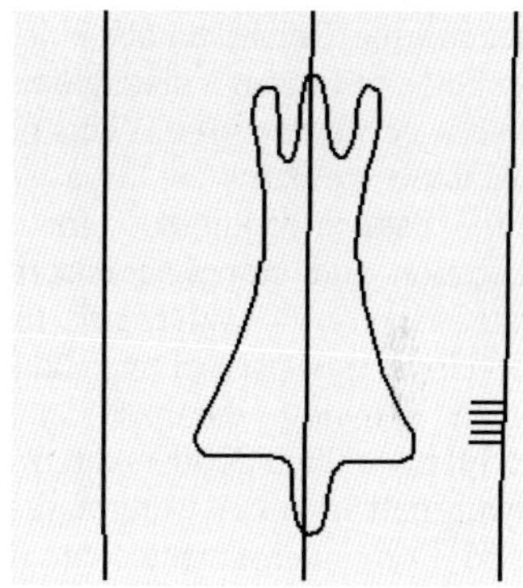

transformation of its control points defines a new Bézier curve, which is just the same transformation of the original curve. So, an extracted contour can be easily adjusted for other watermarks of the same motif.

RELATIONAL DATABASE OF WATERMARKS

A relational data model was chosen for the development of the watermark database. All data are organized in a set of related tables. The database contains two main parts. One deals with the watermark classification according to their textual description. Each watermark motif has a defined place in a hierarchical structure of four levels of watermark subtypes. According to the position in the structure, a unique digital code is generated for each watermark type registered in the database. It is an eleven-digit code consisting of three preceding digits for the watermark type code and four subsequent pairs of digits for subordinated subtype codes. The capacity of this code is enough for coding up to 999 watermark types and up to 99 watermark subtypes in each of four possible subordinated levels. This hierarchical structure is based on the watermark motif classification rules [3].

The second part of the database is used for the registration and management of concrete watermark entities. This part of the database contains a large set of related data required for the classification and complete description of watermarks. All watermark images are stored in this part of the database. Our fast growing database contains currently more than 3,000 watermarks. The main part of these watermarks was captured from manuscripts kept in the library of Klosterneuburg monastery (Austria).

SOFTWARE TOOLS

A large set of software tools is implemented together with the database management in one system. User interaction with this system is realized through a graphical user interface containing a set of menus, buttons, and other controlling items. This graphical user interface was designed following the Microsoft Windows-style.

One module of the system supports the processing and visualization of watermark images and the extraction of contours. The database allows complex textual as well as graphical queries. The software tools include computer-aided measurement of metrical parameters, watermark contour extraction, watermark classification, and automation of other routine jobs. A snapshot from a typical user session is presented in *Figure 4. (pg. 264)*

This snapshot shows three windows: one displaying the watermark image (left side of the screen), one for controlling the watermark contour extraction process (center), and one for the textual watermark description (right side). The system is implemented on a PC platform and runs under Windows 95/98/NT-4.0/2000.

SUMMARY

Methods and tools for digital processing of watermark images and storing them in a relational watermark database were developed and implemented. All methods and tools are integrated in a software system, which allows the user to perform all necessary tasks for watermark classification within one system. The system is fully implemented and has proven its usefulness in a production environment.

ACKNOWLEDGMENTS

This work is supported by the Austrian Science Fund (grant FWF-Projekt P13298-ARS) and INTAS.

REFERENCES

1. Bézier P., *Numerical Control: Mathematics and Applications*, translated by A.R.Forrest and A.F.Pankhurst, John Wiley & Sons, London, 1972

2. Briquet Ch.M. Les Filigranes. *Dictionnaire historique des marques du papier dés leur apparition vers 1282 jusqu'en 1600.* Paris 1907

3. International Association of Paper Historians – IPH, *International Standard for the Registration of Watermarks*, P.F. Tschudin, ed., Rienen, Switzerland, 1997.

4. Karnaukhov V.N., Merzlyakov N.S., Wenger E., Haidinger A., and Lackner F. Digital Analysis of Watermarks of Medieval Manuscripts. In *Computer Optics*, Vol.14-15, 1995, 11-24 (in Russian)

5. Karnaukhov V.N., Wenger E., Merzlyakov N.S., Haidinger A., Lackner F. Thematic processing and retrieving of watermarks. In *Proceedings of SPIE* Vol. 2363, 1995, 32-39

6. Piccard G. *Die Wasserzeichenkartei Piccard im Hauptstaatsarchiv Stuttgart.* Findbuch I-XV, W. Kohlhammer, Stuttgart, 1966-1987

7. Rauber C., Ruanaidh J., Pun T. Secure distribution of watermarked images for a digital library of ancient papers. In *Second ACM international conference on digital libraries*, Philadelphia, ACM, 1997

8. Rauber C., Tschudin P., Startchik S., Pun T. Archival and retrieval of historical watermarks. In *Proceedings of ICIP 1996 - International conference on image processing*, Lausanne. IEEE, 1996

9. Wenger E., Karnaukhov V.N., Haidinger A. Contour Extraction of Watermarks in Old Manuscripts In: Maria Alberta Alberti, Giovanni Gatto and Ivan Jelinek, Eds., *Eurographics '99 Short Papers and Demos,* , Milano, 1999, 236-238

10. Wenger E., Karnaukhov V.N., Haidinger A., Merzlyakov N.S. Digital image analysis for dating of old manuscripts. In: Image Analysis Applications and Computer Graphics. (R. T. Chin, H. H. S. Ip, A. C. Naiman, T.-C. Pong, eds.). In *Lecture Notes in Computer Science 1024*, Springer, Berlin, 1995, 522-523

ABOUT THE AUTHORS

Emanuel Wenger is a researcher at the Commission of Scientific Visualization of the Austrian Academy of Sciences in Vienna. His research interests include computer graphics, digital image processing, and virtual archaeology. He graduated from the University of Technology in Vienna in 1978.
E-mail: *emanuel.wenger@oeaw.ac.at*

Victor N. Karnaukhov is a researcher at the Institute for Information Transmission Problems (IITP) of the Russian Academy of Sciences in Moscow. His research interests include multimedia, digital image processing, and database systems. He graduated at the Moscow State University in 1975. He received his PhD in computer science and image processing from the IITP of the Russian Academy of Sciences in 1986.
E-mail: *victor.karnaukhov@iitp.ru*

Alois Haidinger is a researcher at the Commission of Paleography and Codicology of Medieval Manuscripts in Austria of the Austrian Academy of Sciences. He has published a number of catalogues and articles on medieval manuscripts, mainly from the libraries of Klosterneuburg and Melk, since 1974. He is specially interested in developing methods to use filigranology for dating medieval paper manuscripts.
E-mail: *alois.haidinger@oeaw.ac.at*

Maria Stieglecker is a fellow researcher at the Commission of Paleography of Medieval Manuscripts at the Austrian Academy of Sciences in Vienna. Her research interests include late medieval

manuscripts and their watermarks. She studied history at the University of Vienna and graduated in 1995. In the same year she finished her post-graduate studies at the Institute of Research of Austrian History in Vienna (MAS).
E-mail: *maria.stieglecker@oeaw.ac.at*

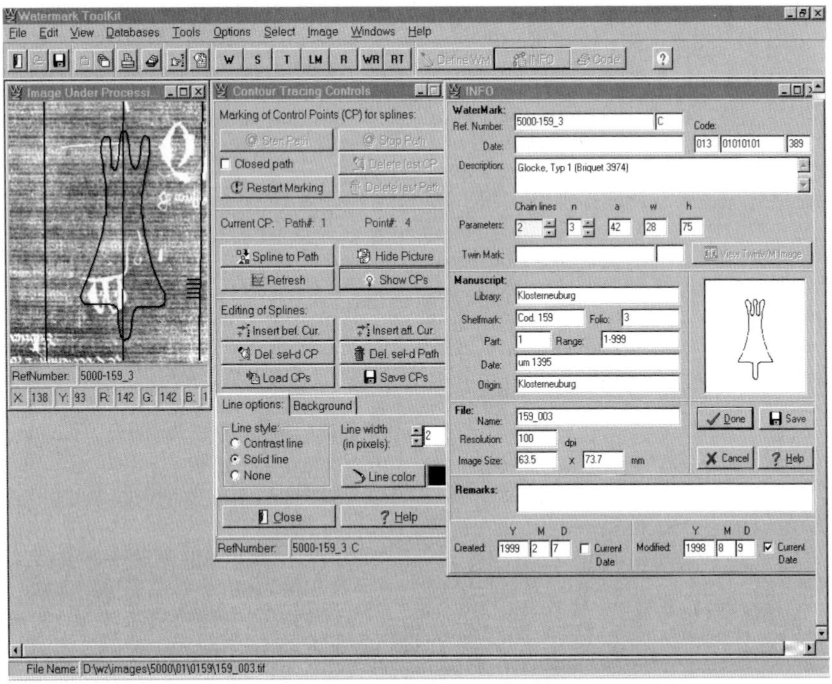

Figure 4: Graphical user interface with three open forms

3D Scanning the Minerva of Arezzo

C. Rocchini (*), P. Cignoni (*), C. Montani (*), P. Pingi (*), R. Scopigno (*)
R. Fontana (#), M. Greco (#), E. Pampaloni (#), L. Pezzati (#)
M. Cygielman (§), R. Giachetti (§), G. Gori (§), M. Miccio (§), R. Pecchioli (§)

(*) Istituto Elaborazione dell'Informazione – CNR, Pisa, Italy
E-mail: roberto.scopigno@cnuce.cnr.it
(#)Istituto Nazionale di Ottica Applicata, Firenze, Italy
E-mail: luca@ino.it
(§) Centro di Restauro Archeologico, Firenze, Italy
E-mail: cygielman@virgilio.it

ABSTRACT
This paper presents the initial results of the "Minerva Project" obtained by means of two different 3D scanning techniques. The Minerva of Arezzo is a bronze statue (Archaeological Museum in Florence) currently under restoration. The "Minerva project" intends to show how 3D techniques can integrate standard diagnostic methods giving useful and powerful tools to the restorers. The main goal of the project is to build complete 3D digital models of the Minerva statue, before, during and after restoration in order to monitor the various phases of the restoration process. The case study is very stimulating, because variations of the shape (removal of plaster, polishing of the corroded bronze surface) and of the colour (removal of the thick paint film on the bronze surface) of the statue are forecasted. In this work we also discuss the experience we get and the problems encountered during the project.

KEYWORDS: digital models, 3D scanning, computer graphics, computed-aided restoration, virtual reality.

INTRODUCTION

This paper reports the first results of an in-progress joint project concerning the development and use of new 3D graphic technologies in the Cultural Heritage (CH) field, specifically oriented to restoration needs and constraints. The restoration of a bronze statue of great historical-artistic interest, the Minerva of Arezzo, gave us the opportunity for testing these technologies.

3D digital models have a wide application field: realization of 3D catalogues and virtual museum, fast prototyping, supporting the restoration project, etc. All these applications require a *3D model*, that is the digital representation of the object involved. Classic 3D modeling tools are often inadequate to reproduce the complex shape of most art works (e.g. sculptures) whereas 3D scanning techniques are sufficiently accurate to provide a high fidelity reproduction of the shape of an object. These techniques have been adopted in a number of recent Cultural Heritage projects, e.g.: the Digital Michelangelo Project of the Stanford University [8], the acquisition of Michelangelo's Pietà in Florence by the IBM T.J. Watson Laboratory [13], and the acquisition of a section of the Coliseum in Rome by an Italian research team [5]. 3D scanning technology evolved in the last few years, but its use in many CH applications has been hampered because of the high cost of 3D scanning devices, the limited diffusion of knowledge about its potentialities and the fact that only few applications are able to manage 3D complex models, thus showing the full capabilities of such data.

The Minerva project goal is therefore to further develop and assess 3D scanning and graphic technologies, with a particular focus on the art works restoration. We also aim to set up a multidisciplinary team of experts in restoration, archaeology, optics and computer science in order to design useful new tools for the restorers.

Figure 1: The Minerva of Arezzo (part of)

THE RESTORATION OF THE MINERVA OF AREZZO

The "Minerva of Arezzo" is a bronze statue discovered in Arezzo in 1541 (see Figure 1) whose origin is still uncertain: it could be either an original Hellenistic bronze dating back to third century B.C., or a variant produced in the Roman Imperial period (first century A.D.). The statue is usually located in the Archaeological Museum in Florence, and is now under restoration at the Archaeological Restoration Centre in Florence. The restoration was considered very urgent because of the highly endangered status of the statue (precarious conditions of the structural wooden elements inside the statue, extensive corrosion of the bronze layer). Besides that, the statue was extensively restored in the past (the main restorations dates back to the sixteenth and the eighteenth centuries), and these actions modified in a significant manner the aspect and integrity of the artwork. As an example, a missing arm (the right one) was replaced in the eighteenth century restoration; plaster was used to join disconnected bronze sections and to fill gaps; finally, a dark greenish paint covers most of the statue surface, giving a uniform aspect to the plaster and the bronze sections but actually covering the original patina. This restoration is very complex because of the heterogeneity of the component materials, so the following actions were planned: the realization of a physical reproduction of the Minerva by casting the statue; an accurate analysis of the statue conditions (performed via: visual inspection; a photographic survey under visible and UV light; an X-ray imaging survey); an accurate clean up of the statue surface, to remove dust and dirty deposits; the removal of the plaster fillings (made by gypsum or glue) and the endangered internal wooden support followed by its replacement with a new structural frame; an accurate polishing of the bronze surface either by mechanical removal of the bronze oxidation or by the use of other innovative technology such as laser-based polishing instruments; finally, the re-assembly of the statue components with techniques and materials more adequate than the ones used in the previous restorations.

The acquisition of a complete digital model of the statue is considered of primary importance for monitoring its status before and after the restoration. Therefore the accuracy of the digital model has to be sufficient to reveal alterations of the statue's shape due to restoration, (e.g. the removal of the plaster fillings or non-original elements); the surface colour alteration due to the removal of the paint film and the metal polishing has to be acquired and all the collected data have to be organised for an easy and interactive visualization, manipulation and analysis on low-cost computers (PC's).

3D SCANNING TECHNOLOGIES DEVELOPED

The Minerva Project gives us the possibility to evaluate and compare two different hardware technologies (a 3D scanner based on structured light and a laser scanner) and to test a suite of software tools for the management of 3D scanning data. Let us describe the characteristics of these components in the following subsections.

A low cost scanner based on structured light

An optical scanner is generally based on a couple of devices: an *emitter*, which projects structured light on the object to be scanned, and a *sensor*, which acquires images of the distorted pattern reflected by the object surface. The 3D geometry is reconstructed by triangulation, given a known location of the emitter-sensor pair.

The scanner developed by C.N.R. [11] has been designed using only consumer technology, thus ensuring affordable hardware costs and enabling easy technological improvements. The accuracy and resolution supported are midway between commercial low cost and high quality laser scanners . The instrument supports the surface color acquisition referred to the statue geometry and ensures easy operability and flexibility.

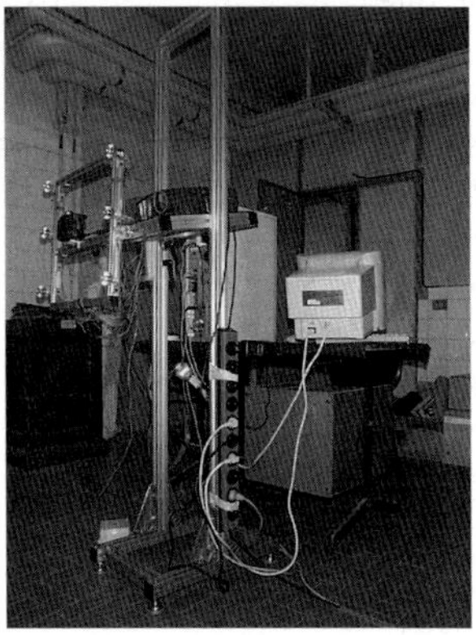

Figure 2: The low cost scanner designed by CNR (Pisa), based on structured light and consumer electronic technology.

In our scanner the emitter is a video projector (DLP technology) and the sensor is a digital still camera. The advantage of using a video projector to produce a structured light pattern is that we can easily experiment and evaluate different patterns. One possibility is to produce an image with a set of thin and tightly packed vertical bands, but this solution gives some problems with objects having a complex topology: the presence of self-occluded parts might cause some bands to disappear or brake, and this makes the reconstruction of band indexing not feasible. The use of a hierarchical pattern where the band spatial frequency is iteratively doubled (binary coded illumination), can solve this problem [7]. Our system uses this approach. A set of colored band pattern is produced by dividing, at every step, each region in two sub-regions, red on the left and blue on the right, separated by a green line (see an example of the recursive process in Figure 3). These images are produced recursively until we reach the projector resolution.

Stripe detection and indexing is implemented using a progressive approach: the new stripes introduced in image *k* are recognized and reconstructed on the base of the knowledge acquired reconstructing images [*1,..k-1*].

A software module running on a standard PC drives the emitter and the sensor devices. Photos are taken to acquire images either of the distorted patterns (from which the geometry is reconstructed by triangulation, producing a range map) or of the object under different illumination conditions (from which the illumination-invariant color, or "albedo", of the object surface is reconstructed). The color images are by definition self-registered with range maps.

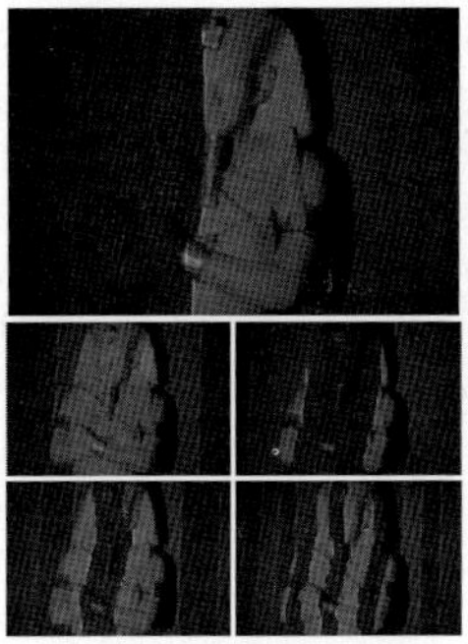

Figure 3: The set of coloured stripe images projected on the surface of the object.

The scanning system, depicted in Figure 2, produces:

- a range map of each selected surface portion, with sample density of approximately 0.7 mm on the vertical axis and 1.4 mm on the horizontal one (using a 1024×768 video projector, located at approximately 1.3 m. from the work of art);
- an illumination-invariant color texture which represents the pictorial detail present on the acquired surface section, reconstructed adopting a technique proposed in [14].

As usual in 3D scanning, complex objects are modeled via the acquisition of a set of partially overlapping range scans. Our proprietary software is used to register and merge all the range maps in a single triangulated mesh, and is described briefly in a following section.

The laser scanner

A laser scanner for Cultural Heritage applications has been realized at the Istituto Nazionale di Ottica Applicata (INOA) in Florence. Among the active non-contact devices for 3D measurements, laser scanner is probably the most versatile one. Its main

advantages are great set-up simplicity, small size and high resolution.
The working principle is based on the optical triangulation: a light source illuminates an object and an image of this light spot is then formed, by means of a lens, on the surface of a linear light sensitive sensor. By measuring the location of the light spot image the distance of the object from the instrument can be determined, provided the baseline and the angles are known [8] .

Figure 4: Experimental setup of the prototype of laser scanner designed by INOA (Florence).

Figure 5: A plaster head with the superposition of the red laser line.

The source employed is a laser diode (Lasiris Inc., $\lambda = 670$ nm, $P = 30$ mW) with a 60° prism head to project a light line, and the sensor is a CCD camera (Jaj Corp., 1300×1030 square pixel 6.7 μm wide, S/N > 56 dB). These two devices are mounted on a motorized stage to allow the scanning of statues or architectural moldings, and the whole system is computer controlled. The experimental setup is shown in Figure 4. In its typical configuration, the laser sheet is horizontal (parallel to the floor), but to facilitate scanning horizontal crevices and extruded parts, the system can be oriented anyhow. The gantry consists of two vertical trusses (2 meter long) on which the motorized stage can vertically translate and rotate around an axis perpendicular to them.
The system has been tested on reference surfaces and corrections for lens distortion were made. The accuracy of location of the image spot, and hence the instrument accuracy, is degraded by laser speckle [1]. The resulting quota resolution is 50 μm and the absolute error is less than 300 μm, with a standoff distance of 50 cm and a scanned area of about 40×40 cm^2.
The control software, realized at INOA, manages both the control (translation stage) and the acquisition systems (grabber and CCD camera). The output is a sequence of profiles gathered in meshes of binary data, i.e. the initial range maps.
The following processing consists of aligning and merging the scans taken from different gantry positions (see following paragraph).

Software tools for 3D scanning

The pipeline of phases of any 3D scanning session is rather complex, involving many steps: **scanning** the object; **aligning** the range maps (that have to lie in the same space); range map **merging** (to build a single, non redundant mesh out of the many, partially overlapping range maps); mesh **editing** (to improve the quality of the reconstructed mesh); mesh **simplification** (to reduce the often huge complexity of the model obtained); finally, mesh **conversion** (to export the mesh produced to the data representation scheme of interest). A comprehensive tutorial of the techniques proposed in literature and most frequently used in applications has been recently proposed in [2]. The CNR project partner has designed a suite of software tools that manages most of these phases:

- **MESHALIGN**: the module allows the registration of multiple range maps (needed to get a complete coverage of the object surface), which by definition represent the distance of the sampled surface points from the sensor location. This means that all these range maps have to be transformed into a unique reference space, such that sections of the range maps that correspond to the same surface zone will overlap. The registration module follows the approach proposed in [9], improved by the adoption of a multiple level of detail representation of the range maps which allows better performances and higher process accuracy.
- **MESHMERGE**: the module allows the reconstruction of a single 3D mesh out of a set of registered range maps. It adopts a new approach [10], which is characterized by a lower space complexity and improved accuracy with respect to the current alternative volumetric reconstruction approaches [4].
- **MESHEDIT**: the module allows performing simple editing actions on the mesh, to improve its quality (e.g. to fill small holes, to remove non-manifold components of dangling edges/faces, to apply smoothing filters, etc).
- **MESHSIMPLIFY**: the module supports the simplification of the [huge] meshes produced by 3D scanning devices, by removing in a controlled manner mesh vertices. The simplification follows the edge collapse approach [6] and has been implemented in an *out-of-core* fashion to allow the management of meshes that could be larger than the core memory of the computer used [3].

RESULTS

The Minerva statue has been scanned using both the two scanner devices presented. In the case of the first scanner (structured light) we took 146 range maps. For each range map we shot: one image of the object under white light plus 10 images to get all the stripe patterns; 6 images for the acquisition of the colour (six different lighting conditions). In total we have 17 images for each range map (shape + colour).

The Minerva has been scanned with the help of a rotating platform. The scanner was positioned at approximately 1.3 meters from the statue, and each range map sampled a section wide approximately 70 cm×50 cm. For each complete rotation of the statue we get 12 range maps, and after completing this piece-wise pseudo-cylindrical scan we raised the height of the scanner to acquire the next section. Due to the required overlap of the range maps, we acquired 8 of these pseudo-cylindrical scans (getting in total 96 range maps). Some more range maps were taken to get some sections of the Minerva surface that were not represented in the previous ones with sufficient accuracy or completeness (e.g. the right protruding arm, the neck, the top of the head, etc). The images shot were around 2,500 for approximately 1,2 GB on disk.

Our software has been used to register and merge all the range maps in a single triangulated mesh. The alignment of this complex set of range maps took approximately 4 weeks (but it is hard to estimate the effective time, because during the aligning session we also spent time to discover bugs or to refine the user interface and the functionalities of the MESHALIGN tool). A complete digital model of the Minerva has been produced, and is shown in Figure 6. The 3D mesh has been reconstructed from the range maps using MESHMERGE and selecting an inter-voxel distance of 0.57 mm. The mesh reconstructed is rather big: 26 millions of triangular faces. To improve usability, this mesh was simplified using our external memory simplifier, MESHSIMPLIFY. The results presented in Figure 6 are simplified models representing the whole statue (with around 1M faces) and the head section (with around 1.1M faces).

The acquisition of the color has given us some problems because our approach assumes a Lambertian reflection model [14, 12]. Conversely, the bronze surface is not Lambertian but highly reflective. For this reason the un-shaded albedo colors returned by our original approach are not completely satisfactory, and we are designing a different approach to cope with this issue.

Further actions planned are the integration of the laser-scanned model with the color data that have been sampled with the structured light scanner and another acquisition of the statue after the restoration.

Figure 6: Digital 3D models of the Minerva's head and full body.

CONCLUSIONS

We have presented the preliminary results of the "Minerva Project", aimed to develop and evaluate 3D scanning technologies in the Cultural Heritage field. A multidisciplinary team has been formed to approach the problems, validate the technologies chosen and interpret the experimental results. A structured-light scanner, a laser scanner and a suite of software tools for the management of 3D scanned data were experimented on the Minerva of Arezzo to realize a digital model of the statue.

The work done up to now is mainly technology-push. Next steps will be to foster these technologies, and to develop specific applications oriented to meet the restoration needs.

ACKNOWLEDGEMENTS
We acknowledge the financial support of the Progetto Finalizzato "Beni Culturali" (Italian National Research Council) and of the Progetto "RIS+" (Tuscany Regional Government).

REFERENCES

1. R. Baribeau, M. Rioux. *Influence of speckle on laser range finders*, Appl. Opt. Vol. 30, No. 20, July 1991.

2. F. Bernardini, H. Rushmeier, *The 3D Model Acquisition Pipeline*, STAR Proceedings, Eurographics 2000 Conference, pp.41-62.

3. P. Cignoni, C. Montani, C. Rocchini, R. Scopigno, *External Memory Management and Simplification of Huge Meshes*, IEI-CNR Tech. Rep., submitted paper, Jan. 2001, pp.10.

4. B. Curless, M. Levoy, *A volumetric method for building complex models from range images*, In SIGGRAPH '96 Proceedings, ACM SIGGRAPH, 1996, pp. 303-312.

5. M. Gaiani,M. Balzani,and F. Uccelli, *Reshaping the Coliseum in Rome*, Computer Graphics Forum, Vol. 19(3),2000, pages 369-378.

6. M. Garland and P.S. Heckbert, *Surface simplification using quadric error metrics*, In SIGGRAPH 97 Conference Proceedings, ACM SIGGRAPH, 1997, pages 209 –216.

7. S. Inokuchi, K. Sato, F. Matsuda, *Range-imaging system for 3D object recognition*, Int. Conf. On Pattern Recognition, Montreal, Canada, 1984, pp.806-808.

8. M. Levoy, K. Pulli, B. Curless, S. Rusinkiewicz, et.al., *The Digital Michelangelo Project: 3D scanning of large statues*, In Comp. Graph. Proc., Annual Conf. Series (Siggraph '00), Addison Wesley, July 24-28 2000, pages 131 –144.

9. K. Pulli, *Multiview registration for large datasets*, Proc Second Int. Conf. on 3D Digital Imaging and Modeling, , Ottawa, Canada, October 1999,pp.160-168.

10. C. Rocchini, P. Cignoni, C. Montani, R. Scopigno, *The Marching Intersections Algorithm for merging range images*, IEI-CNR Tech. Rep., submitted paper, July 2000.

11. C. Rocchini, P. Cignoni, C. Montani, R, Scopigno, *A low cost optical 3D scanner*, Computer Graphics Forum, Vol. 20(3),2001, (in press).

12. C. Rocchini, P. Cignoni, C. Montani and R. Scopigno, *Multiple Textures Stitching and Blending on 3D Objects*, Rendering Techniques ''99, Springer Wien, 1999, pp. 119--130.

13. H. Rushmeier, F. Bernardini, J. Mittleman, G. Taubin, *Acquiring input for rendering at appropriate levels of detail: digitizing a Pieta*, Eurographics Rendering Workshop 1998, G. Drettakis and N. Max (ed.), Springer Wien, 1998.

14. H. Rushmeier and F. Bernardini, *Computing consistent normals and colors from photometric data*, in Proc. of the 2nd Intl. Conf. on 3-D Digital Imaging and Modeling, Ottawa, Canada, October 1999, pp. 99–108.

Objective and Subjective Comparison of Audio Restoration Systems

Sergio Canazza (*), Guido Coraddu (*), Giovanni De Poli (*), Gian Antonio Mian (*)

(*) Dept. of Electronics and Informatics - University of Padova
V. Gradenigo 6/a - 35100 Pd - Italy
{canazza, depoli, mian}@dei.unipd.it

ABSTRACT

This paper presents the results of a perceptual test to evaluate the quality of different audio restoration algorithms based on Short Time Spectral Attenuation method and on Kalman filter. The protocol by International Communication Union (ITU.R P.800) was used. A set of audio stimuli played with different instruments (i.e. different bandwidths, transients, and timbre) was selected and different impulsive and broadband noises were added. Commercial software and proprietary algorithms were used to restore the corrupted stimuli. Twelve subjects were asked to compare the restored and the noise free recordings. Statistical multivariate analysis was performed and accurate characterization of the restoration techniques used was achieved. In order to correlate perceptual results with acoustic characteristics of the different stimuli, a time-frequency analysis was performed. These results permitted to relate the esthetic judgement categories used by listeners with the main features of the restoration techniques used.

KEYWORDS: Audio Restoration, Digital Signal Processing, Perceputal Tests.

INTRODUCTION

The cultural heritage of musical recordings is slowly fading away since the physical media, where the audio signal information is stored, are exposed to deterioration.

The introduction of digital audio carried out remarkable advantages in the field of music preservation and restoration. In fact, the copies digital-to-digital do not suffer of the analog-to-digital re-recordings problems: noise addition, difficulty to fix the 0dB level position, equalization curves choice to set the tape (or vinyl) player. Moreover, in digital domain, it is practicable to carry out the necessary periodic re-recordings in an automatic way. It is also feasible to use algorithms finalized to noise-removal, more sophisticated than analog systems like Dolby B, Dolby C or dbx.

The audio signal is subject to two kinds of technical alterations [9]: unintentional and intentional. The first group is caused by the imperfection of the recording techniques (resulting in various distortions) and misalignment of the recording equipment (resulting in wrong speed or azimuth errors). The group of intentional alterations concern recording equalization, noise reduction systems, and so on.

Furthermore, dirt as well as physical and chemical deterioration have various influences on carrier, ranging from less obtrusive clicks on mechanical carriers to shorter or longer losses of signal, and linear or non-linear distortions. Causes of degrade can have local influence (as scratches and dust) or global influence (as white noise, distortions, wows and flutters). In this scenario, it is getting more and more important to devise different techniques aimed at digital audio restoration. In this sense, the aims of this work are:

- to define a subjective evaluation method of the restoration quality of corrupted audio signal and to apply this method to the comparison of different digital restoration techniques;

- to observe the different listener's expectation, i.e., to understand if the listener (musically-trained or musically-untrained) pays attention to the reduction of noise level, and/or to the artifacts (distortions or lowpass-effects) introduced in the music signal by the restoration process.

In the paper we first shortly overview the most used audio restoration methods. Then we present the setup of the comparison carried out and a time-frequency analysis of the restored stimuli to characterize the algorithms used. A final part is dedicated to relate the perceptive analysis with the results of the acoustic analysis.

AUDIO RESTORATION METHODS: AN OVERVIEW

Time-domain algorithms:

The methods based on Extended Kalman Filter (EKF) theory are based on AR modelling of the data [8]: EKF, in principle, solves the problems of broadband noise filtering, signal parameters tracking and impulsive noise removal [7]. At medium/high signal-to-noise ratios (SNR), the performance of such filter is superior to that of other standard methods like Spectral Attenuation, nevertheless its plain use does not guarantee the best results. One reason for this is that the nonstationarity of audio signals leads to errors in parameter and tracking noise filtering, especially during fast transients. In order to achieve maximum performance from the EKF, it is essential to optimize its implementation [1].

Frequency-domain algorithms:

Adaptations of the spectrally-based techniques applied in speech processing have received most attention. They are based upon spectral weighting, in which individual spectral components are weighted according to expected noise and signal components. Such techniques can be viewed as finite block-size approximations to frequency domain Wiener filtering. As a result of these approximations (necessary to follow the time-varying nature of the useful signal) undesirable distortions can occur, the most notable being known as "musical noise" in which statistical fluctuations in the frequency components of noise lead to random tonal artifacts in the processed signal. Various techniques have been applied to mask or eliminate these distortions. We adopted the Ephraim and Malah suppression rule [5].

Algorithms based on psychoacoustic models:

This task requires to transform the audio signal from an "outer" to an "inner" representation, that is to resort to a representation that takes into account how the human ear perceives the sound. The combination of the psychoacoustic model and Frequency-domain algorithms, permits to define a promising restoration methodology [4].

EXPERIMENT

To single out the best computational methodologies for audio restoration, a perceptual test was made, using commercial software and our implementations of the methods presented in the last section.

Four noise free recordings were intentionally corrupted with different real (i.e., not synthetic) scratches and broadband noises extracted from tapes and vinyl disks. The noise free recordings were selected to cover a wide range of musical genre (Jazz, Western Classical, Pop) and different recording age (from 1940 to 1970); the stimuli were played with different instruments (piano solo, symphonic orchestra, jazz orchestra, voice) in order to analyze the deterioration persistence related to the different physical characteristics of audio signal (bandwidth, transients, timbre).

Finally, the noisy stimuli were restored using commercial software (Dart Pro, CoolEdit Pro, Wavelab) and proprietary software based on Extended Kalman Filter [1], [7] and on a psychoacoustic model [2], [4]. The parameters used to control the different systems were subjectively setup to obtain the best tradeoff between noise-removal and music-signal-preservation.

Twelve subjects were asked to score (from 0 to 5) the restoration quality compared to the original 'noise free' recordings quality. Two subject groups were selected: the first one constituted by researchers (of the Centro di Sonologia Computazionale, University of Padova) and the second one by skilled students. All the subjects were musically trained. The method used to define the evaluation criteria is an extension of the perceptual experiments described in the International Communication Union Protocols [ITU-R BS.1116 and ITU-R BS.708] on digital signal quality.

This protocol is based on "double-blind triple-stimulus with hidden reference" method, which is stable and permits accurate detection of small impairments. In the preferred and most sensitive form of this method, one subject at a time is involved. The known reference ("noise free" stimulus) is always available as stimulus "A". The hidden reference ("noise free" stimulus, a copy of stimulus "A") and the object (restored stimulus) are simultaneously available but are "randomly" assigned to "B" and "C". The subject is asked to assess the impairments on "B" compared to "A", and "C" compared to "A", according to a continuous five-grade impairment scale. As soon as the subject has completed the grading of a trial, it should be possible to proceed directly on to the next trial. The excerpts "A", "B" and "C" may be repeated in any order until the subject makes an assessment. In this way the test procedure is self pacing. The grading scale shall be treated as continuous with "anchors" derived from the ITU-R five-grade impairment scale given in Recommendation ITU-R BS.562 (Imperceptible - Perceptible, but not annoying - Slightly annoying – Annoying - Very annoying).

Different factor analyses were performed on subjects answers to characterize the behavior of the audio restoration algorithms. An acoustic (in time-frequency domain) analysis was carried out on the restored signal to relate the results of factors analysis with the physical characteristics of sound. In this way it was possible to highlight the acoustic parameters that are more important from a perceptual point of view.

TIME-FREQUENCY ANALYSIS

Fig. 1 shows the spectrogram of an excerpt (4 s long) of "Johnny comes lately" (rec. 26 June 1942, Duke Ellington Orchestra, CD Bluebird 5659-RB). Fig. 2 shows the spectrogram corresponding to the addition of broadband noise extracted from a tape. Fig. 3 and fig. 4 give the spectrograms of the restored stimuli using, respectively, Cool Edit PRO and DART PRO. It can be noticed that in figure 4 (restored by DART PRO) there is an evident low-pass effect (around to 6.5 kHz), which gives an audible distortion to the signal. On the contrary, the spectrograms show that Cool Edit PRO (fig. 3) does not remove all the broadband noise, unlike DART (see fig. 4). The measurement of SNR shows that DART improves the value of 10dB. In this case, from perceptual test (see next section) emerges a different listener's expectation, i.e., a cluster of listeners gives importance to the reduction of noise level and seems to prefer low broadband noise (so they prefer DART), and another cluster pays attention to the artifacts (lowpass-effects, in this case) introduced in the music signal by the restoration process (choosing Cool Edit PRO).

The behavior of the other considered methods is intermediate between the one of Cool Edit PRO and DART PRO and will be confirmed by the results of the following section.

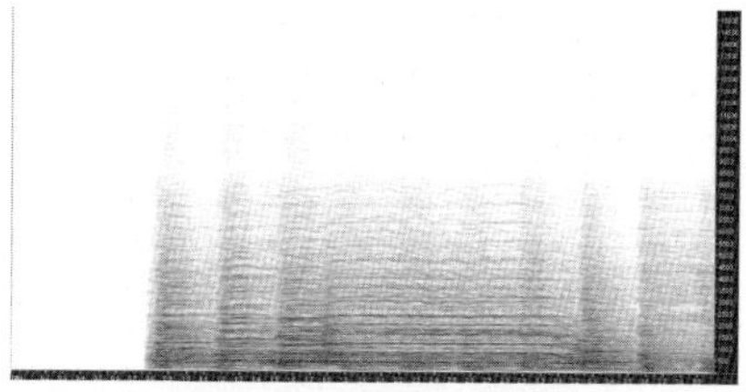

Fig. 1: spectrogram of "Johnny comes lately", from 0 to 4 s ("noise free"). The y-axis (frequency) is scaled from 0 to 15kHz. The x-axis represents the time in s

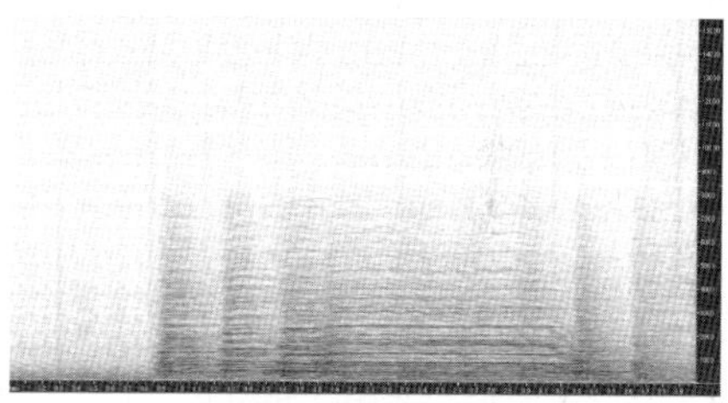

Fig 2: spectrogram of the first 4 seconds of "Johnny comes lately" with broadband noise ("tape") added. The y-axis is scaled from 0 to 15kHz

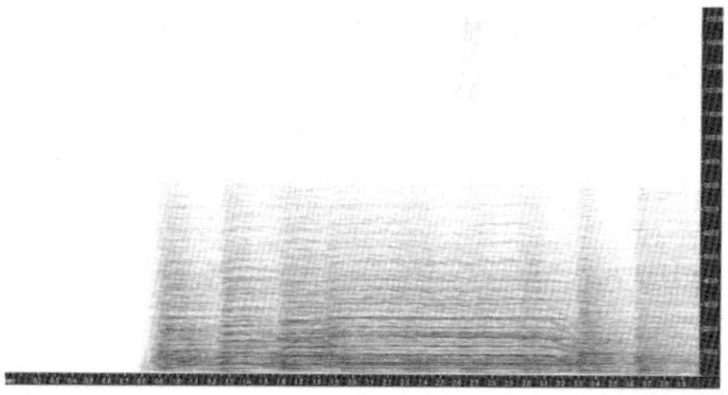

Fig 3: spectrogram of "Johnny comes lately" restored from broadband noise with Cool Edit Pro). The y-axis is scaled from 0 to 15kHz

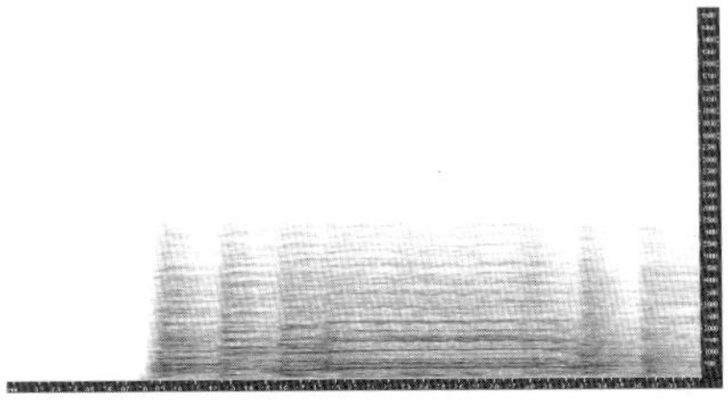

Fig 4: spectrogram of "Johnny comes lately" restored from broadband noise with DART Pro. The y-axis is scaled from 0 to 15kHz

The impulsive disturbances (click and scratches) are totally removed by all systems used. The signal masked by the click is reconstructed by interpolating the missing samples according to different strategies, introducing an evident signal deterioration.

PERCEPTUAL ANALYSIS

A factor analysis was carried out to homogeneously group the subjects, using Principal Component Analysis; the subjects were considered the observable variables. This analysis shows two principal factors that explain, respectively, 55.8% and 9.1% of the global variance. Factor 1 appears to be related to the subjects attention or cooperation during the test. Factor 2 is related to the listeners experience (or competence). Plotting, as in Fig. 5, the factor loadings in a two dimensional space helps to analyze the homogeneity among the subjects. The figure clearly shows that there are two clusters of subjects: the first one containing the researchers and the second one including the students. Note that the first factor has a high value for both groups.

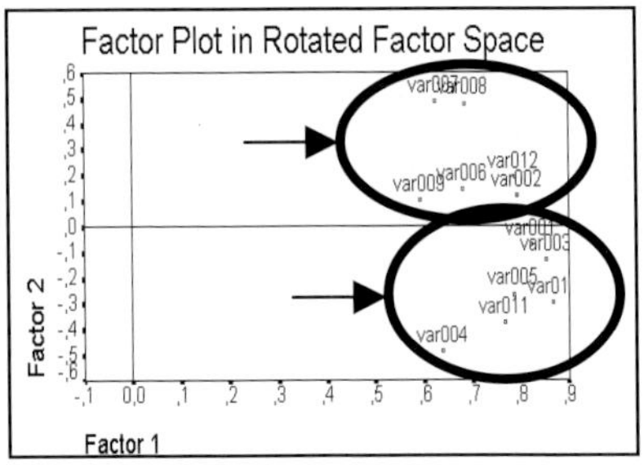

Fig. 5: factor loading plot, taking the subjects as variables. Factor 1 (55.8% of total variance) is related to the subjects attention during the experiments (inattention vs. attention). Factor 2 (9.1%) is related to listeners experience (incompetence vs. competence)

In order to observe the persistence of different noises, another analysis on the data was made transposing the answer matrix and carrying out a factor analysis taking the added disturbances as variables. This time, two main factors emerge: the first one explains 46.7% of total variance and the second one 20.1%. The acoustic analysis relates the first factor with the noise reduction carried out by the restoration methods, and the second factor with signal quality loss. Fig. 6 gives the corresponding factor plot for the different noise types considered: scratch, click and broadband noise. The more persistent deterioration is the impulsive noise (scratches and clicks), which has both a high value of signal quality loss and a low value of noise reduction. The broadband noises used were, respectively, disk noise (D-noise) and tape noise (T-noise) with Signal Noise Ratio 25dB (noise1) and 35dB (noise2).

Finally, the algorithms were considered the observable variables of factor analysis.

Four factors were deduced and Tab. 1 gives the factor loadings obtained by factor analysis (for factors 1, 2, 3 e 4).

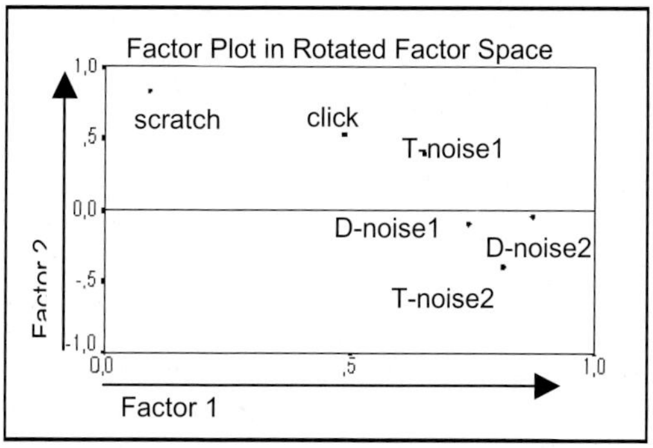

Fig. 6: factor loading plot, taking the different disturbances as variables. Factor 1 (46.7.8% of total variance) is related to the deterioration attenuation. Factor 2 (20.1%) is related to the signal quality loss

Factor 1 (38.5% of total variance) and factor 2 (17.2%) are probably related to click and scratch removal. In fact all the algorithms have similar behavior: removal of impulsive noise that is accompanied by a (perceptually relevant) signal distortion. Factor 3 (14.7%) is related to the distortion caused by broadband noise removal (high vs. low distortion). In fact, Cool Edit has the maximum factor loading (0.85, see tab. 1) and it is the software that introduces the minimum distortion to the signal spectrum. An old (proprietary) algorithm based on Wiener filter (W1*) has the minimum factor loading (-0.28, see tab.1): it is the result of a old experiment finalized to the noise attenuation, without respect to the signal distortion.

Factor 4 (11.1%) is related to broadband noise attenuation (low vs. high attenuation). In this factor DART has a maximum factor loading (0.89, see tab.1): in fact, it is the software that obtains the best SNR.

	Commercial Algorithms			Proprietary Algorithms				
Factor	C	D	WL	P1	P2	W1	W2	K
1	.01	.02	-.01	.0	-.01	.01	.02	.04
2	.03	.02	-.01	.0	-.02	.0	.01	.03
3	.85	.12	.28	.04	-.12	-.28	.45	.33
4	.05	.89	-.25	.13	.28	.29	.18	-.05

Tab. 1: factor loadings for the algorithms used in the perceptual test. Factor 1 (38.5% of total variance) and factor 2 (17.2%) are related to the click and scratch removal; factor 3 (14.7%) is related to the distortion caused by broadband noise removal (high vs. low distortion); factor 4 (11.1%) is related to broadband noise attenuation (low vs. high attenuation). C = Cool Edit Pro, D = Dart Pro, WL = Wavelab, Wn = different implementations of Wiener filtering, Pn = different implementation of a psychoacoustic model, K = Kalman filtering.

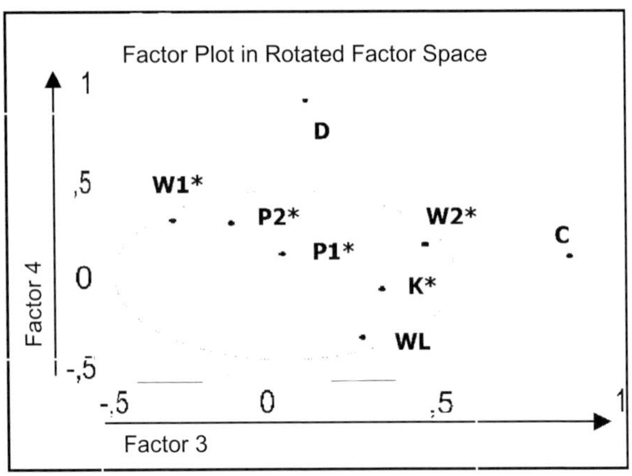

Fig. 7: factor loading plot, taking the different algorithms as variables. Factor 3 (14.7%) is related to the distortion caused by broadband noise removal (high vs. low distortion); factor 4 (11.1%) is related to broadband noise attenuation (low vs. high attenuation). C=Cool Edit Pro, D=Dart Pro, WL=Wavelab, Wn=different implementations of Wiener filtering, Pn=different implementation of a psychoacoustic model, K=Kalman filtering. (*=proprietary algorithms)

Fig. 7 plots the two dimensional space generated by factors 3 and 4. The variables (algorithms) present high correlation index in the first two factors that, consequently, are not plotted.

It can be noticed that, in relation to the distortion caused by broadband noise removal, the best algorithms are those based on Short Time Spectral Attenuation (C and W2*) [6] and on Kalman filter (K*) [1]. In relation to broadband noise attenuation, Dart Professional (D), Wiener filter (W1*), and the algorithm based on psychoacoustic model (P*) ([2], [4]) present a better behavior that is in good accordance with the frequency analysis presented in the last section.

CONCLUSION

The aim of this work was to introduce an evaluation method of the restoration quality of noisy audio signal and to apply it to the comparison of different digital restoration techniques.

A comparison between proprietary algorithms and some commercial products was made on the basis of a perceptual test. The method used to define the evaluation criteria is an extension of perceptual experiments described in Int. Com. Union Prot. [ITU-R BS.1116 and ITU-R BS.708] on digital signal quality. Statistical multivariate analysis was performed and accurate evaluation of the restoration techniques used was achieved. These results permitted to relate the esthetic judgement categories used by listeners with the main features of the restoration techniques.

A factor analysis taking the "subjects" as variables was performed to evaluated if the listening ability was uniformly distributed across all the subjects. Another factor analysis was carried out using the "corrupted musical excerpts" as variables, in order to weight the influence of the disturbances on the signal. From this analysis emerges that the impulsive noise is the disturb most annoying for listeners: moreover, its removal introduces an audible signal quality loss (see fig. 6).

A further factor analysis, using the "restored musical excerpts" as variables, led to classify the effectiveness of the algorithms used with respect to the restoration quality. Each algorithm was classified focusing on the behavior with respect to the different classes of noises and different audio categories.

From the analyses it emerged that it does not exist an optimum audio restoration algorithm, able to remove, in a satisfactory way, all the different disturbance typologies without introducing some quality loss into the audio signal. In particular, some algorithms (e.g., D, W1*, W2* and P2* of fig. 7) are able to completely remove disturbances, but at the expense of signal quality loss (as exemplified in fig. 4). At contrary, other algorithms (e.g., C and K* of fig. 7) are more "respectful" of the useful signal, but do not greatly attenuate noise (e.g. compare fig. 3 and fig. 4).

ACKNOWLEDGMENTS

This work was supported by CNR – "Progetto Finalizzato Beni Culturali".

REFERENCES

1. Bari A., Canazza S., De Poli G., Mian G. A., (1999). "Some key points on restoration of audio documents by the Extended Kalman Filter". Proc. of Diderot Forum, Vienna, 2-4 December 1999, 37-47.

2. Beerends J. G and J. A. Stemerdink, (1992). "A Percentual Audio Quality Measure Based on Psychoacustic Sound Representation". J. Audio Eng. Soc., 40(12), pp. 963-978.

3. Boll S. F. and A.V. Oppenheim, (1979). Suppression of acoustic noise in speech using spectral subtraction. IEEE Trans. Acoustics, Speech and Signal Processing, ASSP-27(2), April 1979.

4. Canazza S., De Poli G., Maesano S., Mian G. A, (1999). "On the performance of a noise reduction technique based on a psychoacoustic model for the restoration of old audio recordings". Proc. of Diderot Forum, Vienna, 2-4 December 1999, pp. 29-35.

5. Ephraim Y. and Malah D., (1984). "Speech enhancement using a minimum mean-square error short-time spectral amplitude estimator". IEEE Trans. Acoustics, Speech and Signal Processing, 21(6) pp. 1109-1121.

6. Godsill S., Rayner P., Cappé O. (1998). "Digital audio restoration". In Applications of digital signal processing to audio and acoustics. Kahrs - Karlheinz Brandeburg (ed.). Kluwer Academic Publishers.

7. Niedzwiecki M. and Cisowski K. (1996). "Adaptive scheme for elimination of broadband noise and impulsive disturbances from AR and ARMA signals". IEEE Trans. Signal Processing, 44(3), pp. 287-300.

8. Paliwal K. K. and A. Basu (1987). "A speech enhancement method based on Kalman filtering". Proc. IEEE International Conference on Acoustics, Speech and Signal Processing, pp. 177-180.

9. Schueller D. (1991). "The ethics of preservation, restoration and re-issues of historical sound". J. Audio Eng. Soc., 39(12), pp. 1014-1016. About the authors

ABOUT THE AUTHORS

Sergio Canazza is researcher at the "Centro di Sonologia Computazionale" (CSC), University of Padova. In 2000 he's been teaching in Master "Transfer of Multimedia Technology in Small and Medium Enterprise system in the field of Cultural Heritage", at the Federico II University of Napoli. Since 1998 to 2001 he's been participating in the research for the CNR "Progetto Finalizzato Beni Culturali" and the "Restauro filologico dei nastri magnetici delle opere di Bruno Maderna" supported by MURST.
E-mail: *canazza@dei.unipd.it*

Giovanni De Poli is Professor in Computer Science at Dept. of Electronics and Informatics, University of Padova. He is director of Centre of Computational Sonology (CSC). Since 1998 to 2001 he's been participating in the research for the CNR "Progetto Finalizzato Beni Culturali".
E-mail: *depoli@dei.unipd.it*

Gian Antonio Mian is Full Professor in Digital Signal Processing at Dept. of Electronics and Informatics, University of Padova. Since 1998 to 2001 he's been director of the CNR contract "Progetto Finalizzato Beni Culturali".
E-mail: *mian@dei.unipd.it*

An Ultrasound Based Technique for the Detection and Classification of Flaws Inside Large Sizes Marble or Stone Structural elements

Mauro Bramanti (*) *and Edoardo Bozzi* (*)

(*) Istituto di Elaborazione della Informazione, Area della Ricerca del CNR, S. Cataldo. 56100 Pisa, Italy
E-mail: bramanti@iei.pi.cnr.it

(*) Istituto di Elaborazione della Informazione, Area della Ricerca del CNR, S. Cataldo. 56100 Pisa, Italy
E-mail: bozzi@iei.pi.cnr.it

ABSTRACT
There is a growing interest in using ultrasound diagnostic techniques for non-destructive testing of marble/stone blocks or structural elements with large sizes, in order to detect internal flaws such as cracks or fissures. Here a specific diagnostic procedure, based on ultrasound transmission measurements, is proposed. A prototype of a measuring set-up, suitable for an in-field use, is described and the experimental results reported.

KEYWORDS: Non-Destructive Evaluation, Ultrasound Testing, Marble/Stone Quality Control

INTRODUCTION

The presence of faults, such as fissures and cracks, inside large sizes structural elements of stone or marble is a real risk factor. Non-destructive ultrasound based diagnostic techniques for the quality control of materials have been studied since many years and several dedicated tools are now available [1]. Most of the currently used ultrasound techniques appears quite suitable for the above types of materials as well. Diagnostic methods based on acoustic waves are already used to assess physical characteristic of underground rocks [2,3], and ultrasound techniques are applied to detect flaws inside concrete and masonry structures [4,5]. Thereore, the development of non-destructive diagnostic procedures for detecting faulty regions inside marble elements up to four meters in length seems be possible and of great interest. A procedure and a prototype tool, specifically designed for in-field use, is here described. Finally, some experimental results are presented and discussed.

THE PROPOSED PROCEDURE

On the basis of propagation experiments of ultrasound bursts through marble blocks of various types and with different internal flaws [6], the following remarks can be made:

i) The value measured for the propagation attenuation and its dependence vs. frequency suggests the use of ultrasonic burst with frequency within 50 -500 kHz range.
ii) The flaws most commonly encountered inside marble blocks are: low level flaws, tipically cracks; medium level flaws, tipically non homogeneous regions due to "filled fault"; high level flaws, tipically empty zones. The presence of flaws of the above said types, affect the propagation attenuation in relation to the frequency of the test signal: generally, the lower the frequency value, the lower the increase in the attenuation caused by a flaw of an assigned type.
On this basis the following test procedure was designed.
A region of the element under test is located, certainly free of flaws, and the propagation attenuation in that region is measured at 500, 100 and 50 kHz, for example, thus obtaining a reference set of values. We shall indicate these values as α^*_{500}, α^*_{100}, α^*_{50}, respectively.
With some guesswork and taking into account the characteristics of the particular element under test, a number of propagation paths are then chosen and the attenuation values (α_{i500}), along these paths, are measured for a burst frequency of 500 kHz .The regions with propagation attenuation values α_{i500} "near" (not greater than about 6 dB, for example) to the reference value α^*_{500} should be considered flaw-free. The paths are then considered with $\alpha_{i500} >> \alpha^*_{500}$ and a subsequent measure of attenuation is made at a frequency of 100 kHz. In the regions where α_{i100} result "near" $\alpha^*_{100,}$ a low level flaw is present, such as a crack. The paths are then considered with $\alpha_{i100} >> \alpha^*_{100}$ and an attenuation measurement at the frequency of 50 kHz is made. If α_{i50} results "near" to α^*_{50} a medium level flaw, such as a filled fault is to be expected. Where $\alpha_{i50} >> \alpha^*_{50}$, a high level flaw, such as void, is present

EXPERIMENTAL RESULTS

Laboratory experiments were first accomplished to determine the variations in the propagation attenuation due to the presence of flaws with different tipology, for the frequency range reported above. In these experiments marble blocks with sizes smaller than those of interest in real cases were used. Flaws of the type described were considered. In some cases, they were artificially created inside the material. On this basis the attenuation threshold levels used for declaring the presence of flaws were derived. Such levels resulted sufficiently indipendent from the type of marble or stone, except for particular cases, *Pietra Serena* for example. A prototype tool for in-field measurements was implemented according to the schematic diagram in Fig. 1. The station is based on the Panametrics high voltage pulser-receiver Model 5058PR. The output pulse is directly applied to the transmitting transducer, while the receiving transducer is connected to the receiver section of the instrument itself. Both the transducers are matched to the element under test by a specifically designed water based matching system. Here, for example, we report the results obtained in testing a *Bianco P* marble block with a width of 1.32 m, a height of 1.50 m and a length of 3.20 m. The test was accomplished by transmitting ultrasound pulses along an array of propagation paths, parallel to the direction of the width, as schematically shown in Fig. 2.

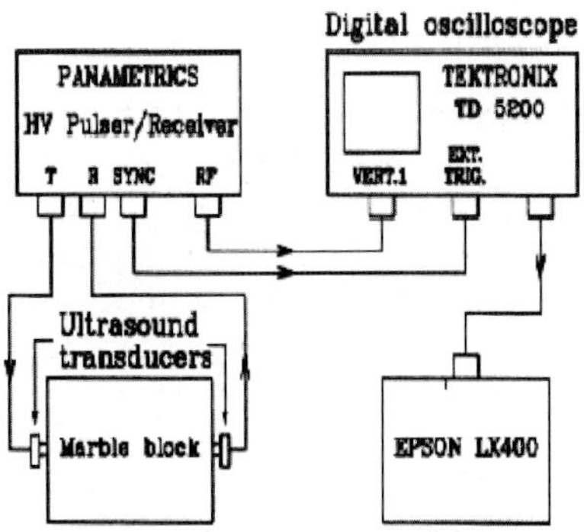

Figure 1: Measurement equipment used in in-field testing

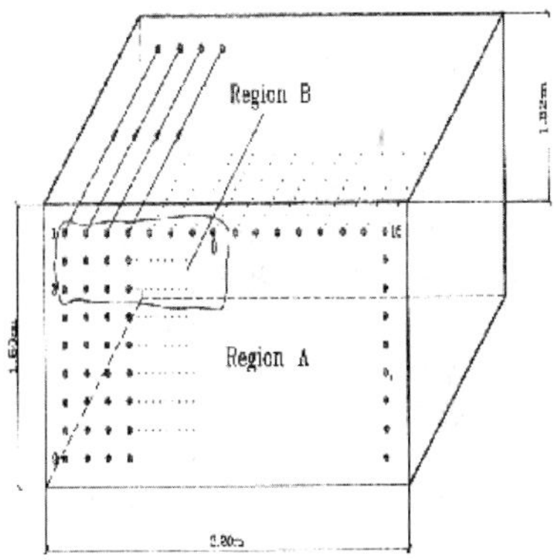

Figure 2: Propagation paths used to test the block

First transmission measurements at a frequency of 500 kHz were made. The propagation conditions were found to be good and sufficiently uniform for all the paths in the region A; on the contrary, for the paths located in the region B no propagation resulted. Then transmission measurements at frequencies of 100 kHz and 50 kHz were made and the results relative to the paths of region A (flaw free region) and region B were compared. An example of the typical signals received at 100 KHz and 50 kHz for paths in region A are shown in Figs. 3 and 4, while the corresponding typical signals for paths in region B are shown in Figs. 5 and 6.

A comparison of the signals in Figures 3 and 5 highlights that the propagation attenuation relative to region B is about 37.3 dB greater than the attenuation relative to the region A.

On the other hand, a comparison of the signals in Figures 4 and 6 shows that the increase in the corresponding attenuation values is limited to 6 dB.

We can thus conclude that region B contains a medium level flaw.

A subsequent direct invasive inspection of the block, which was allowed in this case, really showed the presence of a "filled fault".

CONCLUSIONS

In the technical literature we have found no proposals for the use of ultrasound techniques in non-destructive testing of stone or marble elements. We here proposed a procedure which uses ultrasound pulses within 50 kHz - 500 kHz frequency range for this application. On the basis of propagation tests, the procedure allows one to detect the presence of internal flaws and also to estimate, by comparing propagation measurements at different frequencies, the type of flaw itself.

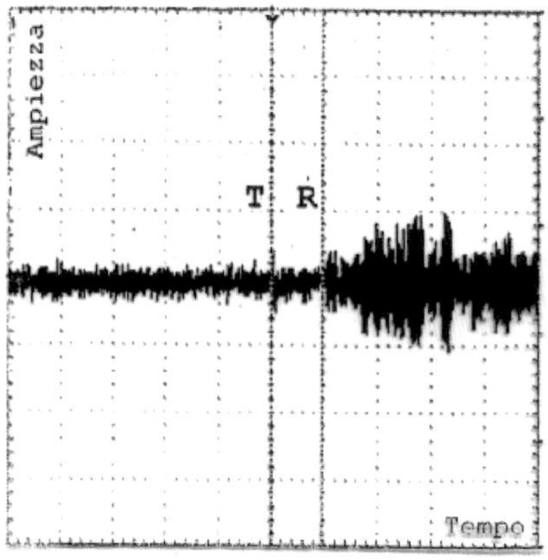

Figure 3: Transmission measures at 100 KHz in region B (medium level flaw)

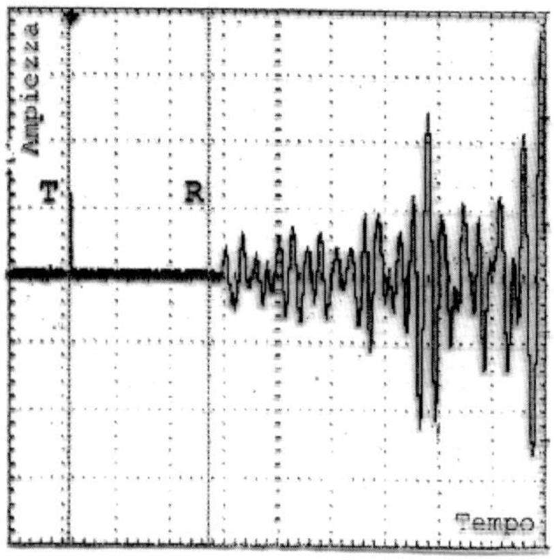

Figure 4: Transmission measures at 50 KHz in region B (medium level flaw)

REFERENCES

1. Bozzi E., Bramanti M., " Propagazione ed Attenuazione di Ultrasuoni in Materiali Lapidei, Alcuni Risultati Sperimentali", AEI Automazione Energia Informazione, Vol. 84, No. 9, Settembre 1997, pp. 59/763, 62/766

2. Krautkramer J. Krautkramer H, " Ultrasonic testing of Materials", Spring-Verlag, 3rd Ed., New York, 1983

3. Mc Cann D. M., Culshaw M. G., Northmore K. " Rock Mass Assessment from Seismic Measurements". Field Testing in Engineering Geology. Geological Society Engineering Geology Special Publication, No. 6; F. G. Bell, M. G. Culshaw, J. C. Cripps, J. R. Coffey (Editors), 1990, pp. 257-266

4. Sandor Popovics, Joseph L. Rose, John S. Popovics, " The Behaviour of Ultrasonics Pulses in Concrete", Cement and Concrete, Vol. 20, No. 2, 1990, pp. 259-270

5. Sansalone M., Lin Y., "Detecting Delaminations in Concrete Slabs with or without Overlays Using the Impact Echo Method", Materials Journal, 1989, pp. 175-184

6. Song, Lin-Ping, " Mapping an Underground Rock Mass by Anisotropic Acoustical Transmission Tomography", Ultrasonics, Vol. 36, No. 10, 1998, pp. 1009-1012

ABOUT THE AUTHORS

Mauro Bramanti graduated in Electronic Engineering at Pisa and since 1965 he was
with CNR, with activities in radio systems, microwaves and NDT for industry and biomedicine.
Director of Research at the IEI-CNR and professor at Pisa University, he is the author of numerous articles and of industrial patents.
Bramanti was awarded the F. Vecchiacchi Prize by the IIC, the 3 M Italia SpA Prize by the AEI and the Philip Morris Prize for Scientific Research and Technology.

Edoardo Bozzi was born in Pisa, Italy, on 1942. He received Laurea degree in Computer Sciences from the University of Pisa on 1975. Researcher at IEI-CNR, his main interest is in designing hardware/software instruments and in developing NDT techniques for industrial applications.

Digital Preservation & Emulation: from theory to practice

Stewart Granger(*)

(*) CAMiLEON UK PROJECT Project Co-ordinator
E-mail: stewartg@dial.pipex.com

ABSTRACT
This paper considers the use of emulation for digital preservation. After describing the theoretical background provided by mathematical logic, the paper moves on to consider practical issues in the use of emulation for digital preservation.

KEYWORDS: emulation, digital preservation, Turing machines, the Church-Turing thesis.

INTRODUCTION: THE PROBLEM OF DIGITAL PRESERVATION

Many libraries and archives are in the process of 'going digital'. The advantages of digital technology are well known and its adoption by libraries and archives seems inevitable, inexorable and well-motivated. Yet the fact remains that several key issues concerning the long term preservation of digital objects remain unsolved. Most writers have tended to identify two key problems: the fragility of digital media (its 'shelf life' compared with, say, non-acidic paper is extremely short) and the rate at which computer hardware and software become obsolete. Some would argue that the first problem is not real - even if we don't know exactly how long, say, CD ROMs will last, it is likely that the drives needed to read them will be obsolete before that date arrives. Many cases have been cited in which valuable data has already been lost because of obsolescence. Digital preservation is a multi-faceted problem of which the technical issues discussed here are but one part. For an overview of some digital preservation issues see *The Task Force Report* [14] and N.Beagrie & Greenstein [12]. These problems have, of course, been exercising the library and archive communities for some time but as yet no one solution or set of solutions has been reached.

To date the most widely advocated solution to these problems has been that of migration. Migration is the process of converting a digital object that runs on one platform so that it will run on another (non-obsolete) platform. But migration has its problems and disadvantages: the process of conversion runs the risk of losing data, the 'look and feel' of a digital object and/or its functionality; it is a time consuming and therefore costly process; it is not obvious how or indeed whether all digital objects *can* be migrated (think of computer programs). For these and other reasons some believe that emulation may, sometimes at least, provide a better solution. In the CAMiLEON project we have reached the conclusion that emulation does have a valuable role to play in digital preservation [3] *Caveat*: this paper explores the possible role of emulation in digital preservation, it does not purport to explain how to write an emulator.

THE NATURE OF EMULATION

Emulation would seem to hold great promise as a method of digital preservation. This is mainly for two reasons: first, it promises a method of preservation that can preserve the functionality and the 'look and feel' of digital objects that migration may not be able to achieve; secondly it could prove to be much more cost effective solution in certain circumstances for the reason that producing *one* emulator could be much cheaper than migrating *every* digital object in an archive. Moreover, there are very strong theoretical underpinnings for emulation and I want to begin by describing some of these.

BACKGROUND THEORY

In the face of the plethora of different computers it would appear be reasonable to conclude that the idea that one machine can *emulate* another machine might appear to be at best contingent, and at worst a highly risky assumption. Yet the fact is that concepts central to the development of computer science and the mathematical theory of computation show this is not the case. Here I want only to say enough to *describe* these results.

.

Turing machines

Possibly the best starting place is with the concept of a *Turing machine* invented by the British mathematician Alan Turing. There are many cosmetic differences in the way Turing machines are described, mine draws on Boolos and Jeffrey [2]. The notion of a Turing machine is abstract and consists of the following:

Represented in the diagram is the notion of a linear tape marked into squares and which is unending in both directions. With a finite number of exceptions, all squares are blank, both initially and at each stage of the computation. If a square is not blank, it has exactly one of a finite number of symbols $S_1 \ldots S_n$ written in it.

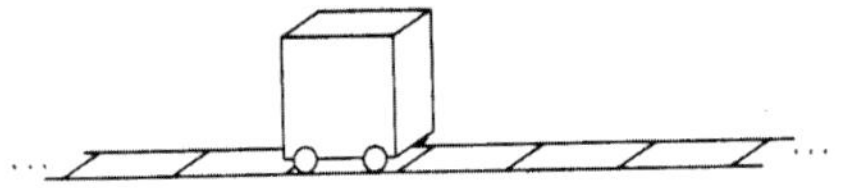

At each stage of the computation, the machine is scanning one square of the tape and is capable of telling what symbol is written on the square. It is capable of erasing the symbol in the square and writing a symbol there. And it is also capable of moving one square to the right or one square to the left.

The machine is designed in such a way that at each stage of the computation it is in one of a finite number of states $q_1 \ldots q_n$. It is irrelevant how these states are realised physically.

The machine is controlled by a set of *instructions*. Each of these instructions is in conditional form. It tells the machine what to do depending on the symbol being scanned is S_0 (blank) or S_1 or $\ldots S_n$. There are n+4 things that can be done:

1. Halt the computation
2. Move one square to the right
3. Move one square to the left
4. Write S_0 in place of whatever is in the scanned square
5. Write S_1 in place of whatever is in the scanned square

-
-
-

(n +4) Write S_n in place of whatever is in the square.

So, depending on what instruction is being carried out and what symbol is being scanned, the machine will

perform one of these $n+4$ acts.
The overall program of instructions that the machine is to follow can be specified in various ways, e.g. by a machine table or by a flow graph.

What can Turing machines do?
Ignoring the issue of speed and efficiency, what can Turing machines do? To the uninitiated the answer might be expected to be 'not very much' but, the answer, amazingly, is that a Turing machine can solve any effectively solvable algorithmic problem! Harel, puts it as follows [8] :

> "….any algorithmic problem for which we can find an algorithm that can be programmed in some programming language, *any* language, running on some computer, *any* computer, even one that has not been built yet but *can* be built, and even one that will require unbounded amounts of time and memory space for ever larger inputs, is also solvable by a Turing machine. This statement is one version of the so-called **Church-Turing Thesis**, after Alonzo Church and Turing who arrived at it independently in the mid 1930s."

The Church-Turing thesis is not mathematically provable (although it is *refutable*) because it employs the imprecise concept of an 'effectively solvable algorithmic problem'. Since it appears to be such a surprising result, why should one believe it?

Evidence for the Church-Turing thesis
Harel again, [9]

> "Ever since the early 1930s researchers have suggested models for the all powerful absolute or **universal** computer. The intention was to try to capture that slippery and elusive notion of 'effective computability', namely the ability to compute mechanically. Long before digital computers were invented, Turing suggested his primitive machines and Church devised a mathematical formalism called the **lambda calculus** (…the basis for the programming language LISP). At about the same time Emil Post defined a certain kind of symbol manipulating **production system**, and Stephen Kleene defined a class of objects called **recursive functions**….All these people tried, and succeeded, in using in using their models to solve many algorithmic problems for which there were known to be 'effectively executable' algorithms. Other people have since proposed numerous different models for the absolute, universal algorithmic device….The crucial fact about all these models is that have *all* been proven equivalent in terms of the class of algorithmic problems they can solve.."

Similarly DeLong [5] writes, "It should be clearly understood that Church's thesis is an empirical thesis, not a mathematical theorem…However the evidence in its favor is so great that it should really be called Church's law since if true it is a natural law." In effect then, the most powerful super-computer can only solve the same class of problems as the simplest home computer. Non-computable (undecidable) problems are solvable on neither, and the computable (decidable) problems are solvable on both.

TOWARDS PRACTICE

It cannot be stressed too highly that the extremely strong theoretical background does not guarantee that emulation is a *practical* digital preservation strategy. Turing machines say nothing about performance or computer peripherals. In addition there is the large question of whether emulation can provide a *cost-effective* digital preservation strategy.

However esoteric most computer users may find the concept of emulation the fact is that it works. Often in fact, emulated systems work better than the original system. Some practical examples can illustrate the general point:

Example 1:

"The 360 …was the first machine that could emulate other computers. The smaller models could emulate the 1401 and the larger ones could emulate the 7094, so that customers could continue to run their old unmodified binary programs while converting to the 360. Some models ran 1401 programs so much faster then the 1401 itself than many customers never converted their programs." [16]

Example 2:

A colleague who works in the CAMiLEON project has been involved with developing an emulator for the George 3 operating system which ran on ICL 1900 series computers. See [17].

Example 3:

We often demonstrate a BBC emulator running on a PC running an early computer game ('Chukie egg'!) and early educational software.

Example 4:

The general point is that emulation is a standard engineering technique and works for good scientific reasons.

APPROACHES TO EMULATION

The specification approach

Through a number of papers and reports Jeff Rothenberg has been perhaps the staunchest advocate of emulation as *the* solution to digital preservation problems [11] It is not the purpose of this paper to describe in detail his proposals; here I only want to mention a key feature of his approach and why this seems unsatisfactory to many people. The key aspect is the idea of developing emulator specifications – that is, a specification such that it *could* be used to develop an actual emulator when the need arose. This seems to be a risky strategy since it will be hard to know that a specification is adequate without actually developing one and testing it. One thing to be borne in mind when considering the practicalities of emulation is that one must deal with the realities of actual hardware and software. It is not unknown for software and hardware to have bugs – which, generally at least, will not be documented. Developing actual emulators is a safer way (the *only* way?) of ensuring that what one hopes will work actually works. For counter arguments to Rothenberg's approach see Bearman [1] and Granger [15].

The virtual architecture approach

IBM and the British Library are involved in a project in which the intention is to design a Universal Virtual Machine (UVM) which is then used for actual emulator development.

This is compatible with the approach we favour since the intention is to use it and then test the actual emulator *before* a given system becomes obsolete. Although it will be interesting to see what comes of this work, we believe this is unnecessarily complicated since a pragmatic solution is already at hand. [7]

The CAMiLEON approach

The two approaches just described so far may be said to be *generic*; that is they attempt to specify a single approach to emulation that will work for all cases. In the long run, when we are ahead in the digital preservation game this may be possible but at the moment, when we are still behind the game, this seems premature. Some work we are undertaking in the project illustrates this: we are attempting to use emulation to rescue the BBC Domesday material [6]. Here we will find it necessary to emulate a very particular video disc player and we have found that it requires research to determine the best way to produce this emulation. The practical realities of computers seem to us to preclude, at least for the moment, a generic, *a-priori*, methodology for emulation.

The CAMiLEON project is building upon work in the CEDARS project [4] which like CEDARS draws upon the Open Archival Information System (OAIS) reference model. [13] The OAIS makes a separation between preservation of digital material and the ability to provide access to its intellectual content. The CEDARS approach emphasises the importance of preserving original byte streams so that providing access to them can be achieved over time. See [10]. Clearly, a preserved byte stream is only valuable if meaningful access to its content can be provided. The CEDARS project shows the importance of identifying the significant properties of digital objects and then choosing an appropriate abstract representation of those properties that enables preservation of those properties. What those significant properties are will depend not only on the nature of the digital object but also on the purpose for which the object is being preserved. Our colleagues in Michigan are conducting some user trials to investigate what properties users find significant with respect to different kinds of digital objects.

The CAMiLEON approach has been described by my colleagues David Holdsworth and Paul Wheatley as follows [17]

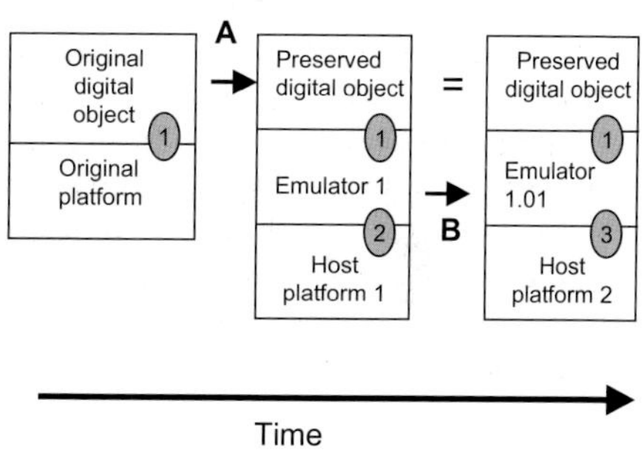

The above diagram attempts to show first the transformation (arrow A) of an original digital object into a preserved digital object, which can then be run under emulation to give adequate access to the significant properties of the original. This step ensures continued access as the original platform becomes obsolete.

Further passage of time leads to the transformation shown by arrow B, in which the obsolescence of platform 1 is handled by updating the emulator to run on a different platform. The large equals sign indicates that the preserved digital object consists of the same byte-stream in each case. The crux of the argument is that the transformations indicated by the two arrows can be implemented economically. Key to this is the design of the abstract interfaces indicated by the numbered circles on the junctions between digital components.

The interface labelled 1 is the API necessary for the successful operation of the original digital object. Emulator 1 recreates this interface in the abstract world of emulation. This interface remains the same forever. The emulator itself is a digital object and relies on interface 2 for its successful operation. A key strategic goal is future-proofing by choice of interface 2 involving only features that are likely to survive the test of time with little modification. Thus interface 3 will be very similar to interface 2, and the emulator for host 2 (emulator 1.01) will be readily derived from emulator 1.

Where emulation is concerned, there is the issue of at what level the emulation should be implemented. Choices will typically be between underlying hardware, operating system or some higher level. The best choice can, we believe, be different for different cases. In some cases emulation at the hardware level still leaves the problem of peripherals. (A problem that we believe Rothenberg considerably underestimates). In other cases it may be a drive that needs to be emulated. Factors influencing the choice here include:

- Complexity (or lack of it) in implementing the emulator
- Availability of documentation
- Mapping of peripherals into easily specified abstractions
- Retention of significant properties

The cost-effectiveness of emulation as a digital preservation strategy will in part depend upon the longevity of a developed emulator and the ease with which it may be updated when necessary. We propose several strategies for giving the source code of an emulator greater longevity:

- mulator code should be produced using standard software engineering techniques. These include the use of good code structure, informative and plentiful commenting, and good documentation.
- mulator code should be written in a subset of C, a subset chosen with a view to ensuring that only those aspects of the language semantics are used that are likely to be found in future programming languages.

Most emulators are likely to incorporate at least some non-standard code (for example for rendering of the emulated machine's display). All non-standard code should be modularised and well documented.

CONCLUDING THOUGHTS: EMULATION IN THE WIDER DIGITAL PRESERVATION CONTEXT

The seminal Task Force Report [14] on digital preservation was published in 1996. Much work in digital preservation has been undertaken since then but many of its key demands have yet to be met. The report spoke of the need for 'deep infrastructure' which included the ideas of certified digital archives and trusted organisations. If (hopefully, *when*) these organisations are created, provided they are set up in an appropriate way (appropriate skill sets, technical capabilities etc), then they could undertake emulation work (developing emulators, maintaining emulators etc) on behalf of many user communities. Emulation, much more than migration, lends itself to a centralised cost model where a few people could do the work for many. This, potentially, has very positive implications for the thorny issue of the costs of digital preservation. It seems to me that the creation of deep infrastructure is the biggest challenge facing those of us working in digital preservation, a challenge that is largely political and not technical. [18]

REFERENCES

1. Bearman, D. "Reality and Chimeras in the Preservation of Electronic Records", *D-LIB Magazine*, April 1999.

2. Boolos, G.S. & Jeffrey, R.C., *Computability and Logic*, Cambridge University Press, 1980

3. The CAMiLEON Project:

www.si.umich.edu/CAMILEON

4. The CEDARS Project: *www.leeds.ac.uk/CEDARS*

6. Delong, H. *A Profile of Mathematical Logic*, Addison Wesley, 1971, p. 195

7. Finny, A. Domesday, at *www.atsf.co.uk/dottext/domesday.html*

8. Gilheany seems to advocate a similar approach. See "Preserving Information Forever and a Call for Emulators" available at: *http://www.archivebuilders.com/aba010.html*

8. Harel, D. *Algorithmics*, Addison Wesley, 1987, p.221

9. Harel, op.cit., p.222

10. Holdsworth and Sergeant, "A blueprint for Representation Information in the OAIS model" available at: *http://www.personal.leeds.ac.uk/~ecldh/cedars/ieee00.html*

11. Jeff Rothenberg, *Avoiding Technological Quicksand: Finding a Viable Technical Foundation for Digital Preservation*,1999. Available at: *http://www.clir.org/pubs/reports/reports.html*

12. N.Beagrie & D.Greenstein, *Digital Collections: A Strategic policy Framework for creating and preserving digital resources,* Arts and Humanities Data Service, 1998. *http://www.ahds.ac.uk%20/manage/framework.htm*

13. The Open Archival Reference Model, **CCSDS 650.0-R-1:** *Reference Model for an Open Archival Information System (OAIS).* Red Book. Issue 1. May 1999. Available at: *http://www.ccsds.org/red_books.html*

14. *Preserving Digital Information: Report of the Task Force on Archiving of Digital Information*, Commission on Preservation and Access and The Research Libraries Group, Inc., May 1, 1996,

http://www.rlg.org/ArchTF/

15. Stewart Granger, "Emulation as a Digital Preservation Strategy", D-Lib, October 2000 *http://www.dlib.org/dlib/october00/granger/10granger.html, October 2000.*

16. Tannenbaum, A.S. *Structured Computer Organization*, 1999, 4th Edition, Prentice Hall, p.22

17. Wheatley, P. and Holdsworth,D. "Emulation, Preservation and Abstraction" available at: *http://129.11.152.25/CAMiLEON/ep5.html*

18. I would like to thank the two reviewers of an earlier version of this paper for their helpful comments.

ABOUT THE AUTHOR

Stewart Granger is the UK project co-ordinator of the CAMiLEON project. He has over 10 years experience in the fields of digital imaging, networking and digital preservation mainly in the context of European projects in the cultural sector.
E-mail: *stewartg@dial.pipex.com*

ZoomageTM : Super High Resolution Imaging & Visualization Tools

Irene Cheng (*), *Anup Basu* (*) *and Afshad Mistri* (#)

(*) TelePhotogenics Inc., #280 SunLife Place, 10123-99st, Edmonton, CANADA T5K 3J1
and Department of Computing Science, University of Alberta
E-mail: irene@telephotogenics.com and anup@cs.ualberta.ca
(#) SGI, Mountain View, California, USA
E-mail: mistri@sgi.com

ABSTRACT
In this paper we describe and demonstrate our technology for creating super high resolution (SHR) & 3D digital content in a variety of applications including museum artifacts and galleries, archeology, anthropology, art design and heritage conservation. Our hardware can capture minute details of static scenes in a composite panoramic format with details surpassing the human eye. We also propose an approach to storing SHR images for efficient retrieval over bandwidth limited networks, such as the Internet. Regions of interest (ROIs) specified by users are stored in multiple levels of detail hierarchy. This hierarchy can be created by analyzing the contrast, texture or other feature changes with varying levels of resolution and creating a hierarchy of ROIs. The depth of the hierarchy for a given ROI is determined by following the changes in the contrast (or other feature) gradient over adjacent levels. An implementation of the browser for Virtual Museums is also presented. In addition to browsing images, our software provides the ability to support "interactive story telling." As viewers interact with a picture using a mouse, the descriptions of objects in the scene appear depending on the location of the mouse and the level of detail in the scene. Finally, the "Zoomage" tools can be used by academic researchers for various applications including object recognition, digital archiving, accurate stereo reconstruction, and 3D visualization.

*Contact address: Dept. of Computing Science, University of Alberta, anup@cs.ualberta.ca.

KEYWORDS : Super high resolution & 3D Imaging, Interactive & Efficient Visualization

ZOOMAGE360 AND ITS APPLICATIONS

TelePhotogenics Inc. (TPGI)'s super high resolution scanning camera, Zoomage360, is specially designed to capture SHR, Super High Quality content for heritage and archeological sites, museums artifacts, dioramas, and priceless art. Figure 1 shows a picture of the SHR capture device, which uses state-of-the-art laser technology. The main advantage of this technology is sheer resolution of the image it can acquire. Zoomage360 can acquire an image up to 5300 x 40,000 pixels for a complete 360-degree panoramic scene. Each pixel location in the captured image has distinct R, G, B values, using accurate calibration of the three-color scanning channels. The technology creates SHR content in a matter of minutes, without any stitching and manipulation of images, for a price only slightly higher than low-resolution software based technologies for panoramic imaging. We have created an easy to use User Interface with a standard USB connection. The user interface also includes computer-controlled gain and shutter adjustments that allow users to take into account the lighting and brightness conditions for a scene. SHR captured content can be used in IMAX theaters (IMAX uses 4000 x 4000 resolution), Silicon Graphics Inc. immersive Reality Centers (RC's) down to a web based application. In short, a complete solution is provided for content acquisition and viewing of SHR content. The 3D scanning option, Zoomage3D, uses a patented technology using a structured laser that is placed on top of the imaging device (Figure 1, top part). The device uses computations based on geometric projections of carefully designed laser patterns, unlike stereo matching based techniques [7-8].

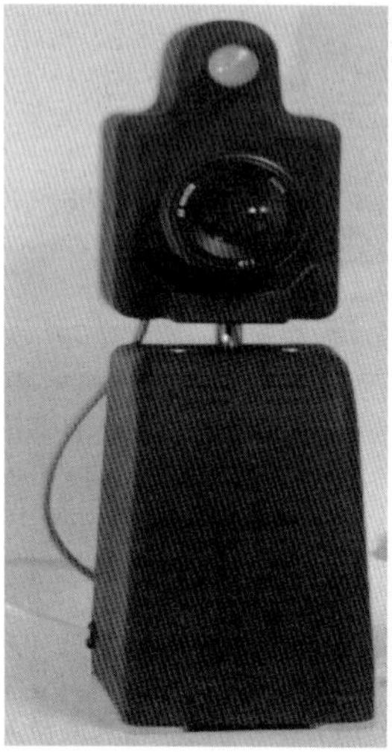

Fig. 1: Zoomage360.

3D SCANNING

In order to scan objects in 3D two scanning modes can be used:

- Static object and rotating camera; and
- Static camera and rotating object.

For the first mode the configuration is as shown in Figure 1. In this mode only one side of an object can be scanned at a time and several segments need to be stitched together for a complete view of all sides of an object. The scanning is performed in two stages: in one stage the object is scanned only for texture content, as shown in Figure 2; in the second stage the same object is scanned with a laser pattern projected on it (Figure 3). The laser pattern can be processed by software that detects a fixed number of laser points for each vertical line (Figure 4).

Fig. 2: Texture scan of a doll.

Fig. 3: Image with laser pattern.

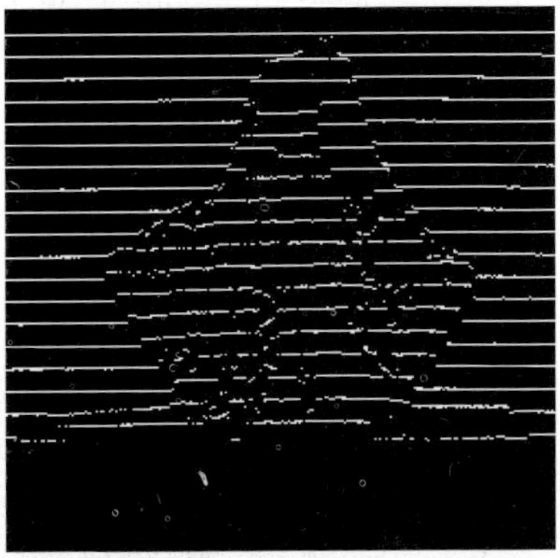

Fig. 4: Detected and segmented laser patterns corresponding to previous image.

It can be observed that the texture image and laser pattern image are captured from identical viewpoints, and the same image sensors are used to record both the texture and the laser data. Thus registration of texture with locations at which depth is estimated is automatic and does not require any special processing.

The background lighting in Figures 3 and 4 is the same indoor illumination. However, during the laser scan it is preferable to turn off the background lighting to make the laser patterns more prominent and easy to detect.

Screenshots of the JAVA3d interface showing different parts of the rotating costume of the doll are shown in Figure 7. Since only a front view of the doll was acquired in the rotating scanner mode, the model was duplicated for a 360 degree rotation mode.

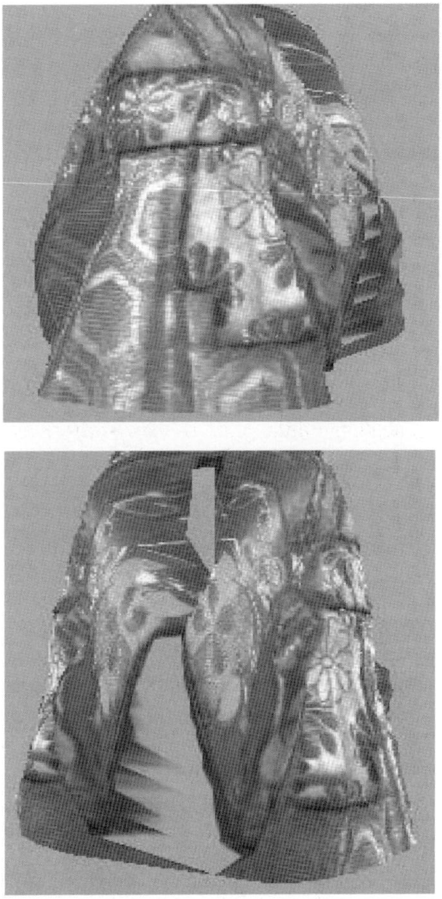

Fig. 7: Screenshots from JAVA3d display of doll.

VIEWING OF SHR IMAGES LEVERAGING COMPUTER HARDWARE AND LARGE-SCALE VIEWING FACILITIES

We have also co-developed another approach for interactive viewing of SHR images - using dedicated hardware, software and viewing facilities. Viewing a 5,000 x 40,000 image on a PC might not be the best way to view SHR content, so we teamed with Silicon Graphics Inc. to put some of their innovations for viewing. Most research institutions and universities own high performance graphics computers made by SGI. Museums, Science Centers and other markets around the world are using SGI Reality Centers as immersive theaters for viewing content.

SGI™ Reality Center™:

SGI Reality Center Group visualization solutions provide collaborative immersive viewing and interaction in real time. Reality Center facilities immerse you in your data so you can perceive it much as you perceive the physical world or in a way that is impossible in the real world.

The Reality Center solutions provide the highest resolution graphics, allowing you to interact with life-size imagery on a large screen and achieve never-before-imagined results via true interaction with large data sets.

Merging the capabilities of TPGI's SHR technology and SGI's powerful visualization solution, we can truly see the future of powerful digital asset acquisition and SHR viewing.

THE ZOOMAGE™ BROWSER

TelePhotogenics Inc.(TPGI)'s Interactive Zoomage™ browser is specially designed to manipulate the display of super high resolution images captured by the TelePhotogenics digital camera, Zoomage360. A custom Zoomage™ browser can be implemented to meet the requirements of different applications. As shown below, the example of a diorama is set up for the Provincial Museum of Alberta. The Zoomage™ browser can be used to obtain a preview of a show on the Internet, or high quality theaters using the SGI Onyx3000 graphics computers in 160 degree (3084 x 1024 resolution) SGI Reality Centers. The focus of our work is not on creating multi-resolution image viewing software [1], but more on software that allows user interactivity and background hardware that allows high resolution content creation. Our software is designed to be compatible with the JPEG, JPEG2000, and the Region of Interest (ROI) processing in MPEG-7 [2-4]. The goal of our software is not only to have efficient data representation, but also to add an element of user interactivity that is often missing from other browsers [5-6].

Problems encountered when displaying Images on the Internet and Information Kiosks:

- Images are of low resolution and thus cannot represent the real quality of objects or products for E-Commerce.
- High-resolution images take too long to transmit.
- Images are scattered across pages or displayed on different screens, which makes it difficult for users to relate various objects to one another. Following one link after another, viewers can easily get lost.

Zoomage™ Approach:

- Provide super high-resolution images, which truly represent the quality of the underlying objects.
- Utilize the Zoomage™ browser, which efficiently transmits high-resolution images.
- Support interactive browsing allowing the viewer to choose different images and descriptions without jumping between screens and pages.

How to use Zoomage™ Interactive browser ?

Operating instructions are also available on the web, and interactively provided on the browser screen.

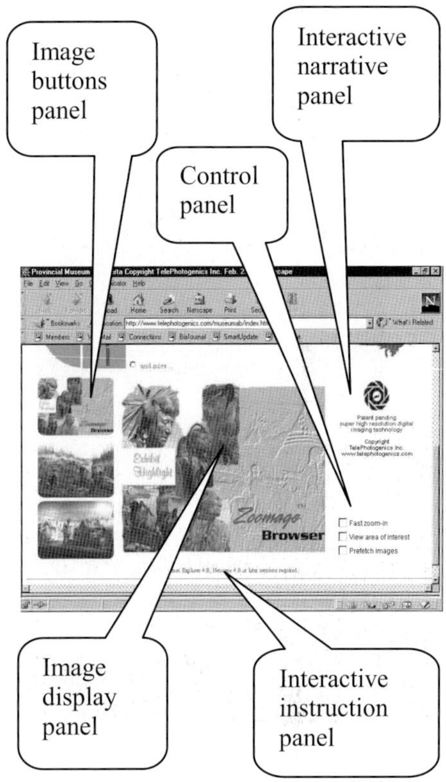

"Interactivity" is an important feature of the Zoomage™ browser. The mouse position on the Image display panel will trigger off the display of the associated description on the Narrative panel. At the same time, the viewer is guided by the hints on the Instruction panel. A zoom-in or zoom-out display can be obtained by a mouse click.

The initial **"Highlight"** screen is another outstanding feature of the Zoomage™ browser. It is a collection of appealing objects selected from individual pictures.

- **Image buttons panel**
  When the mouse moves over a button, the description on the Narrative panel will change. A different topic will be displayed on the Image panel by clicking an image button.

- **Image display panel**

An area of interest can be an object or a scene. It is indicated by a blue oval, which can be turned on and off by checking the **"View area of interest"** box. By clicking inside the area of interest, a zoom-in image will be displayed.

FUTURE WORK

We are currently extending the Zoomage software to handle 3D imaging. As well, applications in other areas including medical, forensics, and computer graphics are being investigated.

REFERENCES

1. Said and W.A. Perlman, "A new fast and efficient image code based on set partitioning in hierarchical trees", IEEE Transactions on Circuits and Systems for Video Technology, 6(3), 1996, 243-250.

2. T. Sikora, "MPEG-4 video and its potential for future multimedia applications", *Proceedings of the*

IEEE International Symposium on Circuits and Systems, New York, 1997.

3. "MPEG-4 overview", ISO/IEC JTC1/SC29/WG11, N2323, July, 1998.

4. "MPEG-7: proposal package description", ISO/IEC JTC1/SC29/WG11, N2364, October, 1998.

5. K. Martinez, J. Cupitt, and D. Saunders, "High resolution colorimetric imaging of paintings", Proceedings SPIE, Jan. 1993, pp 25-36.

6. K. Martinez et al., "Object browsing using the Internet imaging protocol", Proceedings of the 9th World Wide Web Conference (WWW9), May 2000, pp 1-11.

7. S.D. Cochran and G. Medioni, "3D surface description from binocular stereo", IEEE Transactions on PAMI, vol. 14, 1992, pp 981-994.

8. O. Faugeras et al., "3D reconstruction of urban scenes from image sequences", Computer Vision and Image Understanding, vol. 69, 1998, pp 292-309.

3D Reconstruction applied to Virtual Heritage and Cultural Conservation

João G.M. Gonçalves [(*)] *and Vítor Sequeira* [(*)]

[(*)]European Commission, Joint Research Centre, I-21020 Ispra, Italy
E-mail: joao.goncalves@jrc.it, vitor.sequeira@jrc.it URL: http://mortimer.jrc.it/

ABSTRACT
There is increasing interest in creating 3D photo-realistic computer models of monuments or heritage buildings, both indoors and outdoors. Our approach to 3D Reconstruction is to use automated distance measurements together with photographs. Two technologies are used: i) a combined laser - digital camera sensor for data acquisition and ii) mobility for resolving spatial occlusions. After a few scanning operations it is possible to create complete 3D textured models of the building or monument "as-built". The models are visually appealing, Internet compatible and geometrically and dimensionally accurate. The paper describes the main steps in building complete 3D models. Results include the models of the "Cour d'Honneur" and "Salon Delacroix" at the French National Assembly (Palais Bourbon), Paris, and the "Sala dello Scrutinio" at the Doges' Palace, Venice.

KEYWORDS: 3D Reconstruction, 3D Modelling, Virtual Heritage.

INTRODUCTION
It is part of human nature to build realistic representations of the real world. The objectives may be the desire to document information for future generations or to have training material in specific matters. In a world of constant change, there is an increasing interest to use new technologies to create realistic models of monuments, heritage buildings and historic city centres. The objectives are various including accurate documentation, dissemination and use of the material to the entertainment industry (e.g., games, films).

It can be said that an effective representation is the one that best provides "*the feeling of being there*". To achieve this, it is important to provide spatial awareness. The user should not be a passive spectator but rather an actor totally immersed and interacting with the represented world. In recent years, 3D Reconstruction technologies evolved dramatically in generating the 3D representation of real objects, buildings and environments. Progresses in Virtual Reality technologies (e.g., computing power, graphics software, head mounted displays, position sensors) allow for powerful computer interfaces enabling the real-time perception of relative distances and spatial relationships.There are basically two ways of creating the 3D model of a building. The first involves the use of a CAD tool to design the monument into the computer. This approach involves a large amount of manual work, i.e., measuring the correct distances and angles in the real world to have a dimensionally correct model. An alternative approach is the automated generation of a dimensionally correct 3D model, based on the reality "as-

built".
Our approach is to combine distance measurements with photographs. Distances are measured using a laser range scanner, and photographs taken with a digital camera. All system components are commercial.
Figure 1 shows a tripod based 3D acquisition system. It makes a complete 340° by 80° scan with an angular resolution of 0.19° in about two minutes

Figure 1: Portable, tripod based 3D Acquisition system with laser range scanner and digital camera.

3D RECONSTRUCTION

In order to create a complete model of a scene it is necessary to overcome single view occlusions (e.g., objects hiding others). This is achieved by scanning the environment from different viewpoints. The software integrates the different datasets (Figure 3) into a single triangulated model [1]. The triangulation is multi-resolution, in the sense that the density of triangles increases with local spatial detail. As such, there are less triangles in planar surfaces and more, smaller triangles where spatial detail is to be preserved

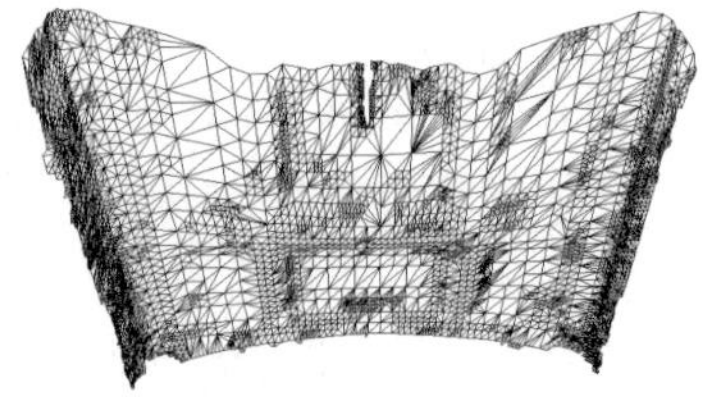

Figure 2: Multi-Resolution triangulation [Salon Delacroix, Palais Bourbon, France]

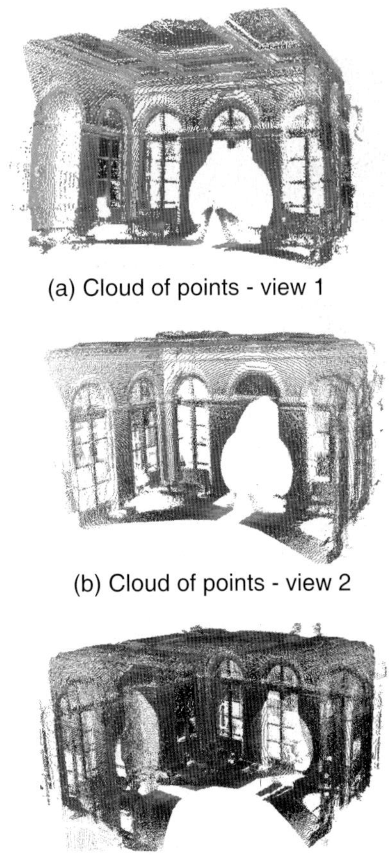

(a) Cloud of points - view 1

(b) Cloud of points - view 2

(c) Combined cloud of points

Figure 3: Integration of two 3D datasets acquired at different capture points [Salon Delacroix, Palais Bourbon, France]

To achieve visually realistic models, digital photographs are used to "paint" the 3D representation of the scanned areas. Each triangle in the model is "painted" with the corresponding area from one or multiple photos [2]. Since pictures may be taken at different spatial resolutions (i.e., zooming factors), it is possible to have different levels of photographic detail. A Z-buffering technique is used to avoid projection mistakes due to occlusions.

To have a complete texture map several images are normally required. As such, a fundamental aspect is colour blending, i.e., the procedure to smoothen out the colour transitions between the different images. Figure is the result of the automated blending of seven images in the model of the Bornholm round church in the Baltic Sea. Coloured lines delimit the parts used in each image.

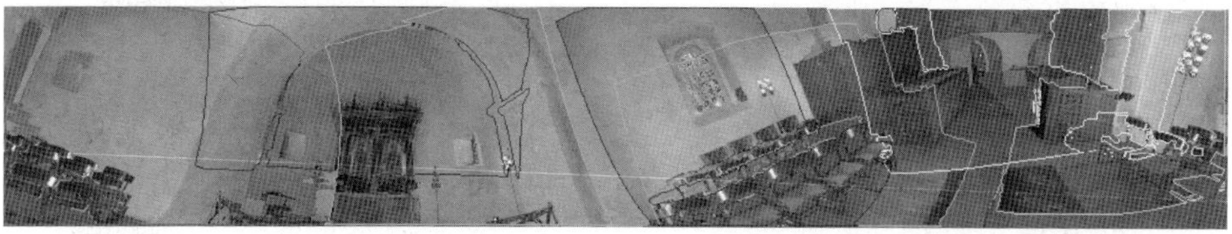

Figure 4: Final texture image obtained from blending of seven digital images [Bornholm round church, Denmark].

FEATURES OF 3D RECONSTRUCTED MODELS

The final 3D models have the following characteristics:

Internet compatible - models are coded into VRML (Virtual Reality Modelling Language) and can thus be viewed with any Internet browser irrespectively of the computing platform or operating system. VRML is a 3D standard for the Internet and MPEG4 communications.

Visually appealing - This derives from the fact that the models are textured with real photos from the scene.

Geometrically and dimensionally accurate - Existing laser scanners acquire 3D points with an accuracy of about 2cm over a distance of more than 100m.

Provide a realistic feeling of "being of there" – The VRML format allows the setting of predefined viewpoints without compromising user's interactivity. It is thus possible to navigate interactively through or around the model and be aware of the existing spatial relationships, as if one is exploring the real environment.

Application specific tools and interfaces - Java scripts/applets can be easily written for easier navigation or for more specific tasks, e.g., measurement of the real dimensions (see Figure 5), surfaces or volumes.

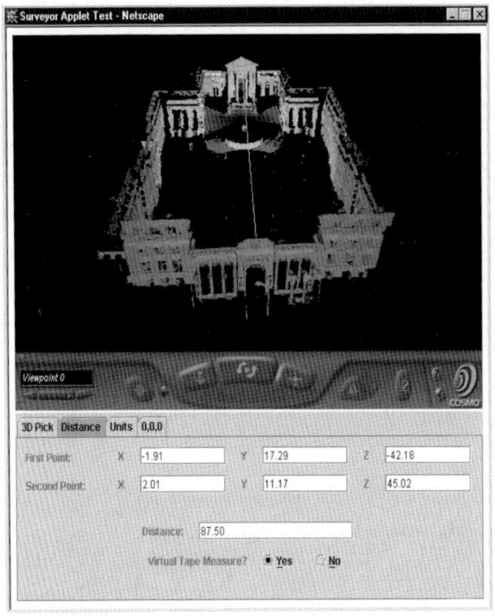

Figure 5: The Surveyor's Java applet provides the distance (in meters) between two interactively selected 3D points [Cour d'Honneur, Palais Bourbon, Paris].

Framework for advanced Human Computer Interface - The 3D models can also be linked to other images/document using hyperlinks. 3D models are thus used as 3D maps. By clicking in a "hot" area, other data can be retrieved (e.g., by clicking on a 3D object, a window with details and close-ups may open). Hyperlinks make navigation more realistic, e.g., the user enters a new room by clicking on a door The insertion of real-time, on-line data is also possible, e.g., live images from a camera can be displayed on a screen inside the model.

Multiple Level of Detail - It is possible to insert or integrate previously acquired detailed models into the overall model (e.g., a highly detailed model of a bas-relief can be inserted in the model of a monument), allowing therefore the easy creation of virtual and dynamic exhibition spaces or museums.

Model size adapted to application - It is possible to adapt the model size to existing transmission or storage requirements by decimating the 3D-triangle mesh and/or re-coding the photographic images.

Contextual Information – Current computing and graphics technology do not allow for large, detailed 3D reconstructed models of a monument including its background. A solution is to use accurate 3D models for the relevant parts and to use panoramic images for distant background scenery providing a realistic spatial context as shown in Figure .

By incorporating panoramas into 3D detailed models there is the opportunity to convey visual quality when at the capture points and spatial quality when moving though the surroundings, usually from one capture point to another. Panoramas are built from still colour images acquired using a digital camera and are also used to texture map the 3D model extracted from the range data.

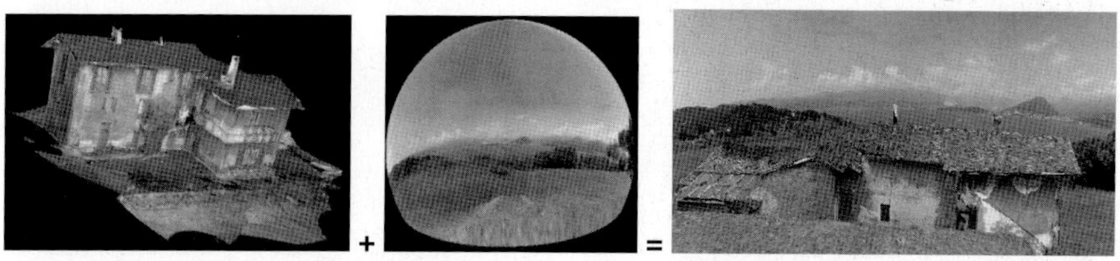

Figure 6: Combination of the background VRML panorama with the 3D model.

SELECTED EXAMPLES

Figure 7: Six snapshots from the 3D model of the Delacroix room, Palais Bourbon, Paris.

Figure 8: Cloister of the São Vicente de Fora monastery, Lisbon, and snapshots of the 3D model of the JRC replica.

Figure 9: VRML model of the "Sala dello Scrutinio", Doges' Palace, Venice, BBC Tomorrow's World, Nov. 2000.

Figure 10: Four snapshots of the 3D Reconstructed model of an Italian farmhouse over Lago Maggiore, Italy.

DISCUSSION AND CURRENT WORK

The paper illustrated a new automated technique for the 3D Reconstruction of monuments "as-built". The main advantage is that it takes reality as the base of modelling.

A main aspect of this technique is that a single acquisition session can produce all sorts of models each one targeted to different applications. Indeed, it is possible to have the high-quality master model, including fine spatial and visual detail (and normally of large dimension), for high quality documentation or scholar applications, and low-size, lower quality model targeted for Internet applications where download times are important. Intermediate quality models are easily derived from the initial one.

It is believed that 3D reconstruction is of particular use in cultural heritage both for indoor and outdoor applications. A few examples were presented. The same technique together with the design of clever Java scripts/applets can be used in archaeology or architectural studies, whenever there is the need to quantify distances and orientations between relevant features. The fact that laser sensors are used for distance measurement makes the data dimensionally and geometrically correct. As an example, the implications of Stonehenge (UK) in astronomy could have been easily found had a geometric and dimensionally correct model existed. 3D models in archaeology can be most beneficial for documenting the findings in different strata, as well as to provide 3D time (and depth) continuity to the objects found. Animations and other forms of data presentation can be easily built. 3D models can also help assembling objects from individual parts.

Our current work concentrates in two aspects: i) increase the visual and dimensional quality of the models; ii) extract structural representations for seismic analysis of heritage buildings and monuments.

A further area of development is the seamless combination of the 3D models with real-time, on-line data originated from the real world. This association of modelled (and yet realistic) data with on-line data helps building powerful Human-Computer Interfaces (HCI). As an example, it is thus possible to visit the ruins of a medieval castle and see in 3D how the castle looked like in ancient times.

ACKNOWLEDGMENTS

The technologies described were all developed at the JRC. Their progress was possible thanks to the participation in the following European projects: RESOLV, InfoBoy, CAMERA. The authors thank E. Wolfart, P. Dias, M. Ruggeri, E. Biotti and E. Bovisio for their help in producing the models shown in the paper.

REFERENCES

1. Sequeira V. et al., "Automated Reconstruction of 3D Models from Real Environments", *ISPRS Journal of Photogrammetry and Remote Sensing (Elsevier)*, (54), pp. 1-22, 1999.
2. Dias P. et al. – "Automatic Registration of Laser Reflectance and Colour Intensity Images for 3D Reconstruction", *8th Int. Symp. on Intelligent Robotic Systems (SIRS'2000)*, July 2000, Univ. Reading, UK.

ABOUT THE AUTHORS

Dr. João G.M. Gonçalves heads the Sensor Based Applications Sector of the

JRC. His interests are in 3D Reconstruction, Augmented Reality and Tele-Presence. The research objective is developing remote verification tools in Nuclear Safeguards and Non-Proliferation.

Dr. Vítor Sequeira is with the JRC and is in charge of developments in 3D Reconstruction, Augmented Reality, Tele-Presence and Internet based control. He works for the Nuclear Safeguards programme.

3D Worlds and Digital Environments

A Content-Centered Methodology for Authoring 3D Interactive Worlds for Cultural Heritage

Augusto Celentano (*) *and Fabio Pittarello* (*)

(*) Dipartimento di Informatica, Università Ca' Foscari, Venezia, Italy
E-mail: {auce,pitt}@dsi.unive.it

ABSTRACT
The application of 3D visualization technologies to cultural heritage has deeply evolved along years, shifting from the role of an ancillary technical solution to the dignity of a new media. However, the responsibility of building a complete 3D world often relies on a single designer who must be expert in different domains: graphics, interaction, and application domain.

In this paper we propose a new *content-centered* methodology for the production of 3D interactive environments for cultural heritage, including a more appropriate definition of subjects and skills involved. The methodology has been applied to the design of guided tours presenting cultural events, in order to test its benefits. The experimental results demonstrate that this methodology enables content experts to produce significant 3D experiences collaborating with a team of computer scientists, without concerning about low level implementation details.

KEYWORDS: 3D worlds, authoring, content-centered methodology, cultural heritage, VRML

INTRODUCTION

Low-level representation and rendering of complex three-dimensional worlds is no longer a troublesome task. There are both a wide number of methods for rendering the scene and standard graphic libraries that communicate efficiently with the underlying hardware. Application developers don't need to worry about underlying hardware, but have a standard way to write their application; this guarantees both rapid development and portability of solutions across different graphic platforms.

Advances in hardware today allow users to explore 3D scenes in real time, enlarging the issues related to 3D representation. This new situation has led human computer interface specialists to study the issues related to interaction in 3D scenes. If the early computer graphics years were characterized by the production of static renderings of scenes, to be used as a complementary support to presentations performed with other media, today 3D graphics have acquired the dignity of a new media, often used autonomously as a new way to communicate, to teach and to play [1].
This change of perspective due to hardware and software advances requires a redefinition of the production methodology, including a revision of the subjects involved in this new media, in order to obtain a full exploitation of its potentialities. In spite of this evolution, the modalities of production of 3D environments often reflect the older model; *authors* (i.e. *3D modelers*) are often still deputed to the whole construction of the 3D interactive worlds, while in many cases such worlds

demand the specific contribution of experts belonging to different areas. For example, a 3D virtual exhibition can be certainly characterized by a three-dimensional building but the main focus is on the layout of the works exposed, that should be placed according to the vision of the art expert that organizes the event.

Therefore a major feature of the production methodology should be the direct involvement of the *specific content expert*: in the exhibition example the focus is on the works of art, so the key figure should be the art expert. Unfortunately in many cases experts are relegated to the role of external consultants, and this doesn't permit a full exploitation of the new 3D interactive media expressiveness.

We propose in this paper a new framework for the production of 3D interactive environments, including a more appropriate definition of subjects and skills involved; the re-definition of *author* will be a key component of this perspective change.

THE CURRENT 3D WORLDS PRODUCTION MODEL

Figure 1 shows the current production model for interactive 3D scenes. It is based on three main roles: the vendor, the author/world builder, and the final user.
The *vendor* implements a 3D browser according to VRML specification, including a number of interface artifacts to perform actions, such as moving or rotating, which are not defined by the VRML standard[1].

The *author/world builder* is the key role in the whole process, since manipulates the 3D geometry, defines the interactive behaviors and organizes often also the hypertextual content. The author operates following the specification of the 3D language and the additional specifications given by the vendor.

The *final user* interacts with the application using the artifacts implemented by the vendor and by the author.

Using this method, the *author/world builder* is directly responsible for designing all the aspects of the application, from the style of presentation to the interaction model. The expert of the content domain gives his/her contribution filtered through the technical sensibility of the author, with no direct control on content organization; for this reason the content expert is not evidenced in the figure.

The author uses a number of specialized tools that require a deep technical knowledge in order to manipulate the 3D primitives, the interaction artifacts and the other media that are part of the interactive world, such as video, text and audio. The technical skill required and the absence of modularity result in 3D scenes which are hardly reusable by other authors because of the complicated low level relations between the various components of the interactive environment.

A COMPREHENSIVE PROPOSAL FOR AUTHORING 3D INTERACTIVE SCENES

The fulcrum of the re-organization of the production environments will be *content*.

[1] We base our analysis on VRML-based 3D worlds due to its wide availability and suitability in a Web environment.

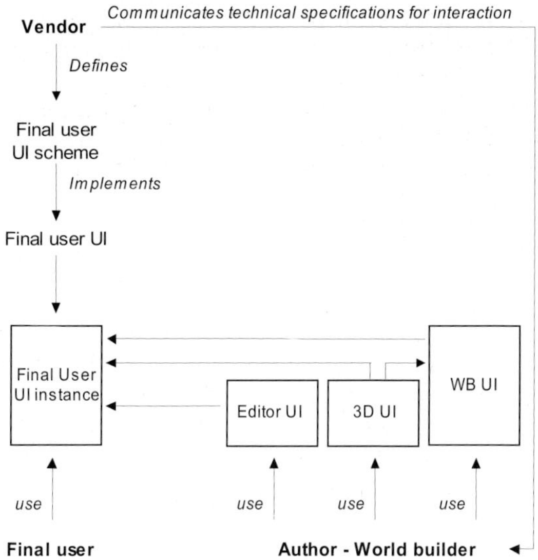

Fig. 1. The current production methodology for 3D interactive worlds

Final user	Author	World builder
familiar with the represented domain	expert in the particular author domain	expert in the geometric building of 3D worlds
able to navigate in 3D scenes using high level interaction objects	able to author the scenes with high level interaction objects	able to build complex behaviors using fine-grained languages that relate geometry to low level interaction objects

Fig. 2. The author's skills: a comparison with the final user and the world builder

We feel that in the recent story of interactive media too much emphasis has often been made on the technological side, even assigning to computer science technicians the role of experts in domains not belonging to their experience.
Our position is that a new range of subjects should be more involved in 3D environments, taking advantage of the 3D media without being constrained by low level technological issues. Our aim is to design a new generation of environments that could be defined *content-centered*, that is environments that take full advantage of interactive technology but, most of all, of the active participation of content experts.

We consider the concept of author in a new way: *authors are users expert in some particular domain (not necessarily in computer science) that want to build an interactive 3D experience for final users, using a given 3D world that they can manipulate through high level interaction objects belonging to their domain.* Figure 2 shows the skills of the re-defined author figure compared to the main roles of the traditional methodology.
The new methodology for designing interaction environments will follow the redefinition of author, including a different organization of phases and development tools matching the skills of the subjects. Figure 3 illustrates the proposed life cycle of the 3D interaction design with the new roles.
The term *interactive 3D experience* is used here to emphasize that the key point of a 3D simulation often is not only the represented geometry, but the whole simulation planned by the author and composed by the geometry, behaviors of objects and, in the case of a multi-user environment, even contributions of other actors in the scene. This is similar to what happens in the real world. For example a certain place, i.e. a square, has some geometric features given by the morphology of the square itself and of the buildings surrounding it, but most of all it is a scene where people can live different experiences.

The subjects involved

The *meta-author* has a deep knowledge of the content domain and didactic skills too; he/she interacts with the *interface designer*, a technician expert in low level interaction mechanisms, whose task is to build interfaces for authors and final users. The *meta-author* communicates with the final user too, in order to focus on didactic aspects of interaction. The focus of the *meta-author* activity is the definition of *typologies* of interactive experiences, resulting in a library of *classes of experiences* that can be instantiated by the authors.
The *author* has a deep knowledge of the content domain, and is concerned mainly with the communication of *contents*. He/she takes advantage of the meta-author design, chooses among the available classes of interactive experiences available the one that fits the application requirements, and fills it with contents. In addition the author searches the iconographic and the bibliographic bases as raw sources for the interactive presentation. Authors may need the support of editors for formatting text and images for the interactive presentation.
The *world builder* builds the animation for the 3D world and the widgets to activate them, following the technical specifications given by the interface designer, while the *3D modeler* is the responsible for the production of static 3D models for the interactive world. The *editors* are support figures that edit graphic and textual materials needed for the interactive worlds according to the author requests.

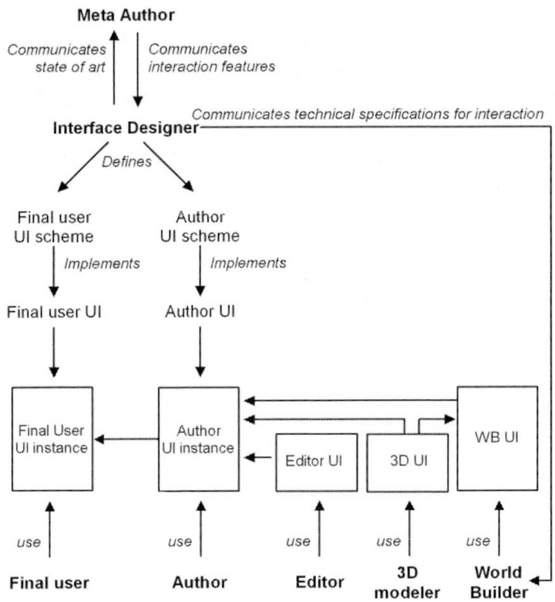

Fig. 3. The life cycle of 3d interaction design

A CONTENT-CENTERED METHODOLOGY

As evidenced from the definitions given above, there is a strong component of content specialists, shifting the attention from technical roles to content, in order to produce more meaningful virtual worlds. Besides, this introduces another fundamental element for the definition of the new methodology: the concept of *class of experience*.

This concept denotes a typology of general enough interactive experience that can be used as a template for particular uses. For example, in a virtual museum context meta-authors may introduce the class *Guided tour* to be subsequently specialized by authors with particular paths, in different museums and showing different artworks, still having the same basic interaction mechanisms: same walk modalities, artwork presentation, information detailing, and so on.

The purpose of this definition is to give a high level metaphor that can be easily understood by the authors. The subsequent user interface is based on the concept of experience, specifying the allowed behaviors for authors and final users. The aim of this re-formulation is to give authors and users a more intuitive metaphor for building worlds, opposed to the old methodology based on the composition of low level objects.

Besides, the concept of class of experience in the context of the new methodology is a consistent step towards modularity and code re-use; actually re-use is not limited to the 3D static models, but thanks to the libraries of classes of experiences even behaviors can be instantiated.

Finally, from the final user's point of view, the concept of experience helps to smooth the learning curve of 3D worlds: final users are presented the same interactive objects (widgets) for performing the same typology of behaviors in different instances of the same class of experiences.

A FOUR PHASES METHODOLOGY

For a better understanding of the role of the subjects involved, we briefly discuss the four main phases of the production methodology. The phases can be conceived as sequential, even if the feedback among them is fundamental, in order to improve the quality of the results, according to the principles of iterative design ([2]).

Conceptual phase

This phase is characterized by the identification of the content and interaction requests, in order to build a conceptual scheme for the author and the final user interfaces. The main role of this phase is interpreted by the meta author; this subject must have a deep knowledge of the contents domain and didactic skills too. The meta-author interacts with the interface designer (the technical counterpart of this phase) to take advantage of the current interaction technology, and communicates with the final user in order to focus on didactic aspects of interaction.
The conceptual phase has, as a result, the production of user interfaces (UI) schemes useful for defining a class of interaction experiences (for example the class *guided tours*). UI schemes are produced both for the final users and the authors, i.e., people instantiating the classes of experience.

Implementation phase

The interface designer, who has a deep knowledge of low level interaction mechanisms, builds the final user interface and the author interface on the basis of the UI schemes produced by the meta-author. The results of this phase are available for the authors who want to build interactive experiences belonging to the specific implemented class.
It is important to note that this implementation phase can be a personalization or a sub-setting of the capabilities of existing tools rather than a complete development of software from scratch.

The interface designer communicates to the world builder the technical specifications in order to coordinate the implementation efforts. The interface designer and the world builder have both technical roles, but the interface designer works at the level of classes of experiences supporting the meta-author, while the world builder works for the specific instance of a particular class, supporting the author.

Content development phase

Authors are the main protagonists of this third phase; they choose among the available classes of interactive experiences and instantiate the one that fits their particular needs. They take advantage of the skills of a number of complementary subjects to build the interactive experience.
In particular they usually need: a writer, for formatting textual information belonging to the interactive experience, since the 3D worlds we are talking about are often complemented by the presence of hypertextual material; a 2D graphic artist, for formatting bi-dimensional graphics; a 3D modeler, for shaping the 3D objects of the interactive worlds

Besides, the author may need to instantiate some particular interactive behavior designed by the meta-author. For example, the author may need to instantiate the behavior *history*, conceived by the meta-author to show the evolution of a venue belonging to the 3D world through time. In this case the author, who is aware of this feature existence through the author interface, communicates to the world builder the particular animation he/she needs. The world builder, following the technical specifications given by the 2D designer, builds the animation and the required widgets.

Final user interaction phase

In this phase, the final user takes advantage of the efforts of all the other people involved, interacting with the contents of the 3D world composed by the author, through the interface implemented by the interface designer.

The final user interaction is monitored in order to improve both the usability of the user interface and, most of all, the effectiveness of content communication.

APPLYING THE METHODOLOGY TO GUIDED TOURS FOR CULTURAL HERITAGE

We applied the methodology to the domain of guided tours for cultural heritage applications in an educational context, referring to interactive experiences designed by one of the authors [3, 4] for Palazzo Grassi, a cultural institution in Venice, Italy. In recent years it has hosted several important exhibitions accompanied by virtual tours in 3D worlds related to the exhibitions themes [5].

Conceptual phase

The theme chosen for the validation of the methodology was general enough to originate a possible class of interactive experiences in 3D worlds.

The meta-author discussed with the interface designer the current state of art of technology, limiting the discussion to VRML worlds. In this reciprocal information exchange the interface designer received information about the kind of content and education techniques of the meta-author; reciprocally, the meta-author received a complete information about the interactive features offered by VRML language that best fit his/her needs for the development of guided tours. The use of VRML derived from the open standard and interoperability requisites of this language and from the availability of a number of tools and run-time engines.

In order to have a wider contribution for the definition of the interface schemes, we chose to write a first formulation of the interaction problem; this description was then given to a number of testers, splitting in this case the responsibility and the point of view of the meta-author for the definition of this crucial passage. The meta-author was then given the task to unify in an unique proposal all the different points of view.

The interaction problem relative to the class *guided tours for cultural heritage applications* was formulated in the following terms:

> *Your task is to prepare a guided tour in a virtual exhibition of some kind for students Start your tour from a given starting location and advance step by step to a sequence of ordered key locations to visit; you have the faculty to go on and to turn back to the previous location. You can reach the beginning and the end of the tour directly, e.g. by clicking over start and end buttons. You can also navigate without constrains using the four main directions.Entering specific locations can automatically activate sounds, spoken comments, animation sequences and information pages. Besides, clicking over selected objects belonging to the scene you may play sounds and animations or display information pages.*

The testers were given a certain amount of time to perform the task; they were asked to list in a written form the verbal sentences they would have used to de-

scribe the actions related to the guided tour, as if they should describe their point of view to a computer science technician for implementation.
This experimental approach produced, as a result, a number of phrases that reflected the subjects' view on the authoring of a guided tour. The results were then given to the meta-author that summarized them in a list of actions such as " set this view as the starting point of the tour", " change location *x*", " start/stop animation when user gets to location *x*", and so on. Some visual schemes on paper were also drawn by the meta-author to have a first rough visualization of the interfaces for the final user and for the author.

Implementation phase

The written sentences and the paper schemes were then used by the 2D designer for the implementation phase. These basic materials were analyzed in order to establish consistency between the features of the final user interface and the author interface.
Then the 2D designer implemented the user and the author interface. For this experiment only the final user interface was fully implemented, while the author interface was given as a partially functional front end.

Content development phase

The author interface resulting from the implementation phase was at this point filled with content to create the *Einstein Tower guided tour*: the 3D model of the building, texts and iconography were used as raw materials to simulate the application of the new methodology.
The simplified interaction mechanisms provided by the author interface were proficiently used by the authors, who found implemented here the interaction modalities described in the first phase (conceptual phase) by the meta-author in the form of verbal action and visual schemes on paper.

The prototype demonstrated to be suitable to produce experiences similar to those ones produced before with the old methodology. This is an important result: in particular the key point is that using the interface resulting from the proposed methodology a new class of authors (as previously defined) was enabled to produce significant 3D experiences collaborating with a pool of computer science experts and other collaborators, but without knowing the peculiarities of VRML language and low level implementation details.

Final user interaction phase

The production of experiences similar to those previously obtained in the first implementations of the 3D worlds for the exhibitions of Palazzo Grassi resulted, not surprisingly, in similar user feedback. Even in this case we registered benefits in terms of simplification of navigation and augmented the usability for the concurrent use of visual and audio information, and hypertext.
Nevertheless it is important to notice that the availability of these tools should take additional benefits even for final users: in particular the definition of classes of experiences and the subsequent definition of a unified interface for each class (like the class *guided tours for cultural applications*) gives final users a standardized way to interact in order to perform similar tasks.

CONTENT CENTERED AND USER CENTERED APPROACH: A COMPARISON

It is worth to compare our approach with the *user centered* approach [6]. The *user centered* design is characterized by a strong attention toward the user; it was born as an alternative approach to the development practice which assumes that the understanding of the user needs is im-

plicit in the design of products. Unfortunately this assumption is misleading: applying this philosophy, only technology guides the development of projects and this often generates products that are not easy to use, unsatisfied customers, difficulties on technical support and training and so on.
Quoting [7], "*User centered design is a method for designing ease of use into the total user experience with products ... It calls for a multidisciplinary team to design everything the user sees and touches and to gather user input and feedback during each stage of the development process ... User centered design focus heavily on understanding users and gathering input and feedback from them throughout the development cycle*".

In other words *user centered* design involves users in all the phases of the development process in order to obtain a better result. Our methodology recognizes the benefits of the user centered approach and goes a step beyond; if according to the user centered methodology users are *consulted* all through the development process, the content centered approach encourages users to *actively participate* in the process of creating the product.

We have emphasized in the previous discussion that the direct involvement of authors, supported by the creation of tools that can be manipulated without a deep knowledge of the computer science world, can greatly improve the quality of contents, especially in computer supported education. That is the reason why in our approach the content expert is not an external consultant, but is the key figure of a content centered methodology, an active part of that multidisciplinary team required by the user centered methodology. We can say that our methodology doesn't want to substitute, but rather to integrate the positive results of the *user centered* methodology with a more active role for users in order to improve the quality of content.

CONCLUSION

In this paper we have discussed the weakness of the current process of building virtual worlds for interactive experiences. We have analyzed the difficulties to exploit the full potential of the 3D media, related to the persistence of an old authoring model characterized by the constant requirement of high level technical skills all through the 3D building process, and to the absence of tools for experts in the content domain to build 3D experiences without being involved in low level technical details;
We then have formulated a new methodology based on an approach that we defined *content-centered*, because of its strong attention towards professional skills of content experts. This approach is characterized by a re-definition of the author and by an enlargement of the skills required by the new potentialities of expression offered by the concurrent advances in hardware and software technologies.

The methodology was then applied to the class of experiences of guided tours for cultural heritage applications in an educational context. The methodology was tested generating a guided tour for an experience previously built with the traditional approach. The experimental results demonstrated that this methodology enables authors to produce significant 3D experiences collaborating with a team of computer science experts and other technicians, but without entering into technical details about 3D modeling language and tools.

REFERENCES

1. M. Gleicher K. P. Herdon, A. van Dam. Workshop on the challenges of 3D interaction. *SIGCHI Bulletin*, 26(4):1-9, October 1994.

2. J. Nielsen. *Usability Engineering*. Academic Press, San Diego CA, 1993.

3. F. Pittarello, *Desktop three-d interfaces for Internet users: efficiency and usability issues*, PhD dissertation, Ca' Foscari University, Venice, Italy, 2001.

4. F. Pittarello, M. Pittarello, G. F. Italiano. Architecture and digital exhibitions –The Einstein Tower world. *Virtual Environments 1998* - Eurographics Workshop Proceedings Series, Springer-Verlag, 1998.

5. *http://www.palazzograssi.it.*

6. D.A. Norman and S.W. Draper. *User Centered System Design*. Lawrence Erlbaum Associates, Hillsdale NJ, 1986.

7. *http://www-3.ibm.com/ibm/easy*

Building “Huys Hengelo” in VRML

Dennis Reidsma (*), *Martijn Kragtwijk* (*) *and Anton Nijholt* (*)

(*) University of Twente, Department of Computer Science, Enschede, the Netherlands
E-mail: anijholt@cs.utwente.nl

ABSTRACT
In this paper we report about our attempts to rebuild a historical building, ‘Huys Hengelo’, its interior, a farm built next to it and other parts of its environment (including a draw-bridge and a gate) using the Virtual Reality Modeling Language (VRML). This castle building played an important role in the history of its region. The main issues we deal with in this paper are: the unreliability of available sources – forcing us to show alternatives rather than ‘the building as it was’, the possibility to allow users to make changes and to experiment with different geographies, animations showing how parts of the wooden buildings were constructed during that time, the interface with the user and, as the project started as a student project on the request of some local historians and architects, some of our experiences with the co-operation between them and computer science students and researchers.

KEYWORDS: virtual reality, source unreliability, historical buildings, archaeological fragments, animation

INTRODUCTION

January 1646 some envoys of the then United Provinces of the Netherlands were on their way to Münster to negotiate a peace treaty between Spain and the Netherlands. They stayed a few nights in Huys Hengelo, a simple castle near the Dutch hamlet Hengelo, not far from the German border. Few is known of this castle itself. Remnants of a simple wooden storage house from the 14th century have been found on the location of Huys Hengelo. This so-called ‘spieker’ has been rebuild a few times until the castle was build. Only a few drawings exist of the castle in its environment, unfortunately historians consider them as unreliable. Around the castle was a canal; there were bridges and a separate farm. In 1821 the castle was demolished and the area was used for industry buildings. During 1993-1995 these building were pulled down and the foundations of the castle were excavated.

In this paper we report about our attempts to rebuild the house (including its interior), the farm that was built next to it and the environment (including a draw-bridge and a gate) using VRML (Virtual Reality Modeling Language). The main issues we deal with are the unreliability of the sources, forcing us to show alternatives rather than ‘the building as it was’, the possibility to allow users to make changes and experiment with different geographies, animations showing how parts of the wooden buildings were constructed in reality, the interface with the user and, as this project started as a student project on request of local historians and architects, some of our experiences with the cooperation between students and between (computer science) students and historians.

Huys Hengelo is one of the projects of

our research group where we experiment with virtual reality environments and especially to add different interaction possibilities for users (or visitors) of these environments. One of the larger projects is the so-called Virtual Music Centre (VMC) in which we have designed interior and exterior of a large existing building according to the design drawings of the architect (see e.g. Nijholt [6]). However, unlike such projects, when we try to model historical buildings, such design drawings are generally not available and sometimes the buildings are only known from descriptions or from paintings.

There have already been lots of projects in which virtual reality techniques are used for historical-preservative documentation of buildings. In the *Eurographics '99* proceedings (Alberti, Gallo, Jelinek [1]), for example, we can find discussions about suitable techniques for restoring buildings like the Camposanto Monumentale of Pisa, the "Kaiserpfalz" (the lost palace of Otto the Great in Magdeburg, Germany), the Danteum in Rome, the Vatican, parts of Pompeii, the Teatro alla Scala di Milano or the Mogao Grottoes near Dunhuang in Northwest China. How to deal with missing data and nevertheless show realistic visualizations is a theme that pops up in many of these projects. Do we have to extrapolate, do we have to show all possible interpretations or should we select one or two most agreed upon interpretations of the experts? How did the building look like, where exactly was it located, what building materials were used? Sites of historical buildings are never undisturbed, excavations may destroy parts of a building's structures, not everything can get uncovered, etc. Simulating archaeological hypotheses is an explicit goal in projects mentioned in Blaise et al. [2] and Roberts et al. [10]. They have a simple manipulation language that is used to describe scripts in which we can define and modify dimensions, positions and orientations of objects and their geographical relationships.

As mentioned in Masuch et al [4], "people tend to take a depicted reconstruction as established scientific truth." For that reason we think that it is important to take a research position that does support discussions about the design decisions in the virtual world rather than aiming at a visualization style that forces experts to agree upon all visible details.

BACKGROUND OF THE "HUYS HENGELO" PROJECT

A Short History

In the years between 1993 and 1995 a large factory was taken down in Hengelo in the east of the Netherlands. It had already been known for a long time that the factory had been built on the site of a historically important building (Huys Hengelo), so the opportunity was seized to excavate the foundations of the house and the remnants of the moat and the gardens. After the factory has been taken down, a residential quarter will be built on and around this site. Since Huys Hengelo has been central to local history for hundreds of years, and the site has been the location of different kinds of houses and buildings for some thousand years, it was decided that the history of Huys Hengelo would be given a place in the new quarter. Part of these plans was the project described in this paper, which had as goal the modeling of a virtual version of the house.

The building site of Huys Hengelo has housed some kind of building at least as early as the 13th century (Reynders [8]). The first building was probably some kind of storage shed (a so-called 'spieker'). Somewhere after 1500 the spieker was probably extended to house the owner of the land around the place when he had been hunting there. In later years the building was subsequently extended several times and even partly rebuilt until it became a large mansion. The owners of the mansion were from a few important and influential families in that part of the Netherlands, including landgraves and counts and such. Some owners used it as a lodging for special occasions, other owners lived there all the time. All in all, the place has been central to local politics for a very long time. Sometimes, as we mentioned, it even comes up in stories of national politics, as when in 1646 a delegation from the Netherlands, headed for the ratification of the Covenant of Münster stayed there for some time. The presence of Huys Hengelo caused the nearby village Hengelo to grow until it became one of the mayor cities in the east of the Netherlands. In the year 1821 however the house was torn down, after it had fallen in disuse.

Compared with Huys Hengelo, much more is known about the farm that was built next to it (Nijhuis [7]). It existed until 1916 and a few photographs have been taken. Furthermore, the archives of Hengelo contain a detailed description of a renovation of the farm. Back then, it was a building of only 5×10 meters. In Hesselink [3] an expert explains from recollection how this type of farms used to be built until the beginning of the 20th century. This expert has worked on one of the last farms that were completely built in this traditional way. Horses dragged the timber to the site, and many craftsmen were involved in building the wooden framework. No other sources containing this knowledge are known. In a later section we will show how we visualize this expert knowledge and make it accessible to a wider audience.

Unreliable Sources and How to Make Decisions

The project to rebuild Huys Hengelo, the spieker and the direct environment, including the farm, was completed in cooperation between students in computer science and experts on architecture and archaeology. Of course the students hardly knew anything about those issues, so they had a lot explained to them by the experts. Even then however, they usually didn't know what decisions to make, what to model, which sources to believe (the most important sources contradicted each other vehemently) and what to make of the two existing drawings (very much unalike and considered uninformed and incorrect by all experts in any case). Therefore the experts were included in every step of the building process. They were so helpful as to provide the students with constant feedback. But the experts were even more than average illiterate on the subject of computers. This meant that they had difficulty understanding the implications of many design decisions until they could see actual implementations of parts of the house. Furthermore the experts couldn't agree on many subjects. This all meant that, though the experts were very enthusiastic and willing to do almost anything to help, the students had to make many decisions on what was to be used in modeling the house on their own. A next section about handling lack of information addresses part of these problems.

BUILDING 'HUYS' AND ENVIRONMENT

The Actual 'Huys

In Figure 1 we have a drawing made in 1860 by Jacobus Craandijk, based on stories of a child who used to go there often.

Figure 1: Huys Hengelo by Jacobus Craandijk, 1860

An other drawing, made in 1700 by Stellingwerf (1666-1727), shows a slightly different house. But there are also floor maps and a map of the site available. Experts disagree which drawings are correct and give different interpretations to the other sources, descriptions and excavations. Some drawings are known to be made when the building didn't exist any more. In Figure 2 we have the virtual reality version that we came up with after studying the sources (Nijhof [5], Reynders [8,9]) including an inventory from which an interior could be reconstructed. Here we will not dwell on the design and details of implementation (in VRML) of the Huys Hengelo building.

Figure 2: Virtual 'Huys Hengelo' and Environment

Handling a Lack of Information

In this project many things had to be modeled based on partial information. It might not be known whether a stone fragment had been part of a window or a door, how large some element had been, where the gate to the island had been or what kind of glass the house used to have in the windows. Besides, there had been many versions of the house. Some stone fragments were probably from an earlier version of the house. And the experts often couldn't agree on interpretations. So it was decided to make it possible to visualize all kinds of different possibilities. That way a user could assemble the house on several properties, giving a version from a certain era or presenting the views of a certain expert. However, the viewers of the model were supposed to be computer illiterates, so it would never do just to model all separate parts and let the user assemble them, placing them together in a coordinate system. Instead a console was made in Java, which presented the user with simple properties that could be switched on or off. The different options were parameterized: the VRML-modeler only had to name the objects and their different versions and state a few properties like the number of versions of the object and whether the object could be switched off. For example: when the house of 1850 was presented, the windows could be set to a version with square glass panes and a version with diamond-shaped glass panes but the windows could not be switched off. However, the shutters for example could be switched off, when it was not certain whether there had been any shutters. Certain configurations of everything could be saved to a file for later reviewing.

MODELLING HOUSE PARTS AND ARCHAEOLOGICAL FRAGMENTS USING TOOLS

Building the Walls and Windows of a House

When modeling a house in 3D, it is necessary to build walls and to place doors and windows in them. And sometimes those doors and windows need to be moved later in the modeling process. However, most modeling programs don't give the possibility to build a wall with some general properties and then define a *hole* in it, which can be moved around. Therefore a small and simple tool was developed in which a wall with holes in it could be defined. The holes could be manipulated even later in the process, when all kinds of other objects had been added to the model.

The Modeling of Archaeological Fragments

Part of the modeling process consisted of modeling the many fragments that had been found at the excavation site. Those fragments ranged from parts of windowsills and steps to sandstone ornaments. We wanted to be able to model many of them in a fast and flexible way. This meant that simply modeling them by hand was not an option, since that takes too much time. 3D-scanners can be very fast and accurate but are far too expensive for this kind of project.

Furthermore, a model as produced by such a scanner is not easily modified. For example when the user wants to produce a complete windowsill, extrapolating from a small fragment, this would still need a lot of work. To achieve this, a handful of simple tools were developed in the project. The tools are all based on the concept of an *extrusion*, with which it is very easy to model things like ornamental vases or pillars. One simple way to define an extrusion is by a polygon base, a 3D path and a scaling function. The resulting extrusion is then the volume that can be obtained by moving the polygon along the path, scaling it along the way according to the scaling function (see Figure 3).

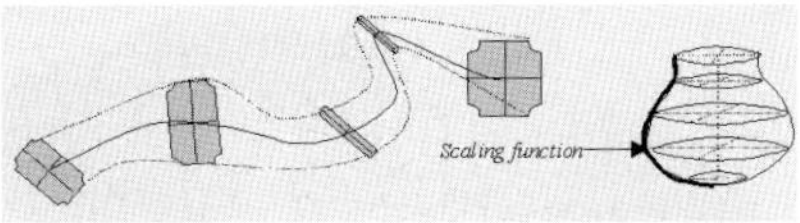

Figure 3: Modeling Vases and Pillars

The *polygon editor* makes it possible to define the polygon that is to serve as the base for the extrusion. The user can load an image into the editor and then draw the points of the polygon's outline over this image. The *line editor* makes it possible to define a two-dimensional path for the extrusion. Though this path could be three-dimensional, in this project only paths in one plane were needed. The *extrusion scaler* finally can be used to graphically draw the scaling function using a photo of a pillar base or ornamental vase and tracing the outline. Figure 4 shows how this works.

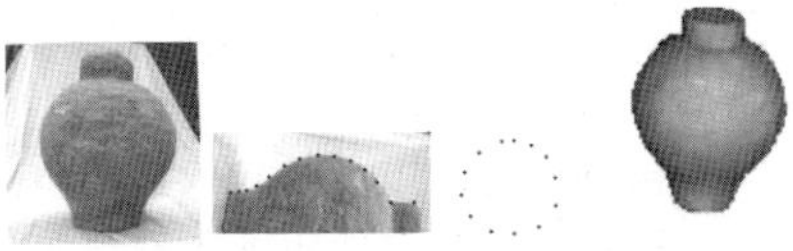

Figure 4: Different Phases in Modeling: Photo of Vase, Scaling Function, Base Polygon and Virtual Vase

It may be clear that such simple tools offer the speed as well as the flexibility in modeling the fragments. The user can make side and front view photos of the fragments and then use them to quickly model the fragments in the editors. When that is done the resulting objects can easily be extrapolated into complete windowsills, pillars and staircases and such, for example by drawing a longer path in the line editor (see Figure 5 for an example).

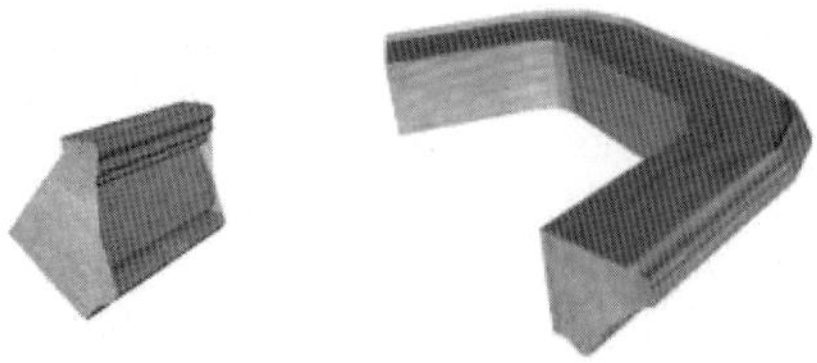

Figure 5: Extrapolating a Stone Fragment to a Complete Window Sill

MODELLING THE TWENTE FARM

As we mentioned in a previous paragraph, there was also a small farmhouse next to Huys Hengelo, about which a lot more was known than about the Huys itself. Using this information, we have constructed an interactive application that takes the user on a guided tour around the building site of a traditional Twente farm like the one next to Huys Hengelo.

Figure 6: First Phase of the Building Process, only the Foundation is Visible

Figure 7: The Skeleton of the Farm, Almost Completed

Figure 8: Almost Finished: the Construction of the Roof and the Walls is Shown

All information in this application is taken from Hesselink [3] and Nijhuis [7]. The user is able to walk through different phases in the construction of the farm, ranging from the stones that are put in the ground to form the foundation (Figure 6) via the skeleton (Figure 7) to the construction of the roof (Figure 8). The building process is made clear through animations, and the user can get more information on building elements by clicking on them.

The building of the farm is divided into ca. 10 phases that can be viewed by the user in any order. Each building phase consists of a 3D scene, accompanied by a short explanatory text. VRML allows the user to walk through the scenes using the mouse. Animations that show how the wooden building elements are put together can be started on the user's demand. A lot of effort has been put in showing a truthful, detailed view of the wood joints that were used, all according to [3]. If the user clicks on a building element, a short explanation of that object appears below the 3D scene. Moreover, a separate 3D scene (containing only this object) is shown, and the user can rotate the object with the mouse to view it from all sides (see Figure 9).

Figure 9: Interface to the 3D Building Site

CONCLUSIONS

Modeling the Huys itself resulted in three things. The first result is the virtual model of the house, which can for example easily be incorporated in some kind of museum exhibition showing also photos, drawings and building fragments. The second result is a set of simple tools that make it very easy to model stone fragments in VRML and to manipulate them into complete pillars, ornaments and other objects. The third result is the interface that makes it possible to present the uncertainty inherent in this kind of project to the user in a flexible way.

With the farm we have presented an application to demonstrate how these farms used to be build in our region. This application helps to preserve this valuable knowledge and pass it on to our descendants, through a series of interactive VRML scenes. It requires no specialist technical knowledge, and is therefore very suitable to teach school children about how people from their region lived 200 years ago.

In this project we have chosen VRML. It is available to anyone and in a very short time it can be used to prototype first versions of buildings and make experts aware of possibilities of this technology. Moreover, in interaction with experts it became clear which tools had to be developed in order to attack similar problems in future projects. Although it is quite obvious that the game industry is able to produce editors and environments with which much more beautiful graphics can be obtained, it is also clear that these tools are not yet available for virtual reality modelling of unique buildings in the current early situation and in a situation where we try to find tool requirements in interaction with experts.

ACKNOWLEDGMENTS

We gratefully acknowledge the work of some other students that helped in this project: Maartje van Hardeveld, Ilja Clabbers and Alexander Broersen. The 'Wetenschapswinkel' of the University of Twente initiated the contacts with H. Reynders and E. Jans who taught us about historical and architectural issues. They don't bear responsibility for the final outcome of the project.

REFERENCES

1. Alberti, M.A., Gallo, G. & Jelinek, I. (eds.). *EuroGraphics 1999*. Short Papers and Demos. 20th Annual Conf. of the European Association for Computer Graphics, Milano, 7-11 September, 1999.

2. Blaise, J.-Y, Drap, P.& Florenzano, M. A distributed interface for archaeological restitutions. In: Proceedings *Nîmes 1998: Virtual Reality Applications*, I3 Conference, LLIA Nr. 134-135-136, May-June-July-August 1998, 168-175.

3. Hesselink van der Riet, M.L.M. *De Bouw van de Twentse Boerderij,*

Stichting Matrijs, Utrecht, 1984.

4. Masuch, M., Freudenberg, B., Ludowici, B., Kreiker, S. & Strothotte, T. Virtual reconstruction of Medieval architecture. In Alberti et al. [1], 87-89.

5. Nijhof, H. De tekening van Stellingwerf. *Oald Hengel*, February 1999, 8-9.

6. Nijholt, A. Virtual Cultural Environments: Exploration of Information. In: *Cybernetics and Systems 2000*. Proc. *EMCSR 2000* (15th *European Meeting on Cybernetics and Systems Research*), Vol. 2, R. Trappl (ed.), Vienna: Austrian Society for Cybernetic Studies, 2000, 530-535.

7. Nijhuis, G.A.B. Reconstructie schoolmeestershuis en schuur bij Huys Hengelo. *Jaarboek Twente*, 1995.

8. Reynders, H. *Historie en opgraving Huys Hengelo*, 1996.

9. Reynders, H. De bouwfases van Huys Hengelo. *Oald Hengel*, February 1999, 5-7.

10. Roberts, J.C. & Ryan, N. Alternative archaeological representations within virtual worlds. In: R. Bowden (ed.), Proc. of the 4th *UK Virtual Reality Specialist Interest Group Conference*, Brunel University, November 1997, 179-188.

ABOUT THE AUTHORS

Dennis Reidsma and **Martijn Kragtwijk** were M. Sc. students when this project was started. Presently both of them have temporary research positions at the Department of Computer Science of the University of Twente, where they work on natural language processing and virtual reality, respectively.

Anton Nijholt is full professor of computer science at the University of Twente. At this university he is coordinating the research and education in the areas of human-computer interaction, speech and natural language processing and virtual realit. Among other things he is interested in research issues that deal with interacting embodied agents in virtual environments.
E-mail: *anijholt@cs.utwente.nl.*

Ancient Olympia as a Three Dimensional Museum Experience

Cliff Ogleby (*), *Sarah Kenderdine* (#)

(*) Department of Geomatics, The University of Melbourne, Victoria, 3010, Australia
Email: c.ogleby@eng.unimelb.edu.au
Powerhouse Museum, Sydney, 2007, Australia
Email: sarahk@phm.gov.au

ABSTRACT
This paper will present the techniques used in the production of a 3D narrated tour of the ancient precinct of Olympia set in 200 BC; and detail the approaches undertaken to digitise the Statue of Zeus using an hand held laser scanner, together with the processing of this data into a high resolution model suitable for interactive display. The web site will be briefly described with focus directed to a statistical analysis of its usage.

KEYWORDS: Olympia, virtual reality, reconstruction, website, stereo, 3D, laser scanning, cultural heritage, panoramic, high resolution, zoomable

INTRODUCTION

The advent of the Sydney 2000 Olympic games in Australia was an occasion for celebrations and festivities of many different kinds, as well as the sporting event itself. In conjunction with the athletic events there were cultural events, ranging from classical and contemporary dance through art displays and including the *1000 Years of the Olympic Games – Treasures of Ancient Greece* exhibition at the Museum of Applied Arts and Sciences. This Exhibition showcased 54 antiquities loaned from the Hellenic Ministry of Culture that depicted Olympic sporting events, and incorporated two virtual reality components.

The subject of this paper is the preparation of the virtual reality data used in the two interactive exhibits, and the associated website. The works included a digital reconstruction of the site of the ancient games, Olympia, and a high-resolution digital model of a statue from the Museum of Antiquities in Athens. The works are analyzed below in terms of both quantitative and qualitative survey. The website will be briefly described with focus directed to a statistical analysis of its use (a full analysis and design specification of the website can be read in Kenderdine 2001).

PROJECT ELEMENT A: ANCIENT OLYMPIA

Olympia is located between the Cladeus and Alpheios rivers on the western side of the Pelloponese. It is an archaeological site, where most of the monuments lie scattered as a result of two earthquakes and numerous floods over the centuries. The foundations that survive from many of the buildings date to different periods, from the Archaic, Classical, Hellenistic and Roman times.

Olympia became the primary site for the games held in honour of Zeus, and with the rise of Zeus through the pantheon of Greek gods the games at Olympia increased in importance. The Games were a time of peace throughout the Hellenic world and an occasion for displays of sportsmanship and athleticism. The games were also a time for business deals to be struck, war booty to be displayed, and state and city honour to be upheld. Following the Roman rule of the Hellenic world, and the rise of Christianity, the games diminished and eventually the site was abandoned.

Excavations at Olympia were begun in May 1829 by French archaeologists, who transferred many of the finds to the Louvre where they are still being exhibited. When the Greek government was informed of the looting of artefacts, the excavation was stopped. Excavations started again 45 years later by German archaeologists, which continues to this day. Much of the material published as a result of the excavations formed the basis of the reconstructions used in this project (Adler, F. et al, 1892-1897/1966).

ARCHITECTURAL AND SCULPTURAL ELEMENTS

Olympia was first and foremost a place of worship, and a place for the celebration of the games held in honor of the Gods. Many of the structures on site were temples, with a few administrative and special accommodation buildings. In 200BC, there were buildings at Olympia from most of the main architectural periods, dating back to around 700BC. The date of 200BC was chosen as the epoch was indicative of the evolution of the site, but without the more recent Roman additions.

Information on the architecture was obtained mostly from the publications of the this German material, and from Pausanias, a travel writer of the first Century AD (Levi, 1979). For a full set of reference please refer to the website bibliography and associated resource materials (*http://www.phm.gov.au/ancient_greek_olympics*).

The reconstruction of ancient Olympia has been a long and involved process. At all times effort was made to create models and materials that were not only visually appealing but also historically correct. A multitude of published sources were consulted in order to determine the architecture of the missing buildings, the form and colours of the decoration, the statues, the terrain features and vegetation. All of the architectural elements have been built as close to their 'original' descriptions as possible.

The original building models were created in Bentley System's MicroStation CAD package, primarily because of the level of experience with this package and its suitability for precise dimension control. This caused many difficulties when translating data into the 3D modelling package used, the only common workable export format was DXF which not only gave considerable errors in relative position of elements but also created very large polygon counts for even simple objects. Much of the building elements were re-created in 3D Studio Max from scratch.

All of the structures are in the locations derived from the archaeology of the site, and are true to scale in all dimensions. The heights of the buildings were derived from the architectural remains, and were extracted from the publications. The materials and textures were composed of colours derived from the archaeological and historical record, as well as using contemporary materials typical of the period. The descriptions of the buildings and statues by Pausanias were also used as the basis for much of the material mapping, and for the style and location of the statuary.

Figure 1: Postcards from Olympia

All of the animations were created using 3D Studio Max. Consideration was given to using a variety of other packages however the 'render-farm' facility of 3D Studio Max. offered substantial (essential) timesavings. The render-farm finally consisted of approximately 60 machines, ranging from very fast Pentium PIII to PII computers. Whilst the render-farm required considerable nurturing the process greatly facilitated the delivery of the exhibition in a very short time period.

The rendering consisted of 14 animation sequences timed to fit the narration, created as single left and right channel frames for later composition and post production, in excess of 40,000 frames.

Presentation and delivery

The 3D experience offered to visitors was developed in Macromedia Director using a 'stage' that was 1600x600 pixels. The two separate 768x576 (PAL) animations sequences that made up each tour path were composited into one 1536x576 movie, ensuring perfect synchronisation between the left and right channels. These animation sequences were then used in Director, where the interactivity and additional material was added. Visitors to Ancient Olympia would be guided through the site along pre-determined paths, accompanied by a narrative explaining the monuments with anecdotes of life in during the games.

Figure 2: A decision point screen from the narrated tour interactive exhibition

Two rear-projection capable projectors were used to overlap the left and right channel images onto a 4m wide screen in a 5m square room. A custom built console was designed to facilitate user interaction at the decision points, and mounted on a stand approximately 3m from the screen.

The computer used to deliver the movies was a 2x800MHz Pentium III machine, with 500Mbyte of

RAM, fast disk access and a Matrox G400 graphics card. This card has a twin screen output of 800x600 resolution, which was used to drive the two projectors.

Figure 3: The navigation map with legend

PROJECT ELEMENT B: THE STATUE OF ZEUS-POSEIDON

One of the most significant sculptures in the National Archaeological Museum in Athens is the statue of Zeus from Artemision (Artemesium), also considered to be perhaps a statue of Poseidon. This bronze sculpture is slightly larger than life size, and was found in 1926 in the sea off Cape Artemision. It is one of the few surviving examples of Early Classical statuary, and represents the deity as a perfect human form. It is presented in the Museum in Athens on a pedestal, in a sunlit room with other statuary around the walls.

In order to develop a high resolution digital 'copy' of the statue of Zeus it was decided to use state-of-the-art reverse engineering techniques. This approach would enable a dense three-dimensional coordinate data set to be acquired within the minimum time possible.

A Modelmaker Laser Scanner mounted on a Faro 3D Coordinate Measurement Arm was used to create high resolution point clouds of the surface of the statue. A purpose built scaffold was fabricated in Athens to give the facility to elevate the scanner in order to access the top of the statue whilst also ensuring a stable fixture. The entire sculpture was digitised over several daytime sessions in multiple parts. There were considerable restrictions on this part of the project (including initially no photography, essential for material maps), no method or re-orienting the scanning system datum, and a very short working period in the Museum. These hurdles created complications in the data stitching process, as each scan was effectively an independent data set with no points in common other than overlapping surfaces.

Figure 4: Scanning the Statue of Zeus/Poseidon

Presentation and delivery

The Zeus model was presented in a separate 3D virtual reality booth at the Exhibition. It was displayed using shuttered glasses technology (they were suspended from the roof), driven by proprietary software running on Intergraph hardware. The use of shuttered glasses necessitates the presence of a museum attendant; this option was not pursued for the reconstruction of the city that used polarised projection. The model was

rendered in real-time, and was displayed at life size on one screen and as a close-up of the head on another.
A limited amount of interactivity was provided where users could zoom in and out and rotate the model about its vertical axis.

Figure 5: The Statue of Zeus/Poseidon virtual model

ANALYSIS
The response to the Olympia model in the Exhibition was very enthusiastic. A survey based on responses from 302 visitors was undertaken by the Powerhouse Evaluation and Audience Research Group. It identified that 37% of the visitors preferred the virtual exhibits in preference to the objects themselves. Also that the group was equally represented through age groups, and that most had exceptionally high levels of formal education.

Particularly notable during the Exhibition itself was the large number of senior or retired citizens who readily engaged with the Olympia model. With good content the technology did not appear to present difficulty of use.

When the interactive components were designed it was expected that visitors would spend perhaps 5 minutes (of the possible 20 minutes) in the Virtual Tour booth, and then move. From observation, a large proportion of the visitors stayed for the entire duration of the tour, some even waiting for the tour to re-start so as to see the part they may have missed. The Virtual Olympia component occasionally caused a blockage in the flow of visitors through the exhibition (although this problem of congestion was common to the whole exhibition); if the exercise was to be repeated then more space would need to be allocated and perhaps a better designed booth to ensure that lingerers would not cause obstructions.

Consideration was also given during the exhibition to duplicating the Virtual Olympia tour as a 2d display for another lecture theatre in the Museum for the large number of school children who tended to congregate in the booth. The survey identified the huge demand for a product to takeaway from the Exhibition in the form of CD ROM.

> '*Outstanding – Virtual reality tape descriptions were exceedingly accurate. Best however was the 3D Olympia - unique! You should market it to schools!*' (respondent to the survey).

PROJECT ELEMENT C: THE WEBSITE
The website consisted of six content layers. These contained the following information.

1. **The Exhibition**: a series of zoomable panoramas of the Exhibition at the Powerhouse Museum augmented by 3D object movies and high-resolution zoomable images of objects, and supplemented with object information as text. (see illustration on next page).

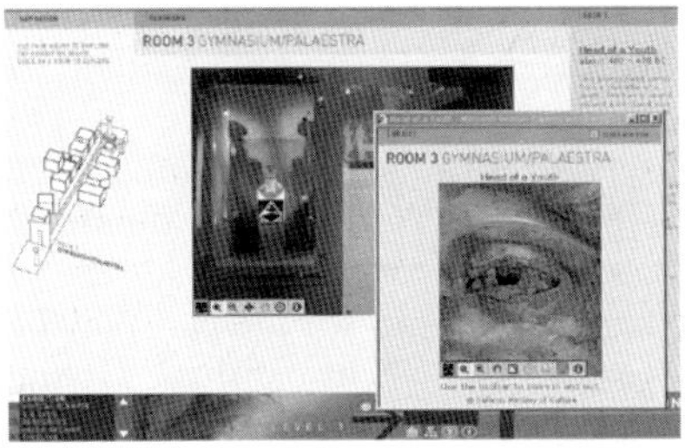

Figure 6: The Web Based Exhibition Tour

Virtual Olympia: this section contains a rich collection of materials separated into 4 sub-sections including:
Archaeological and geographic maps of Olympia;
A panoramic tour of the digital reconstruction of Olympia presented as 360° zoomable panoramas (viewable at full screen) with audio files. The option to download individual buildings to compliment scholarly research is also available. Essays are available as PDF files;
29 zoomable 360° panoramas of the archaeological site at Olympia taken in March 2000 (viewable at full screen). Essays are available as PDF files;
The statues from the Temple of Zeus shown as a 360° panorama of the main room at the Museum of Olympia, and a series of zoomable object movies. The metopes are further supplemented by reconstructed line drawings and colour interpretations using archaeological and historical records as the sources for reconstruction.

3D Zeus: a web version of the laser scan of the Zeus of Artemision (using the Zoom Viewer plugin), with the option to view as a 3D anaglyph through downloadable anaglyph glasses.

Resources - this section contains:
14 researched PDF articles;
Education programmes;
Glossary of terms linked to the PDF articles;
A nodal genealogy of the gods;
Supplementary programme of events to compliment the Exhibition at the Powerhouse Museum;
All audio files are listed.

Image archive: due to the extensive visual components in the website this section offers a quick entry to all the individual panoramas, zoomable images, object movies and jpegs. Selected buildings from the digital reconstruction of Olympia are available here, for research purposes, to download and print.

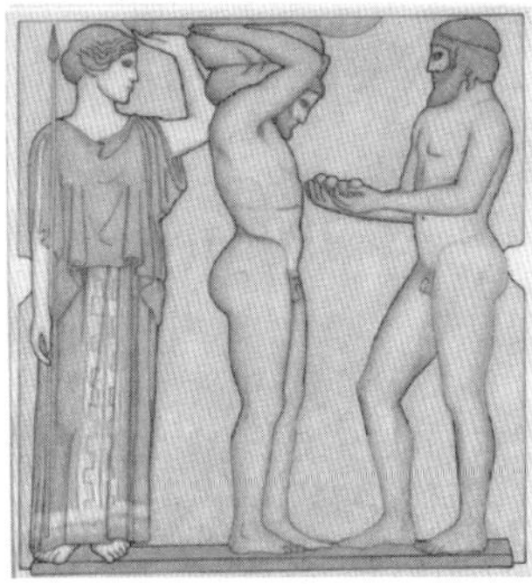

Figure 7: Artist's Impression of a Metope

Timeline: a timeline of events at Olympia over a 1000 years period is organised into:
Events specific to the Games at Olympia (events added after 200 BC not included);
Events relevant to Olympia itself (building programme until end of Hellenistic period only);
Events, political, around the Mediterranean world – after 200 BC only events directly relevant to Greece or Olympia included: politics, arts & sciences.

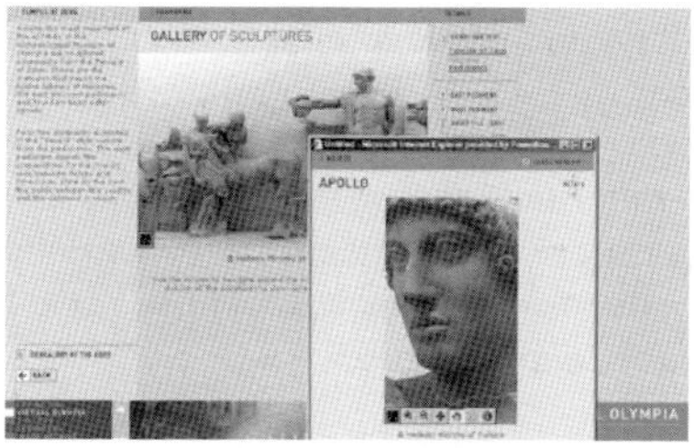

Figure 8: The Metopes as LivePicture Objects

THE ANALYSIS

Statistical data has so far been analysed for the four-month period of the Exhibition — from the beginning of August 2000 until the end of November 2000. An examination of the period shows a predictable variation in the data display. The most active day of use was during the Olympic period and on September 16, 2000
with 92,352 hits. Weekdays rather than weekends showed a larger number of hits although it was after-hours that numbers were marginally higher. This contradicts our initial expectations that the high bandwidth required to access the site would preclude extensive home use.

What is potentially of more interest is the length of stay by each visitor. 45% of visitors accessed just the first two pages of the website. There was a corresponding drop off as the number of pages increases. However, a relatively high proportion (8% of the visitors) was staying over 20 minutes to access 43% of the total page views. This last statistic would suggest that those visitors who were interested in the site found rich and rewarding content. At one point in the 4 month period over 18% of the visitors were staying over 20 minutes. In web lingo this has been referred to web "stick-ability" (and hopefully not a measure of slow content download!).

CONCLUSION

The overall project has won many awards in the short time since its launch. The most important of these include a Virtual System and Multimedia award for the Best Virtual World Heritage reconstruction; the Best of the Best website for the Australian Interactive Multimedia Industry Awards; the Best Interface Design at AIMIA; Hitwise Top 10 for Education and Research 2000; and the website was a finalist in the British Association for Film and Television Interactive Entertainment Awards 2000.

The versatility of use for the Olympia and Zeus models, and the rich website content demonstrate that rigorous scholarship is a worthwhile investment and can command a significant audience. The repurposing of the underlying content has enabled the works to be developed for different platforms and delivery mechanisms. The works have now been repurposed for a CD ROM that will be distributed to schools in Australia courtesy of Intel Corporation.

ACKNOWLEGEMENTS

This project was made possible through the generous provision of the Hellenic Ministry of Culture (http://www.culture.gr). A number of people in the Ministry worked hard on our behalf to make the data collection process possible. Sponsorship to complete the work was generously provided by the Intel Corporation (http://www.intel.com). Thanks also to the Australian Ambassador in Athens, Ross Burns, and consul Deborah George, to Jan Casson-Medhurst at the Australian Archaeological Institute at Athens, Judith Swaddling at the British

Museum for her time in reference to the scale model of Olympia housed at the Museum. The website was conceived and design by the Powerhouse Museum (http://www.phm.gov.au) with implementation by Massive Interactive. The photographic work at the archaeological site was undertaken by Peter Murphy, Sydney. The overlays for the metopes were draw under the direction of Kate Da Costa by a meticulous and hardworking David Loong.

REFERENCES:

1. Adler, F., Curtius, E., Dörpfeld, W., Graef, P., Partsch, J. & Weil, R. 1897/1966 *Olympia I: Topographie und Geschichte*, Amsterdam

2. Adler, R., Borrmann, R., Dörpfeld, W., Graeber, F. & Graef, P. 1892/1966 *Olympia II: Die Baudenkmäler von Olympia*, Amsterdam

3. Kenderdine, S., 2001, '1000 years of the Olympic games: treasures from ancient Greece. Digital reconstruction at the home of the Gods'. Paper for WWW 10, Hong Kong, 1-5 May 2001. CD ROM.

4. Levi, P. 1979 (trans.) Pausanias *Guide to Greece vol 2: Southern Greece*, London

ABOUT THE AUTHORS

Cliff Ogleby is a Senior Lecturer in the Department of Geomatics at the University of Melbourne, with 20 years experience in applying information technology in the disciplines of archaeology and cultural heritage management. He is Secretary General of CIPA-I^{2}DOC, the liaison committee for the documentation of cultural heritage between the International Society for Photogrammetry and Remote Sensing (ISPRS) and the International Council on Monuments and Sites (ICOMOS). He works in visualization and virtual reality, as well as site documentation.

Sarah Kenderdine is a maritime archaeologist and museum curator. Sarah has been involved in the development of large-scale internet based museum web projects and portals since 1994; and is currently involved in digital reconstruction and visualization projects. She was Project Manager and Creative Director for the *1000 years of the Olympic games: treasures of ancient Greece* Exhibition website (http://www.phm.gov.au/ancient_greek_Olympics) and CD-ROM. She is currently Project Manager of Special Projects at the Powerhouse Museum in Sydney.

From dust to stardust: a Collaborative 3D Virtual Museum of Computer Science

Thimoty Barbieri(*), *Franca Garzotto*(*), *Giovanni Beltrame*(*), *Luca Ceresoli*(*), *Marco Gritti*(*), *Daniele Misani*(*)

(*) Politecnico di Milano, Hoc-Hypermedia Open Center,
Department of Electronics and Information, Milano, Italy
E-mail: {barbieri,garzotto}@elet.polimi.it

ABSTRACT

One of the outcomes of Moore's Law - according to which the exponential growth of technical advances in computer science is pushing more and more of our computers into obsolescence - is the birth of museums and exhibitions about "ancient" computer science, referring to facts that occurred no more than 50 years ago. The italian computer science virtual museum collects personalities, facts and events of relevance related to the development of Computer Science in Italy during the past century. The museum can be visited online, as a traditional web site, but provides also a three-dimensional virtual reality interface, that can allow people connecting from different parts of the Internet to visit it together. In addition, it contains simple interactive games that illustrate basic principles of computer science, and gives and effective demonstration of how data was handled at the time, presenting in 'real size' a working model of a mainframe of the time. All of this in a sci-fi setting, in which visitors can fly and jump from floating platform to floating platform, in a scenario set in a starry space sky.

This paper will describe the features of the 3D portion of the virtual museum, and the way it enhances navigation inside the museum contents, providing additional fun and entertainment.

THE VIRTUAL MUSEUM OF COMPUTER SCIENCE IN ITALY

The Virtual Museum of Computer Science in Italy started from a project which consisted in gathering documents, material, and facts about the early days of Computer Science in Italy. The results of the research have been collected in a CD-ROM, containing pictures of the first machinery and computing devices of the 50s-70s in use in industries and academies. Along with information about how these devices were operating at the time, and which were the tasks they were used for, the CD provided interviews, both in textual and in video format, to the personalities which contributed most to the development of the field in Italy. In the interviews, curiosities and facts are put in evidence, linking to related information and pictures.

This material has been converted to an on-line, web-based virtual museum, augmented with an entertaining and innovative 3D interface. The 3D interface allows the users to navigate in the museum in a collaborative fashion, because all people connected to the site can see each other, talk to each other, and go together to look for information. Moreover, this interface offers some simple games, that can be played together with other visitors, and that are inspired to the contents of the museum. In this short paper we will describe the features that the 3D part of the museum offers, and the criteria with which we built this innovative interface.

The technology that was used, WebTalk-I, was developed internally at the Politecnico di Milano, and was effectively used to develop other examples of collaborative access interfaces to Virtual Museums of different kinds.

STRUCTURE OF THE 3D INTERFACE

When entering the world, every user can choose its avatar, that is the shape he will assume inside the 3D Virtual Museum, to be seen by other visitors. There are different kind of robots or human being shapes that the virtual visitor can pick. The 3D museum is organized in pods, which are floating in space, against a starry black sky, which sets the virtual museum in a science-fiction, Star-Wars like environment.

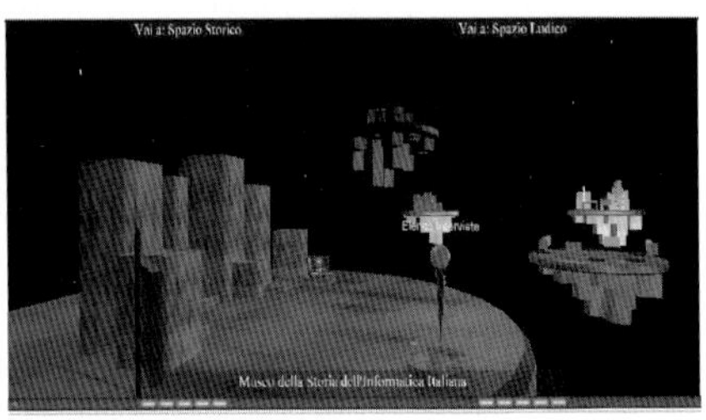

Fig. 1 - A breath-taking view of the floating pods in the starry space. The pods host different 3D objects and information related to the Computer Science museum.

Each floating pod contains a collaborative game, or a collection of items referring to a particular topic of the museum. From the 3D pods it is possible to open the regular web pages and visit the virtual museum web site, with pictures, textual information, and streaming video interviews.

All the pods share the same basic structure:

- an informational billboard, that can be clicked to get help about the game on the pod, or to browse the web page presenting the related information
- a landmark, that is a 3D icon which allows users to distinguish easily one pod from another
- 3D objects that can be examined, or with which users can play, to learn more about early computer science in Italy.

When the users first enter the Virtual Museum, they start from the JumpStation Pod. The JumpStation is the main hub of the museum, containing all teleporters to jump to every other pod. Three teleporters - gathered in the Playground Area of the museum - lead to gaming pods, the other three teleporters lead to the Information Access Area, which provides access to the traditional museum website. By entering one of the teleporters, the users can explore the pods by themselves, or they can decide to follow each other in space to play together, or to go together to a particular page of the museum website.

The most common scenario is that users follow a virtual guide, an expert of the museum that leads other users to the discover of the games and of the contents of the museum, explaining, teaching and discussing via the chat window available under the 3D representation.

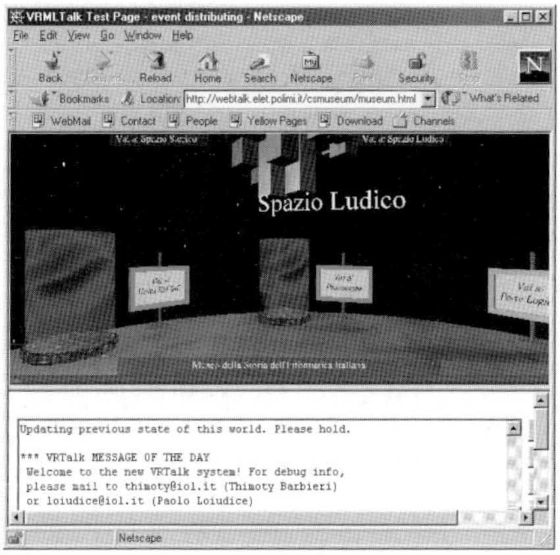

Fig. 2 - The JumpStation Pod provides the users with teleporters to quickly move from one part to the other of the museum. Users can interact via the chat window provided under the 3D view. The system works in a regular browser.

The gaming pods provide a Logic Circuit quiz in which users learn what are logic ports and boolean logic, playing together with input switches to AND, OR and NOT ports. In another pod, there is a simulation of a CPU interacting with registers and a ALU. All these fundamentals notions are explained via direct interaction with the game, and by information accessible from the 3D billboard.

Another pod provides a reconstruction in real size of a mainframe, and a huge stack of punched cards - the equivalent of a 1.44 MB diskette - to convey to the user the feeling of the dimensions of these devices at the time. The computer can be operated to read the punched cards in, and provides a link to the website page describing it. Via the information access pods it is possible to go to an interview, or to particular device or machinery.

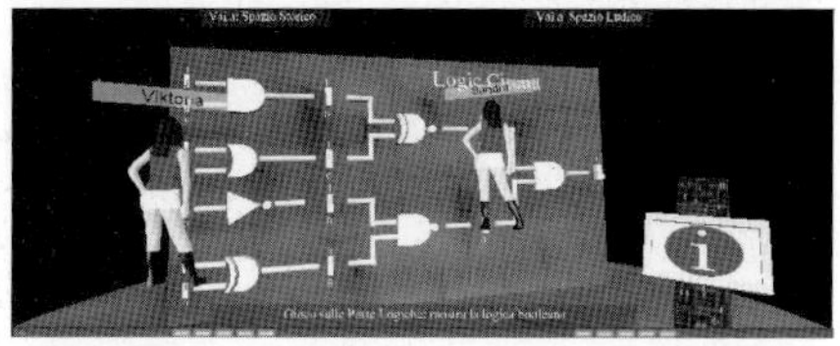

Fig. 3 - Three users (two in view, the third looking at them) are playing together on the logic circuit pod

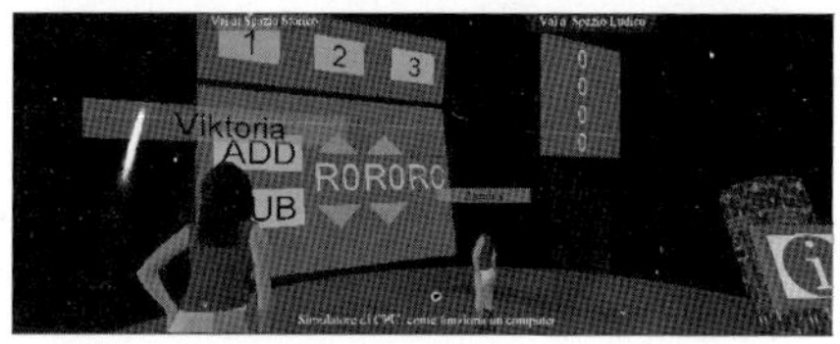

Fig. 4 - The users are exploring the CPU/ALU game

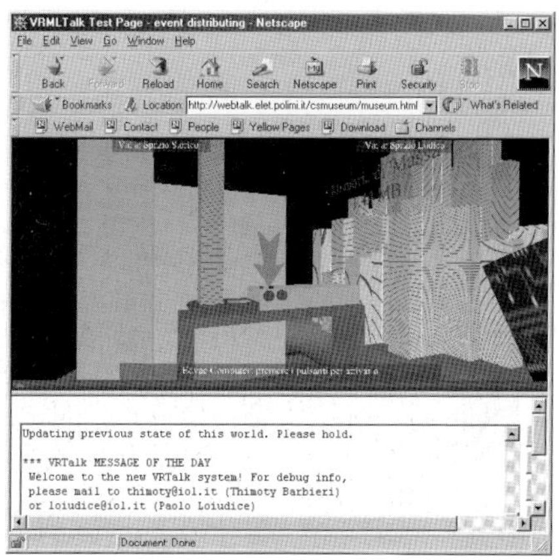

Fig. 5 -The big EDVAC unit is reading the huge stack (equivalent to 1.44MB floppy) of punched cards

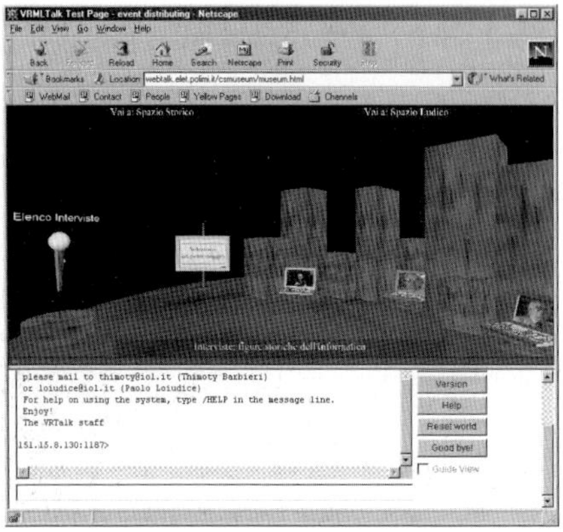

Fig. 6 - The pod providing collaborative access to the interviews. Each laptop links to an interview. The pod has also a landmark (the microphone) to allow the users to easily distinguish this pod from the others.

PREPARING FOR A NEW TECHNOLOGY

The Computer Science Virtual Museum is basically an experiment on how to organize complex contents for collaborative exploration, using 3D interfaces. The museum itself is not online, due to the instability of the technology supporting it. Soon a new framework, WebTalk-II, already in an advanced status of development, will be distributed to the public to address this issue. There is however another project similar to this one, Leonardo Virtuale, which was made available to the public on the web site of the Italian National Science Museum, and that gave rise to interesting results in terms of response from the general public.

While the WebTalk-II framework is being finished in our R&D Labs, the team continues in its experiments around designing and trying collaborative 3D access interfaces, in order to understand better how do they work and which is the best way to tackle the problem at the design stage. A lot of work is being put into this subject at the moment.

CONCLUSIONS

In our constant research about how to provide ways to collaborate while accessing information via the Internet, we are getting more and more convinced that museums provide a formidable application for 3D collaborative access structures. This Computer Science museum, with its innovative access systems, is more entertaining, more engaging, and more fun to visit than its CD or conventional website counterpart.

As in the near future we will continue developing our technology (applying it not only to museums or cultural heritage related applications), we will make it stabler and easier to use. This way of navigating and interacting with others may become common practice in on-line cultural heritage applications, where the complexity and the richness of information calls for ways of guiding each other to the discover of the contents which are provided.

REFERENCES

1. WebTalk-I Web Site, available at *http://webtalk.elet.polimi.it*

2. The Italian Computer Science Virtual Museum - *http://webtalk.elet.polimi.it/csmuseum/museum.html*

3. Virtual Leonardo at The Italian Science Museum - *http://www.museoscienza.org*

4. Barbieri T., Paolini P. – *WebTalk: a 3D Collaborative Environment to Access the Web* in Proceedings EUROGRAPHICS '99, Short Papers pg. 111-113, September 99

5. Paolini P., Barbieri T., et al. – *Visiting a Museum Together: how to share a visit to a virtual world*, in Proceedings Museums&Web '99, New Orleans (USA), March 99, pg. 27-35

6. Barbieri T., Paolini P. – *Cooperative Visits to WWW Museum Sites a Year Later: Evaluating the effect*, in Proceeding Museums&Web2000, Minneapolis (USA), April 2000

7. Barbieri T. – *Networked Virtual Environments for the Web: The WebTalk-I and WebTalk-II Architectures*, in Proceedings IEEE for Computer Multimedia & Expo 2000 (ICME), New York, USA, July 2000

8. Barbieri T., Paolini P. - *Cooperation Metaphors for Virtual Museums*, in proceedings Museum&Web 2001, Seattle (USA), March 2001

9. C. Gutwin, S. Greenberg, *A Framework of Awareness for Small Groups in Shared-Workspace Groupware*. Technical Report 99-1, Department of Computer Science, University of Saskatchewan, Canada, 1999

10. Bridges, H.A. and Charitos, D., *The architectural design in virtual environments*, R. Junge (ed) CAAD Futures'97, Kluwer Academic Publishers, Dordrecht, 1997

11. S. Benford, D. Snowdon, et al. *Visualising and Populating the Web: Collaborative Virtual Environments for Browsing, Searching and Inhabiting Workspace*, in proceedings JENC8

12. C. Cerulli, *Exploiting the Potential of 3D navigable Virtual Exhibition Spaces,* in proceedings Museum&Web 99, New Orleans

13. Charitos, D., Rutherford, P., *Ways of aiding navigation within VRML Worlds,* Proceedings of the 6th EuropIA Conference, Edimburgh, 1997

14. Campbell, D., *Design in Virtual Environments Using Architectural Metaphors: a HIT La Gallery.* March Dissertation, University of Washington, 1996

15. Charitos, D., *Defining existential space in virtual environments*, Proceedings of Virtual Reality Worlds '96, February '96, IDG Magazines, Stuttgart, 1996

16. B. Rohel, et al. *Late Night VRML 2.0 with Java*, Ziff Davis, 1997

ABOUT THE AUTHORS

Thimoty Barbieri is pursuing his PhD in Computer Engineering at Politecnico di Milano. He is active in the Computer Graphics field, and has published several works on 3D collaborative systems.
E-mail: *barbieri@elet.polimi.it*

Franca Garzotto is associate professor at Politecnico dI Milano
E-mail: *garzotto@elet.polimi.it*

Giovanni Beltrami, Luca Ceresoli, Marco Gritti and Daniele Misani are graduating in Computer Engineering at Politecnico di Milano.

Exhibiting in Shared Virtual Environments: Realistic 3D Museum Representation over the Web

Tasos Harisis (*), George Koutalieris (*), Constantinos Ksourias (#), Fotis Karantzis (#), Christos Douskos (*)

(*) Systema Informatics S.A., Athens, Greece
E-mail: {tasos, george.koutalieris, christos.douskos}@systema.gr
(#) Silicon Technologies S.A., Athens, Greece
E-mail: anastilosis@systema.gr

ABSTRACT

The work proposed in this paper is part of the development of an integrated environment for the exhibition of digitised museum artefacts. This environment will enable the realistic representation of the virtualised (i.e. digitised 3-D surrogates of physical objects) exhibits and the Museum's interiors. The system is currently under development and will enable geographically dispersed users to visit this virtual museum at any time inside a shared digital environment where they can meet, associate and interact with each other and the exhibits over the World Wide Web. The system will be the basis of an integrated learning environment and this system aspect is also investigated. Other services like general info and guidance for the visitors will also be implemented. This work is currently funded by the project "ANASTILOSIS" [1] of the Greek Ministry for Development.

SYSTEMA INFORMATICS S.A

KEYWORDS: 3D Virtual, Environments, VRML, Virtual Museums, Digital Object Libraries, Online Communities.

INTRODUCTION

Virtual Archaeology was first proposed by Paul Reilly [2] to refer to the use of 3d computer models of ancient buildings and artefacts. The key concept is "Virtual" and signifies an allusion to a model, a replica or the notion that something can act as a surrogate or replacement for an original. Virtual Reality is being used as a generic word to refer to the growing range of dynamic-interactive visualisation [3],[4]. Virtual Reality is such a hot concept that many people tend to use it even when its use is logically inappropriate. It should be defined as those environments where the human operator is transported into a new interactive environment by means of devices that display signals to the operator. This process is called "User Immersion" and is fundamental to all Virtual Reality systems.

The work presented in this paper, is based on the technological advances in the field of Virtual Environments and the potential of accessing this type of computer interfaces over the Internet. A Virtual Environment (VE) is a computer system, which generates a 3-dimensional (3D) virtual environment, with which the user can interact, in such a way that he receives real time feedback. If multiple users use the VE and they are able to interact with each other the above definition would be extended to multi-user or Shared VE (SVE).

SVE technology has been comprehensively considered in order to support the activities of the Benaki Museum in Athens, Greece. This technology will be used for the enhancement and dissemination of the planned actions related to the opening of the new Islamic Wing of this important Greek museum. Prior to this opening a number of related events have been scheduled and the proposed system will also be used to inform the public as well as provide access to simulations of the actual exhibits to a vast number of visitors over the Web.

MUSLIM STUDIES IN GREECE: THE BENAKI MUSEUM ISLAMIC BRANCH

Even though it plays a broader role, which serves more than one social need, the Benaki Museum is perhaps unique among the complex of Greek museums in having a composite structure. The scholarly study and the cataloguing of its Islamic collections were put in hand with the more particular and immediate objective of promoting Islamic studies in Greece. The primary focus of this scheme is not only to disseminate Islamic artistic creation but also for the Museum to function as a Centre for the Study of Islamic Civilization.

Despite the cultural proximities between Greek and Islamic civilizations, and despite the presence of Muslim minorities across Greece, Islamic studies in Greece remain at a premature stage. The Benaki Museum has organized a series of important events, which will precede the opening of the Museum Islamic branch in order to sensitize the public and provide access both to the actual and the digitally replicated exhibits as well as to historical and other sources of information related with them.

On February 2002 the Benaki Museum, in association with the Metropolitan Museum of Art, New York and the Corning Museum of Glass, Corning New York, is planning a major exhibition that will explore the most significant aspects of the production of glass from Islamic lands and its influence on Europe. The exhibition is entitled 'Glass of the Sultans' and will include over 150 of the most culturally impressive objects in glass, ranging from the seventh to the nineteenth century originating from twenty different countries. Another important event that will be shown at the Benaki Museum from November 2001 to January 2002 and is closely related to it in subject matter and historical sequence is the exhibition "Ancient Glass from the Holy Land", organised by the Israel Antiquities Authorities. This exhibition will trace the origin and development of glassmaking in the periods immediately preceding the Islamic era.

E-ISLAM: A VIRTUALISED SPACE

Two major considerations govern the e-Islam project in regard to the final system's enforcement. The first is the use of e-Islam environment for disseminating the Museum's Islamic collection to the public, notably actual museum visitors and virtual visitors of the e-Islam web environment. The second major consideration is the use of the e-Islam environment as an educational tool, for the study of the Islamic culture. Both of these issues are the conceptual bedrock for the design considerations and the derived system architecture. Key system features and components are described in following.

The Digital Objects Library

Central to the e-Islam environment will be the Digital Objects Library (**DOL**) system component. This will hold not only the 3D digitised exhibits but also metadata and other linked information that fully qualify and interpret the archived exhibits. The **DOL** system will be the core system both for the presentation of the e-Islam artefacts and the related educational material. Certain rules will be embedded within this system and different user profiles will be documented and supported. The **DOL** will structure the data information according to standardised information architecture.

The purpose of the information architecture is to represent the riches and variety of library information, using the building blocks of the **DOL** system. From a technical view, the digital library is built up from simple components, notably **digital objects** [5], [6]. A digital object is a way of structuring information in digital form, some of which may be **metadata**, and includes a unique identifier, called a **handle**. However, the information in the digital library is far from simple. A single work may have many parts, a complex internal structure, and one or more arbitrary relationships to other works. To represent the complexity of information in the digital library, several digital objects may be grouped together. This is called a **set of digital objects** [7]. All digital objects have the same basic form, but the structure of a set of digital objects depends upon the information it represents.

The different types of material in a digital library, information can be divided into **categories**, e.g.: text with XML mark-up, VRML objects, supporting software tools, or digitised multimedia material. Within each category, rules and conventions describe how to organize the information as sets of digital objects. For example, specific rules will describe how to represent a VRML quality digitised exhibit or a higher quality textured version of the same exhibit in Cult3D format. For each category, the rules describe the digital objects that are used to represent material in the library, how each is represented, how they are grouped as a set of digital objects, the internal structure of each digital object, the associated metadata, and the conventions for naming the digital objects.

A user interface that is aware of the rules and conventions applying to certain categories of information is able to interpret the structure of the set of digital objects. Complex information can be presented without the user having any knowledge of the complexity. Since the user interface recognizes how material is represented, it can provide unsophisticated users with flexible access to rich and complicated information [8].

For the implementation of the **DOL** subsystem certain relevant standards will be followed in regard to the metadata that will be used for the annotation and management of the digital objects. The Consortium for the Interchange of Museum Information (CIMI) [9] has made some important advancements in the field of standards for museum resource description. Phase 1 of the CIMI Dublin Core Testbed Project [10] was undertaken in 1998 with the goal of testing "assumptions related to the flexibility and simplicity of the Dublin Core element set, and its suitability and readiness for deployment". Seventeen CIMI member organizations worked to create object-level descriptions using the Dublin Core standard, and identified issues surrounding the functionality of Dublin Core for resource discovery on the Internet. One of the problematic topics that was raised as a result of this project was "characterizing resources as either item-level or collection-level -- i.e., determining the unit of analysis for description such as with an exhibition, a collage of photographs, or other aggregated objects" (Section 5.2 Issues). Phase 2 of the Dublin Core Testbed Project [11], which began in 1999, includes the publication of a "Guide to Best Practice" for museums using Dublin Core, and an "examination of Resource Description Framework (RDF) as an effective method for enabling interoperability between applications that exchange meta-data [12].

The Educational Aspect

The enhancement of the **DOL** with courseware directly related to the digitised exhibits, defines a concrete educational target for the e-Islam environment. The courses provided will be in accordance to the Greek Muslim Studies Curriculum and could be expandable with additional courses and supporting material in the future. In this way the e-Islam environment will be the basis of a unified learning environment where the learning experience will exceed the domain knowledge of the Museum's Islamic collection and will be used for the needs that are directly implied by the Greek Muslim Studies Curriculum.

In order for this educational dimension of the system to succeed, certain learning scenarios will be developed in cooperation with educators and cognition specialists. These learning scenarios will be developed early on in order to take full profit of the e-Islam infrastructure, approach and unique content. These will also form the basis for the educational subsystem to build upon.

A special purpose subsystem - the Courseware Manager Subsystem (CMS)- will be developed inside the e-Islam environment for the presentation and management of the embedded courseware. CMS will be directly linked to the **DOL** and will use the profile information to present the users with personalised views of the available data. At this early stage CMS will be the sole subsystem for the support of learning inside the e-Islam environment, however extensions to this subsystem are already planned in order to fully support the educational procedures thus leading us eventually to a complete Learning Management System that will fully support the Centre for the Study of Islamic Civilisation. The overall architecture indicating the entry points and different presentation layers of the e-Islam environment is depicted in Figure 1.

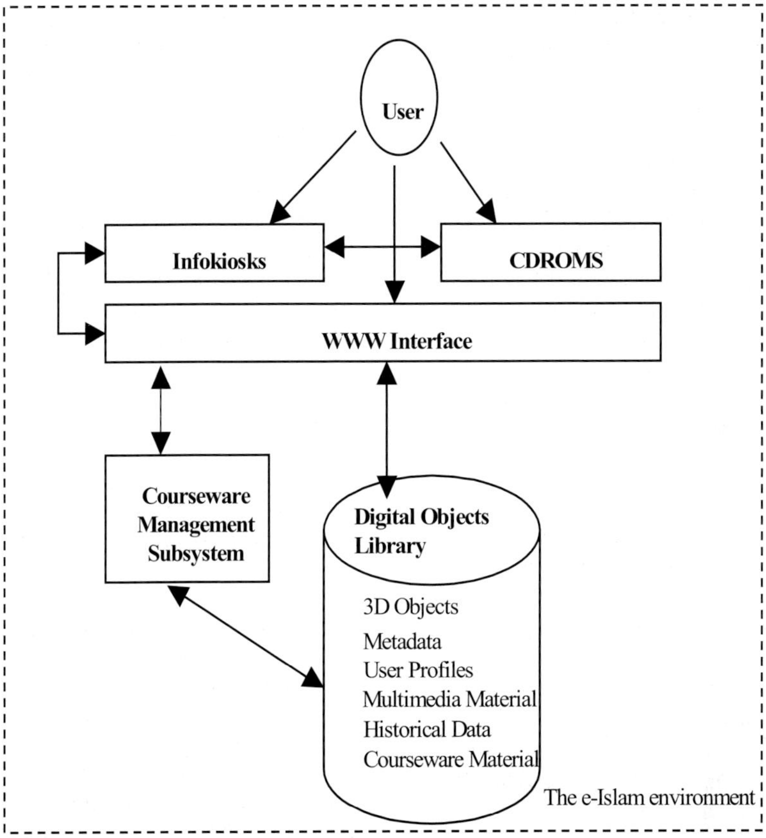

Figure 1: The e-Islam entry points

CLUSTERING OF 3D OBJECTS

The integration of visualisation technologies into a new advanced digital library model that could support development and access to cultural resources raises a number of technical and non-technical issues. Identifying current trends in what the research community is working on as well as to what new technologies have to offer for the digital library, is essential to any development effort in this field.

Technical considerations for the collection and final selection of Islamic art objects according to clustering disciplines include monitoring of strategic areas for research and development at European and international level, deliberation of key technological challenges and priorities in the immediate and longer term and close examination of applications and service areas that will support a continuing leading role in the digital environment for cultural and memory institutions.

The underlying technological challenges have, therefore, to deal with:

- multilinguality of the content and in making it available to and presenting it to users;
- interoperability and comparability between sources;
- the diversity of the content over time;
- ensuring developments are user-centred, based around emerging and distinct communities of users
- quality of service and quality of information, including branding.

Moving on to the non-technical considerations, a number of important questions are emerged in the context of digital libraries for memory institutions, most notably including:

- What can add value in context of the past and the digital future?
- What do people (users) do with new technologies and how to capture that?
- How to address the booming online communities?

As far as copyright issues are concerned, heavily protected data can have implications in interactivity and performance. These implications are not beneficial to either content owners or to its users. Tension between copyright and business models can also be a burden to creativity. The issue is to find the right balances between protection of rights, free use and commercial exploitation.

The considerations described above, define a framework of rules according to which the Islamic art objects will be clustered (selected and deployed) within the e-Islam environment. The aspect of digital objects being used as learning material with supplementary courseware within the same environment, adds further significance to the clustering and selection of the exhibits to be included [13].

DIGITISATION TECHNOLOGIES AND TECHNIQUES

There are currently a number of available technologies that allow the viewing and interaction of 3D objects on the Internet. As far as the E-Islam environment is concerned, two technologies are going to be utilized, namely VRML and CULT3D.

VRML (Virtual Reality Modeling Language) provides the technology that integrates three dimensions, two dimensions, text, and multimedia into a coherent model for the World Wide Web. When these media types are combined with scripting languages and Internet capabilities, an entirely new genre of interactive applications are possible [14].

Cult3D is an interactive 3D solution that allows designers to easily build and display high-quality interactive 3D graphics for Web sites. Cult3D is a multi-platform 3D object-rendering engine. It is 100% software. No additional hardware-such as 3D accelerator cards-is required [15].

Data Capture

The digitization of real objects can be achieved using various technological approaches. Different methods are currently available for achieving the digitization of real objects. The most common used methods are either the use of a 3D scanner or the use of software that can generate 3D models from 2D photographs.

Using a scanner has proven to be the most efficient way in order to produce good quality 3D objects from real objects. However, the cost of obtaining a 3D scanner is high and the time and effort required to produce the desired results depends on various parameters like, lighting conditions of the environment, coloring of the scanned object, transparent areas, size of the object,
shape complexity in three dimensions and hidden curvatures. For the E-Islam environment the Minolta VIVID 700 3D laser scanner is to be used [16].

Several software vendors have produced software that allows the creation of 3D models from photographs. For the purpose of the E-Islam environment, two software packages have been tested. Image Modeler from REALVIZ [17] and Softface 3D by EPTRON [18]. Further more the use of Shapesnatcher by Eyetronics [19] will be investigated. A major drawback of this approach is the amount of time required to achieve the desired results.

Online Communities

Blaxxun Interactive's Community Platform is going to be investigated during the development of the E-Islam environment. Blaxxun Community Platform is a client-server architecture that supports VRML-based, interactive networked SVEs. Several applications and multi-user environments have been developed on the Blaxxun platform, and a particular focus is on product promotion. [20] [21]. Other systems that utilize SVEs for the support of online communities are particularly impressive [22].

CONCLUSIONS – FURTHER DEVELOPMENT

In this paper we presented the concepts, considerations and key issues involved in the construction of a user- friendly and widely accessible virtual museum – the e-Islam environment - that will also form the basis for a more generic educational environment. e-Islam will not only replicate physical structures in detail over the Internet but will also provide information and courseware for different types of users. In achieving the desired outcomes, different information technologies have been evaluated and integrated.

ACKNOWLEDGEMENTS

This work was performed by the project team which also included the following additional members: Evaggelos Athanasiou, Katerina Galanopoulou, Alexandra Karanika, (Systema Informatics S.A), Aristeidis Antonakis, Christos Georgiou, Eleni Katsouli, Georgios Mpalas, Giannis Roussopoulos, (Silicon Technolgies S.A.), Giannis Georgakakos, Despoina Mihaloglou (Informap S.A). This project was partly funded by the project "ANASTILOSIS" [1] of the Greek Ministry for Development. The authors would also like to acknowledge the staff at the Benaki Museum, Athens Greece for their invaluable contribution to this effort.

REFERENCES

1. Greek General Secretarial of Research & Technology, PENED 99, " ANASTILOSIS", a collaborative project between Systema Informatics and Silicon Technologies SA, Greece.

2. Reilly P, Rahtz S, (1990), "In Archaeology And The Information Age London", Pp.312-322, Routledge

3. GILLINGS M., (1999), "Virtual Archaeology: Great Discoveries

Brought To Life Through Virtual Reality", Thames And Hudson

4. LLORET, T, (1999), "Multi-Dimensional GIS: Exploratory Approaches To Spatial And Temporal Relationships Within Archaeological Stratigraphyh. Analecta Praehistorica Leidensia", No. 28 (2), Pp. 207-316

5. Kahn R, Wilensky R, "A Framework for Distributed Digital Object Services", May 1995, http://WWW.CNRI.Reston.VA.US/ home/cstr/arch/k-w.html

6. Arms W Y, "Key Concepts in the Architecture of the Digital Library", D-Lib Magazine, July 1995, http://www.dlib.org/dlib/July95/07arms.html

7. Arms W Y, Blanchi C, Edward A, "An Architecture for Information in Digital Libraries Overly Corporation for National Research Initiatives Reston", Virginia {warms, cblanchi, eoverly}@cnri.reston.va.us D-Lib Magazine, February 1997 ISSN 1082-9873

8. Arms W Y, Blanchi C, Edward A, "An Architecture for Information in Digital Libraries Overly Corporation for National Research Initiatives Reston", Virginia {warms, cblanchi, eoverly}@cnri.reston.va.us D-Lib Magazine, February 1997 ISSN 1082-9873

9. Consortium for the Computer Interchange of Museum Information (CIMI), July 26, 2000 *http://www.cimi.org*

10. CIMI Dublin Core Metadata Project Phase 1 Final Report. "Consortium for the Computer Interchange of Museum Information", July 26, 2000 *http://www.cimi.org/documents/meta_phase1_final_report.html*

11. CIMI Dublin Core Metadata Testbed Phase II, July 26, 2000, "Consortium for the Computer Interchange of Museum Information", *http://www.cimi.org/documents/meta_011899_pdII_final.html*

12. Dunn H, "Collection Level Description - the Museum Perspective", D-Lib Magazine September ISSN 1082-98732000, Volume 6 Number 9

13. EUROPEANCOMMISSION, DG INFOSO, "Digital Libraries Research for Access to Cultural and Scientific Resources, Report of the Experts Meeting", Luxembourg, 24 March 2000

14. The WEB3D Consortium, VRML 2.0 Specification, *http://www.web3d.org*

15. Cycore, Cult3D, *http://www.cycore.com*

16. Minolta3D, the Minolta VIVID 700 3D laser scanner, *http://www.minolta3D.com*

17. [17] RealViz, ImageModeler, *http://www.realviz.com*

18. Eptron, SoftFace3D, *http://www.eptron.es*

19. Eyetronics, ShapeSnatcher v.1.1, *http://www.eyetronics.com*

20. Colony Cities, *www.cybertown.com*

21. Blaxxun Interactive, Community Platform, *http://www.blaxxun.com*

22. Singhal S, Zyda M , (1999), “Networked Virtual Environments-Design and Implementation”, Addison Wesley

ABOUT THE AUTHORS

Dr. Anastasios (Tasos) Harissis holds a degree in Electrical Engineering from the National Technical University of Athens, Greece, an MSc in Microelectronics Systems Design and PhD in Electrical Engineering from Brunel University, U.K. Dr. Harissis is the Development Director of Systema Informatics and has more than 15 years of experience in Information Systems' analysis, design and development.
E-mail: *tasos@systema.gr*

George Koutalieris is a Software Engineer graduated from the Computer and Informatics Engineering Department, University of Patras, Greece. He currently holds the position of Information Systems Manager at Systema Informatics S.A. and has participated in various research & development projects in the fields of educational/training computer technologies and networked virtual environments among others.
E-mail: *george.koutalieris@systema.gr*

Constantinos Ksourias holds a degree of Macroeconomics & Statistics from the University of Piraeus. Mr. Ksourias has over 14 years of experience in the IT market and in particular in issues involving
management, sales and networking.
E-mail: *systema@systema.gr*

Fotis Katrantzis holds a Diploma in Electrical Engineering-National Technical University of Athens (1985). His specializations include Industrial Marketing,
GIS applications, and 3Dmodeling.
E-mail: *systema@systema.gr*

Christos Douskos is a post-graduate in analytical philosophy (specialising in cognitive fields), has an interest in the subject of man-machine interaction, and is currently working as an educational multimedia developer
for Systema Informatics.
E-mail: *christos.douskos@systema.gr*

Virtual Walking Tour of Ryoanji Temple using Locomotion Interface

Tsutomu Miyasato[(*)]

[(*)]ATR Media Integration & Communications Research Laboratories, Kyoto, Japan
E-mail: miyasato@mic.atr.co.jp

ABSTRACT
This paper describes an application of VR (Virtual Reality) system that combines locomotion interfaces which we have developed up to now from the viewpoint of communications that allow both shared experiences and shared emotions. In the context of communication, in order to mutually understand the emotions and the excitement of that moment, it is not enough to pass on information only through words or images -- the environment surrounding that experience and the physical sensations are also important. To accomplish this, we have been developing "Locomotion Interface," as a kind of communication equipment that can give the user the feeling that they are actually walking and talking with a conversation partner who is in fact in a remote location. We have recreated Japanese Ryoanji Temple on a computer through VR technologies, and, using equipment that simulates the sensation of walking, have made it possible for users to experience a realistic "walking tour" of the historical and valuable temple built 550 years ago.

KEYWORDS: VR, locomotion interface, walking simulator

INTRODUCTION

Immersive virtual reality is the sensory experience in which a participant feels him/herself immersed in a computer-generated world, including avatars of other participants and, potentially, subsets of his/her own physical environment. Using current VR technology, many research and development projects are attempting to create high-fidelity simulated real worlds or to enhance the reality of the real world in a virtual world. Most of these projects, however, focus on the "*Visual Reality*" of a simulated world. Humans also have other sensations, i.e., auditory, tactile, somatic and so on. These sensations are a very important part of the whole human sensory mechanism, but current VR technology is insufficient in stimulating them. To cope with this problem, we are developing a VR device that can stimulate these "non-visual" sensations.

Some VRML viewers, called a "walkthrough simulator," allow a user to visually move around in a virtual world on the screen. The user merely controls the direction and speed of the viewing point by using a mouse. We believe it is much more important to actually walk than to change the viewing point using a joystick or a mouse. Using a mouse or joystick cannot give the user the sensation of walking, because this is just like driving a car without realistic feedback for locomotion.

Walking is the most basic way of moving around for humans, and plays an important role in recognizing the environment. Humans can sense complicated impressions about the environment, i.e., the feeling of distance, the feeling of spending time, and the feeling of space, from locomotion. As the saying goes, people won't get lost where they actually walked. This is one of our main reasons for developing the virtual locomotion interface (VLI) in VR technology. VLI is a control technique for allowing a person to move in a natural way over long distances in the virtual environment, while remaining within a limited small space.

VLI can be used for: training, evaluating architectures, communications between people at different locations, and entertainment. We have proposed the locomotion interfaces called ATLAS [1] and GSS [2], which allow a user to get a true feel, even when walking on a remote, uneven ground. As an application of ATLAS and GSS, we created "Virtual Walking Tour of Ryoanji Temple."

LOCOMOTION INTERFACES

ATLAS (ATR Locomotion Interface for Active Self Motion) [1]

In the ATLAS system, an active treadmill is installed on a motion platform with 3-axis rotation, and there is a function that freely tilts the walking surface (Fig. 1). ATLAS can cancel out the forward motion of a person walking via an actively operating treadmill. In order to achieve this canceling-out function, a method is required to detect the walking movements of the user. Because it uses a non-contact, non-restrictive sensor, ATLAS offers the advantage of not interfering with the natural walking motion of the human user. Furthermore, by rotating or tilting the entire treadmill with the motion platform, it is possible to give a sensation of turning, or to generate or recreate the slope of a graded path.

In the ATLAS system, a CCD camera fixed at the front of the treadmill ahead of the user's position records an image of the front user's feet; through image processing, walking speed is detected by obtaining a movement pattern at the front of the feet. An ultra-red filter is fixed to the CCD camera, and an ultra-red floodlight is positioned on approximately the same axis. In this way, the user only needs to affix a reflective marker to the front of his shoes; there is no need to wear or carry any other special equipment or sensors. The available walking area of ATLAS is 145cm (L) x 55cm (W), and the treadmill is 4.0 m/s.

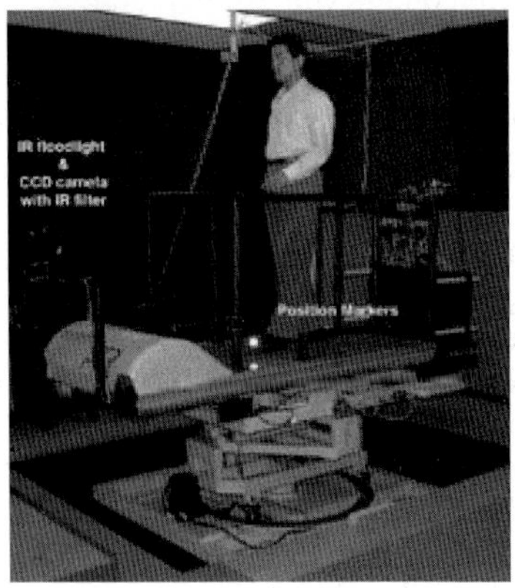

Figure 1: Locomotion interface: ATLAS [1]

GSS (Ground Surface Simulator) [2]

Figure. 2 shows an external view of the GSS. Because the GSS can recreate an uneven walking surface, the user can experience the sensation of walking on uneven terrain in a remote location, or on an uneven surface that has been generated virtually. The GSS has the form of a treadmill, but with the following significant features:

1) Based on walking movements and ground formation

information stored in a computer, the pattern of irregularities on the ground is recreated on the walking surface of the belt.

2) As in the case of ATLAS, the user need only walk normally on the belt, and his position is constantly maintained at the belt's center. This is accomplished with measurements of the user's walking movements and automatic control of the belt speed.

With these two functions, GSS can provide the user with the sensation of walking on uneven terrain.

The size of the walking surface that recreates the irregularities is 150cm (L) x 60cm (W). Underneath the surface of the walking belt, there are six panels with corresponding actuators for the irregularity simulation function. The panels are positioned in strips (25cm wide) lying at right angles to the direction of the belt's movement. The movement of the simulated irregularities is one-dimensional, in the direction of the walking movement, and the maximum height of the irregularities is 15cm.

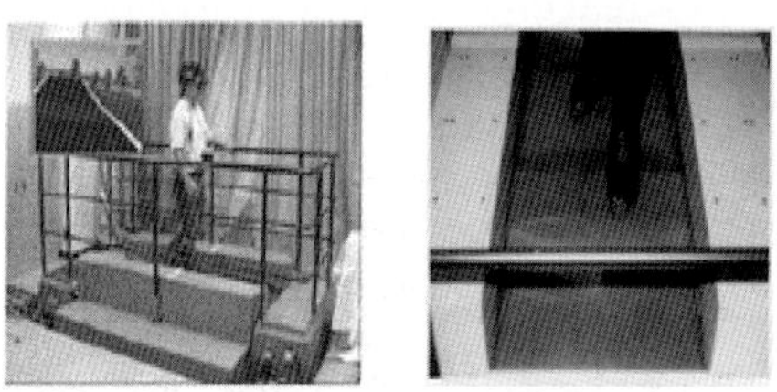

Figure 2: Locomotion interface: GSS [2]

WALKING TOUR OF VIRTUAL RYOANJI TEMPLE USING LOCOMOTION INTERFACE

Ryoanji temple was built in 1450 in Japan, and is most famous for its stone garden called *Sekitei*. *Sekitei* is made from only white sand and fifteen stones, with no trees or grass. Each visitor finds enjoyment in his or her own personal interpretation of the garden. Since people, however, are not allowed to enter the Sekitei, they can not see the room from inside the garden. They can not feel the space of the garden by moving around, either.

In the virtual walking tour of Ryoanji temple using the locomotion interfaces (ATLAS and GSS), the temple recreated by CG is projected on a large screen or on a HMD worn on the head of the user. Visual information alone, however, is insufficient to provide a spatial awareness of the size of the garden of the temple or the positional relationship between objects (stones). We think that actual movements of one's own body is essential to spatial awareness. A walking simulator therefore becomes an important element of the "virtual tour," in order to provide this spatial awareness in virtual space or when experiencing unfamiliar locations.

By using such locomotion interface, unlike visual tours in which the user moves around inside the virtual space by manipulating a control pad, it is easier for the user to actually feel how large Ryoanji temple is. At the same time, the sense of reality is increased, making it seem as though the user was actually in the temple. In addition, by connecting several walking simulators as shown in Figure 3 through telecommunication lines, people in various remote locations can enjoy talking to one another while taking "virtual group tours."

Figure 3: Virtual group tour using ATLAS and GSS

Along with developing the two types of locomotion interfaces described earlier, the authors are also conducting a test production of the mobile equipment, called a Tele-GSS, which becomes the user's counterpart and moves around in a remote location as a reflection of the user's movements. The Tele-GSS is equipped with a sensor that measures the irregularities in the ground; as it moves, it sends information on the measured irregularities to the locomotion interface. In this way, as the changes in the ground are recreated on the walking surface of the GSS belt, the user on the GSS is able to experience the sensation of the irregularities in the surface of a remote location. The Tele-GSS has a TV phone function, one of the goals of which is to create a channel for conversation and at the same time take the place of the person speaking in a remote location to bring out a feeling of actually being there. ATLAS also can control tele-GSS. Figure 4 shows the Tele-GSS controlled by a walker on the ATLAS in a remote location. The walker with an HMD (Head Mounted Display) walks on the ATLAS while watching the image sent from the Tele-GSS. As shown in Figure 5, by remotely controlling the Tele-GSS in the real Ryoanji Temple, a person in a remote location could take a walk with a friend. By sending information regarding the movement of the Tele-GSS to the locomotion interfaces, both the friend in the Tele-GSS side and the walker on GSS or ATLAS can feel the same sensation of moving together in Ryoanji Temple.

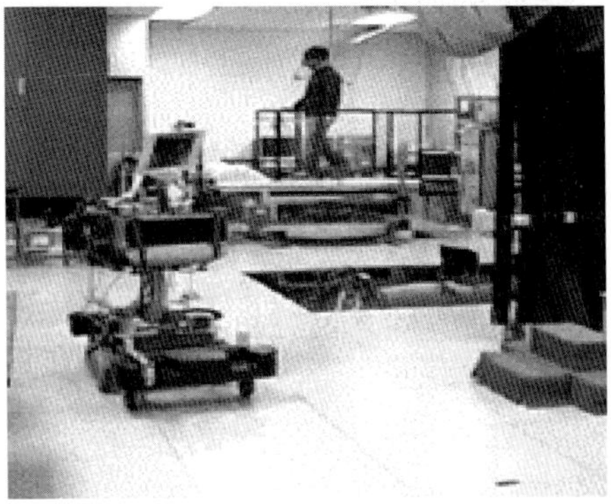

Figure 4: Tele-GSS controlled by a walker on ATLAS

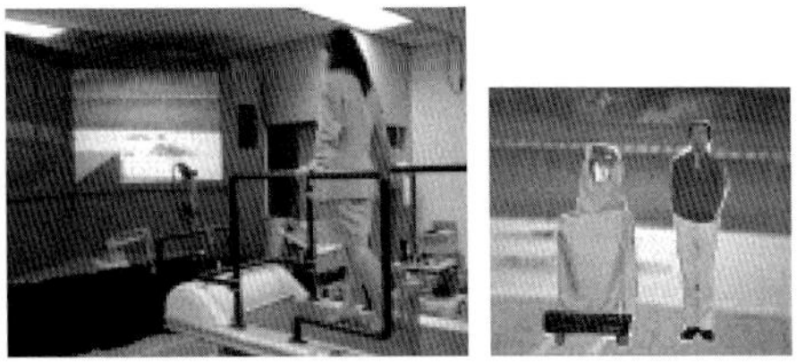

Figure 5: Taking a walk with a friend in a real environment by remotely controlling Tele-GSS

CONCLUSION

In this paper, we presented Virtual-walking system as a part of the research being carried out to creating the optimal environment for communications between people using VR technologies.

The system applies a locomotion interface and remote-controlled moving equipment as means of communication.

The system makes it possible to stimulate dynamic physical sensations, such as the sensation of walking or the sensation of moving and thus provides a realistic environment in which the persons involved can share physical experience.

REFERENCES

1. Noma H. and Miyasato T. Design for Locomotion Interface in a Large Scale Virtual Environment, ATLAS:ATR Locomotion Interface for Active Self Motion. In *ASME-DSC-Vol. 64*, 1998, 111-118.

2. Noma H., Sugihara T., Miyasato T. Development of Ground Surface Simulator for Tel-E-Merge System. In *IEEE VR2000, New Brunswick, USA*, March 18-22, 2000, 217-224.

ABOUT THE AUTHORS

Tsutomu Miyasato received the Ph. D. degree from Tokyo Institute of Technology in 1991. Since 1978, he was with the Research and Development Laboratories of the Kokusai Denshin Denwa (KDD) Co., Ltd., Tokyo, Japan. Since 1993, he is with ATR (Advanced Telecommunications Research). He is currently engaged in virtual reality techniques and human interface in virtual environments. He is a visiting professor of Tokyo Institute of Technology.
E-mail: *miyasato@mic.atr.co.jp* or *T.Miyasato@alum.mit.ed*

Photographic Virtual Reality and Architectural Heritage

Roberto Paiano (*)

(*) SETLAB-Software Engineering and Telemedia Laboratory, University of Lecce, Lecce, Italy

E-mail: rpaiano@ingle01.unile.it

ABSTRACT

Multimedia applications concerning Cultural Heritage represent a great opportunity to spread the knowledge about the history, culture and tourist environment of a specific site. But this kind of applications requires a large effort in term of costs in order to guarantee the high quality needed. This requirement has also a remarkable impact on the technology used to realize the multimedia contents as well as on the lifecycle and the methodology adopted to develop the application software.

The application described in this paper is related to the architectural heritage, which, due to the specifically visual impact of sculpture and architecture, needs a special photographic technique to be displayed in a virtual tour.

We describe, moreover, how, by using a HDM methodology, we developed a general-purpose run-time engine reducing the costs needed to realize new applications.

KEYWORDS: HDM, photographic virtual reality, architectural heritage, virtual tour

INTRODUCTION

In order to develop a multimedia application able to completely present a particular architectural heritage, unlike applications concerning other forms of cultural heritage, the designer must take into account that the traditional multimedia usable units, such as images, sounds, texts and videos, allow the final user to have only a partial outlook of such heritage without appreciating either the harmonious structure of its halls and cloisters or the surrounding natural environment.

Furthermore, there are two metaphor of visit:

- The user is guided through a static visit using image and text;
- The user can look round a building having the sensation to be there.

Both these kinds of visit are effective but in the second way the common visitors are more involved in the tour.

To satisfy these requirements it is important to realize a virtual tour by using photos at 360° edited using tools such as QuickTime VR.

The use of photographic virtual reality is necessary but it is not enough. The visitor needs also an in-depth study of some aspects of the heritage such as paintings and sculptures, as well as the particular stone used to build the architectural heritage and other examples. So, it is important for a designer to model a traditional hypermedia application that provides also a virtual tour.

Hypermedia applications were, at the very beginning, hand-coded pages: the information handled by each page was structured “ad-hoc” and the links among pages were set “manually”. This production method was acceptable when only a few pages had to be produced, linked and maintained. It became rapidly unmanageable when applications of several hundreds or thousands of pages with complex interactive objects had to be considered. In particular, two related problems became rapidly relevant (and still are very relevant): how to ensure the “usability” of modern large hypermedia-applications [1-3], and how to improve the efficiency of the production-maintenance process for the same applications.

In good hypermedia applications, in fact, the reader should be able to effectively exploit the potential information managed by the application: i.e. he/she is able to quickly locate the objects of his/her interest, to understand the inner structure of the objects, to easily move (navigate) from one object to another, to easily accomplish complex access sessions. Several factors concur to the achievement of usability, and one of the most important factors is to have a good structuring of the information objects and a good structuring of the navigation patterns. Several authors and several communities have recently proposed the adoption of design models [4-8], and design patterns [9-12] in order to improve the quality and usability of hypermedia applications, at least for those aspects concerning structure and navigation. Design models provide, in fact, the primitives that allow structuring the information objects and the corresponding navigation patterns along regular and systematic features, improving consistency, predictability (for the user), robustness of the design, and therefore improving usability. The ancestor of these models can be traced to HDM [13] that since its initial definition has undergone a number of changes and improvements. Almost all the other design models that have been proposed for the Web over the years have features similar to those of HDM.

To come to the point, the adoption of a model (like HDM) to design the internal structure and the navigational features of hypermedia applications is desirable for almost three reasons:

- resulting applications are usable;
- the production process can be decomposed in sub-problems easy to manage;
- the application model can be directly “executed” by a suitable “interpreter” to create the application pages in a way that is independent from the specific application.

The remaining part of this paper is divided into: section 2, where we describe the technology used for photographic virtual reality; section 3, where we examine the main concepts concerning the HDM interpreter which SETLAB developed; in section 4, we describe the application to the Olivetani monastery built around the interpreter; in section 5, we draw the conclusions; section 6 is for references and bibliography.

THE PHOTOGRAPHIC VIRTUAL REALITY: TECHNOLOGY ASPECTS

There are several technologies to realize 360° photos to be used in photographic virtual reality and the most known is the IPIX technology [17] that has a complete system.

The IPIX system realizes the full 360°images by capturing two opposing photographs, called "hemispheres," with

a fisheye lens in vertical direction (for example from the top to the bottom); this photos are later linked together by means of an IPIX tool and can be delivered in a Java environment.
This system is cheap and easy to use and we used IPIX for some Web applications describing room interior of buildings.
In fact, this kind of systems isn't suited for external environments where the variation of light could create high contrast between objects. This problem affects also large interior environments when it is impossible to have a uniform light.
The Olivetani monastery and the adjoining church aren't in a situation of uniform light due to the architecture and to the particular stone and sculptures that turn out dark in the sunlight.
For this reason we adopted a different system, more expensive in term of costs and manpower, but that guaranteed high quality results.
The system is based on a traditional camera mounted on a tripod. A steel hemisphere is fixed horizontally on top of it, reflecting the 360° environment. The camera photographs an image, obtained with the use of mirror effects, inside the hemisphere.
Several photos are taken for each node with a different aperture of diaphragm to best capture all the elements of the image. Then, the little spots on the film are digitalized by using a special scanner at 2600 dpi and the results were life-size images. In the editing phase all the photos taken for the node are combined to reach the best results of light for all the elements. In the last step images are linked together with the use of QuickTime VR.
This system was adopted to realize the virtual tour of two important sites in Italy: the Cathedral of Milan and the Basilica of St. Peter in Vatican (Rome). This activity is more expensive than the IPIX solution in term of 1:4 requiring the lights expert, the photographer and the director to tell the story.

THE HDM INTERPRETER

A short introduction to HDM is essential for the comprehension of the following part of the paper.
In brief, for HDM a multimedia application consists in a Hyperbase, an Access Layer and a Visual Layout. The Hyperbase contains the actual information objects and their basic connections; the Access Layer provides access paths to the objects of the Hyperbase, the Visual Layout description is made up of page templates used to create the application.
The Hyperbase consists of Entities grouped in Entity Types. All the entities belonging to the same type have similar structures and similar basic connections (e.g. "Artist" and "Masterpiece" in an application on painters and paintings). In HDM, entities may have an inner structure: in general they consist of nodes arranged in some way (hierarchy is the most commonly used topology). Structural Links are used to "keep together" the different nodes belonging to the same entity type, and also to provide "navigation" across the nodes.
Application Links, grouped in Link Types, are used to connect different entities, belonging to different types or to the same type (such as the link "Has-created" in the painters-paintings example, connecting an "Artist" to one or more "Masterpiece"). The definition of the Entity Types and Link Types is called, in HDM, "Hyperbase Schema in-the-large".
The "Hyperbase Schema in-the-small",

instead, details the inner structure of the nodes, i.e. the smaller structuring units permitted for entities, and the inner structure of links. In brief "Slots" are the smallest (atomic) pieces of information defined in-the-small, Frame are aggregate of slots and/or other frames. A Frame is associated to a Node and to one or more Link instances. The interested reader is deferred to the bibliography for more details [13,14].
The Access Layer of an application is organized into Collections (e.g. "all the Artists in alphabetical order" or "Artists in the Cubist movement"). A Collection groups together entities (of one or more types) and/or other Collections (nested Collections). A Collection is not simply a set: it must have a Topology, i.e. a way to organize its members (Sequential, Tree and Lattice topologies are the most commonly used), and it must have a Center (with a Frame associated to it), i.e. an additional information structure that helps the user to make an effective use of the Collection itself. The center may be used to "introduce" the collection, to "synthesize" its content, and to reach the collection members. A Collection introduces new connections within the application, called Collection Links. Collection Links are of two different categories: Index Links and Guided Tour (or Topology) Links. Index Links connect the center of the collection to its members and vice versa; Guided Tour Links connect the members together, according to the topology of the collection. "NEXT" and "PREVIOUS", for example, are typical navigation primitives for Sequential Guided Tours. Again the reader is deferred to the bibliography for more details.
According to the previous description a database can be used to store the application components described by the model, then a run-time engine can be used to extract that components from the database and to display it to the reader as an interactive multimedia presentation.

THE APPLICATION: THE OLIVETANI MONASTERY IN LECCE

Using the same run-time engine already used for other hypermedia applications [18] we have developed an application to present the Olivetani Monastery, now seat of the Department of Cultural Heritage of the University of Lecce.
The monastery was built at the end of XII century by the Norman Count Tancredi di Altavilla according to the style of the great coeval buildings realized in Sicily by the Normans. During the time, in the XVII and XVIII centuries, the monastery was enriched with the typical elements of the baroque of Lecce, famous in the world for the embroideries of its sculptures, which exploit the properties of the local stone particularly soft and easy to work.
The monastery combines some XII century-style elements, such as the dome similar to the one of the Holy Sepulchre in Jerusalem, with baroque elements in a splendid harmonious building.
The application provides also some monographs meant to present the Salento region in terms of culture, history and natural heritage so as to attract a high level tourism.
According with HDM the design model of the application is shown in fig.1.

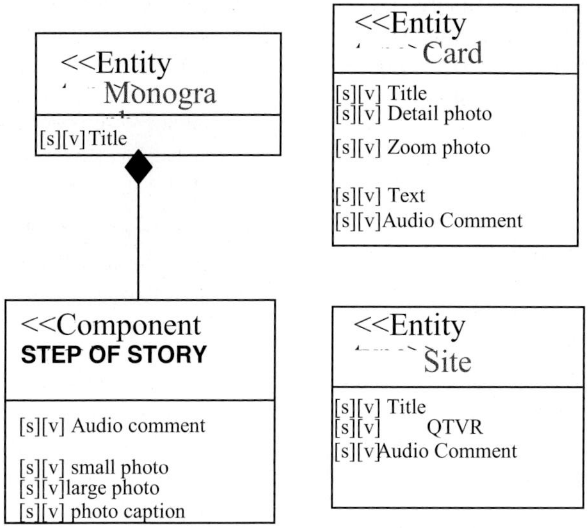

Figure 1:The Olivetani Hyperbase

The entity type Monograph is used to study in depth arguments related to architectural elements, to the history and to the symbolism. Each monograph is composed by steps, each of one supported by an image and audio comment. The monograph can be appreciated in two ways: as an automatic or a manual guided tour. The instances are: the Monastery from its origin to present time, the Tancredi's period, the Olivetani Order and its history in Lecce, Water as a source of spiritual life and the Baroque.
The entity type Card allows the description of the architectural details of the monastery by using about 30 instances.
The entity type Site is used to realize the photographic virtual tour in association with an audio comment.
Fig.2 shows the semantic links that represents the navigation capability provided to users. The hyperbase of these applications is very simple but there are more complex applications where the navigation links must be designed carefully.

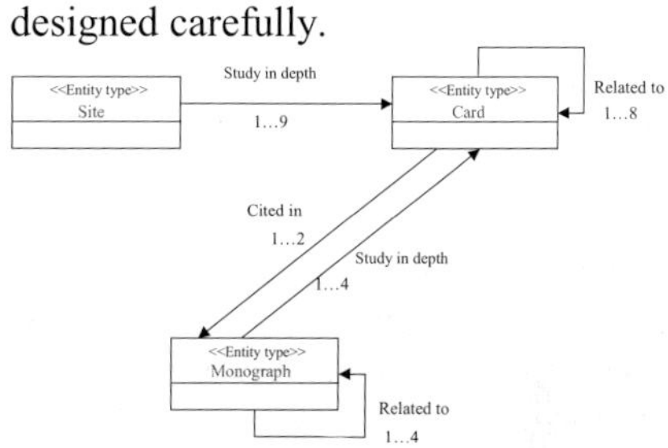

Figure 2: Navigation Links

The structure of the hyperbase with its entity types and the navigation links, according to HDM, allow the designer to develop a run-time engine independently from the specific application.
So, the designer can devote himself to designing the hyperbase in the large (definition of entity types and links) and in the small (definition of multimedia unit for each entity types named slots) without worrying about the software development.
The application design is completed by

the definition of Access Structure, i.e. the best way to allow the user to enjoy the application information. (Monograph collection, Virtual Tour, Cards Index, Site collection..)
Fig.3 shows the Home Page of the application, in Fig.4 we see a node of the Virtual Tour with the position on the map in the left side and in Fig.5 we show the monograph about Monastery from its origins to present time with its component in the left side.

CONCLUSIONS

The development of high quality multimedia applications requires a large effort in term of costs and work mainly related to the realization of contents and to the software approach.
The software costs are strongly reduced by the construction of a reusable engine for hypermedia applications based on the HDM model that is the logical extension to the model-based approach. The reusable engine, in fact, is able to implement hypermedia titles in an independent way from the specific application. The generality of the approach is based on the ability of HDM to describe a wide range of complex hypermedia applications. So, it is possible to concentrate the efforts on the development of high quality multimedia contents, which are, together with the usability granted by the methodology, the key to success of an application aimed at spreading the Cultural Heritage.

ACKNOWLEDGMENTS

The author wishes to thank IMT Company and SETLAB people.

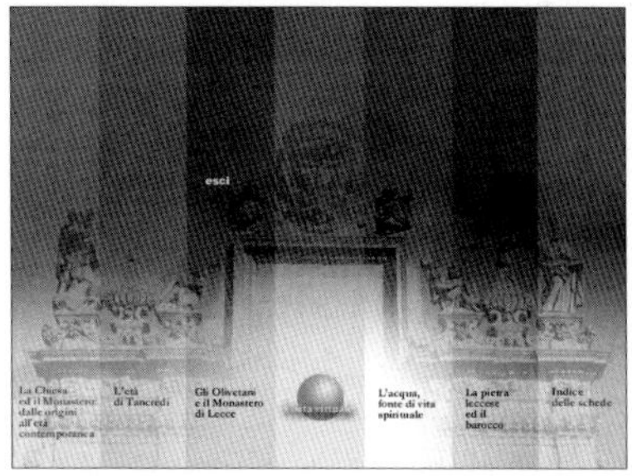

Figure 3: The Olivetani Home Page

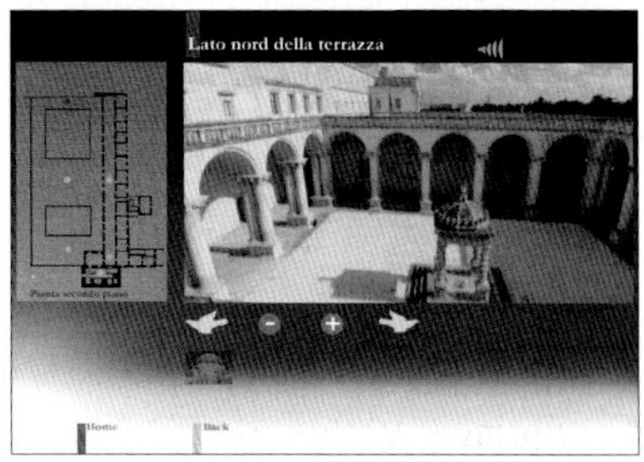

Figure 4: The Olivetani Virtual Tour

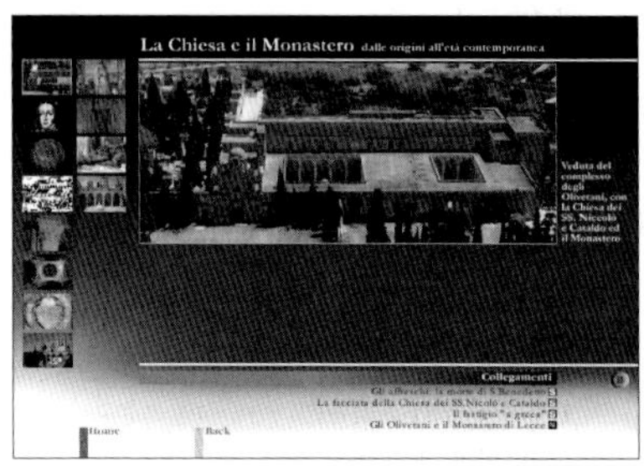

Figure 5: The Olivetani Monograph

REFERENCES

1. F.Garzotto, M.Matera, "A Systematic Method for Hypermedia Usability Inspection", In *The New Review of Hypermedia and Multimedia,* 3, 1997

2. J, Nielsen,R.L.Mack, "Usability Inspection Methods", John Wiley & Sons, New York 1994.

3. F.Garzotto, M.Matera and P.Paolini, "Abstract Tasks for Hypermedia Usability Evaluation", Dept. of Electronics and Information, Polytechnic of Milan, Technical Report No.03-99, 1999.

4. F.Garzotto, L.Mainetti and P.Paolini, "Hypermedia Application Design: A Structured Approach", In J.W.Schuler, N.Hannemann and N.Streitz Eds., "Designing User Interfaces for Hypermedia", Springer Verlag, 1995.

5. D.Schwabe, G.Rossi, The Object-Oriented Hypermedia Design Model, Communications of the ACM, Vol. 38, N. 8, pp.45-46, Aug. 1995

6. T.Isakowitz, E.A.Stohr, P.Balasubramanian: "RMM: a methodology for structured hypermedia design", Communications of the ACM Vol.38, No.8, Aug. 1995, pp. 33-44.

7. G.Mecca, P.Atzeni, A.Masci, P.Merialdo, G.Sindoni: "The Araneus Web-Base Management System". SIGMOD Conference 1998. Pp. 544-546.

8. Fraternali P., Paolini P.-" The Autoweb System"; submitted for publication

9. E. Gamma, R. Helm, R. Johnson and J. Vlissides. Design Patterns. Elements of Reusable Object-Oriented Software. Addison-Wesley, 1996

10. F. Garzotto, P.Paolini, D. Bolchini, S. Valenti "Modeling-by-patterns of web applications "In World-Wide Web and Conceptual Modeling - Proc. the international workshop WWWCM'99 - to be held in conjunction with the 18th International Conference on Conceptual Modeling (ER'99), Paris, France, Stephen W. Liddle (ed.), Springer Verlag - to appear

11. M. Nanard, J. Nanard and P. Kahn. "Pushing Reuse in Hypermedia Design: Golden Rules, Design Patterns and Constructive Templates". In Proc. of the ACM International Conference on Hypertext '98, ACM Press, 1998, pp. 11-20.

12. G. Rossi, D. Schwabe and F. Lyardet, "Improving Web Information Systems with Design Patterns". In Proc. of the 8th International World Wide Web Conference, Toronto (CA), May 1999, Elsevier Science, 1999

13. F.Garzotto, P.Paolini and D.Schwabe, "HDM - A Model Based Approach to Hypermedia Application Design", ACM Transactions on Information Systems, 11, 1 (Jan. 1993), 1-26

14. Franca Garzotto, Luca Mainetti, Paolo Paolini, "Adding Multimedia Collections to the Dexter Model". ECHT 1994: 70-80

15. M. A. Bochicchio,R. Paiano, P. Paolini " JWeb: an HDM Environment for fast development of Web Applications" - ", Proceedings of Multimedia Computing and Systems 1999 (IEEE ICMCS '99), Vol.2 pp.809-813

16. M. A. Bochicchio, P. Paolini, "An HDM Interpreter for On-Line Tutorials", Proceedings of MultiMedia Modeling 1998 (MMM '98), pp.184-190, Ed. N. Magnenat-Thalmann and D. Thalman, IEEE Computer Society, Los Alamitos, Ca, USA, 1998.

17. www.ipix-eu.com

18. M. A. Bochicchio,R. Paiano, " Software Reuse in Hypermedia Applications" – Procedings of IRMA 2000 - Multimedia Computing Track - May 2000 Anchorage - Alaska - USA , pp.826-829.

ABOUT THE AUTHOR

Roberto Paiano received the DrEng degree in electronic engineering from the University of Bologna, Italy. He worked in IBM Italy until the 1996. Since the 1997 he has been at the Department of Engineering, University of Lecce, where he's assistant professor. His research interests include the methodology of Web application design and distance learning tools. He is member of the IEEE and the IEEE Computer Society.
E-mail: *roberto.paiano@unile.it*

Virtual Historical Heritage: Yuso Monastery

María Aróstegui[*], *José Ignacio Barbero*[*], *Carolina Cruz Neira*[#]

[*] Labein, Mechanics Unit, Bilbao, Vizcaya. Spain.
E-mail: arostegui@labein.es
E-mail: jose@labein.es
[#] VRAC; Virtual Reality Application Center, Iowa,USA.
E-mail: carolina@vrac.iastate.edu

ABSTRACT
Due to the maturity acquired by the new technologies, the research work developed within the project of the Virtual Reconstruction of the monastery of San Millán de la Cogolla, is focused on the use of VR techniques in very different environments, such as historical documentation or artistic purposes.

This project consists of the development of a VR application with spreading aims. The research work about cultural innovation is one of the main goals.

KEYWORDS: Virtual historical heritage, cultural innovation.

INTRODUCTION

The computer graphics application that recreates the monastery of Yuso, is born as the collaboration of two research centers: Foundation LABEIN (Bilbao, Spain), and the North American Research Center VRAC (Virtual Reality Applications Center, in ISU, Iowa State University, USA).

The aim of this research work developed within the project of the virtual reconstruction of the monastery, is the deepening in using tools of virtual reality and some photo-realistic techniques. Other clear objective is using these technologies on historical documentation, or on industrial, and artistic environments.

On the other hand, one of the main characteristics of the project of San Millán is the easy portability of the computer graphics application. What is more, due to the conceptualization of the application like a class (using VRJuggler), or in any case like a code programmed with an oriented objects philosophy, the application can be used with all kind of input/output devices, being this way useful with most of the commercial devices. So is that, it can be used with trackers, joysticks, mice and/or keyboard (like input devices), and with DIGS / SSVR systems or Head Mounted Displays (like output devices).

In this paper several topics will be studied; in the first place, the developed tasks along the different stages of the project, will be analyzed. Besides, future implementations will be valued. And to sum up, some of the consequences and the most immediate conclusions derived of the carried out work, will be exposed.

STAGES OF THE PROJECT OF SAN MILLÁN.

First of all, it would be convenient defining what is meant by Virtual Reality. One of the best definitions comes from the "The Silicon Mirage: The art and Science of Virtual Reality"[1].

"Virtual Reality is a way for humans to visualize, manipulate and interact with computers extremely complex data"

The **visualization** part concerns the virtual world generating, so is that, visual, auditory and other sense outputs within the virtual world. The **programming** part refers to the way of manipulating objects within the virtual world, moving from the virtual world, or interacting with the virtual world.

This project has had until the moment two different stages, due to the nature and objectives of the project, that coincide with the **visualization** and **programming** parts. These ones are going to be detailed immediately after.

Modeling and Image Treatment Stage

The first stage of the project consisted of the 3D modeling of the architectural complex, as well as textures mapping to obtain a loyal real scale reconstruction of the current monastery of Yuso in San Millán of the Cogolla.

Describing the scene within the virtual world, does not only mean describing graphical primitives and geometric objects, but lights and cameras. The Yuso computer graphics application, use a tree data structure allowing new kinds of nodes, such as attribute-definition nodes or matrix transformation nodes.

The real-time model of the monastery used in is defined in terms not only of geometry, but also of structure, or hierarchy. The Hierarchy is used to organize the components of the database. The Hierarchy is defined by means of nodes, the database items, connected in one or more tree structures.

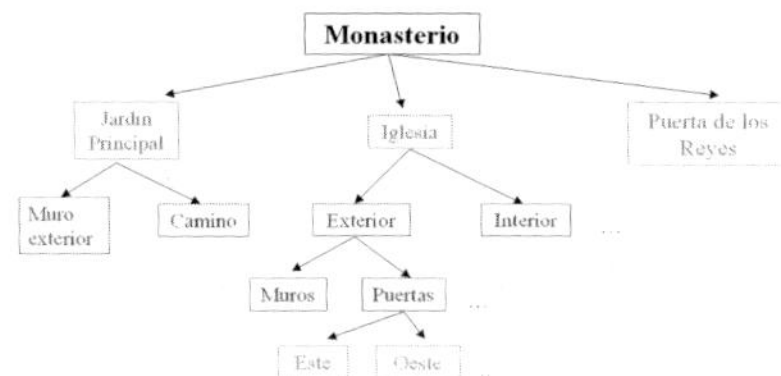

Figure 1- Example of the hierarchical structure of the database.

A scene graph holds the data that defines a virtual world. The scene graph includes low-level definitions of geometrical objects and their appearance, higher-level definitions of spatial information, such as specifying the positions, animations, and matrix transformation of objects, as well as additional application-specific data.

The scene graphs couples perfectly with the object-oriented paradigm supported by languages such as C++. This way we can define classes to manipulate encapsulation to build up scenes of great complexity. We can add animation nodes in the scene graph that provide the needed parameters to change the scene, and even to redisplay it.

Depending on the attributes and the geometry created for each node, they can be classified in two different ways:

(A) Paying attention to the animation properties inherent in the nodes, they can be divided into:

*SCS, Static Coordinate System,

*DCS, Dynamic Coordinate System

(We defined DCS, when animation properties are attached to that node, and SCS in the contrary case.)

(B) Because of the geometry used to defined them, each node might be:

Geometric nodes; Vertexes, Edges, Facets, Objects, Groups, LOD(Level Of Detail),etc.

Attribute nodes; Transformation nodes, Sound beads, Color nodes, etc.

On the other hand, the Yuso monastery database is built using approx.: 67.000 triangles, 513 group nodes, 816 object nodes, and more than 100 level of detail nodes.

Level of Detail are sets of models that represent the same object with varying degrees of complexity. The real-time system selects one of the LODs to display, depending on the distance from the eyepoint to the LOD and the number of polygons the real-time system can process.

If the eyepoint is far away from an object, the object is displayed as a low LOD, containing relatively few polygons. As well the eyepoint moves towards the object, the real-time system replaces it with increasingly complex LODs. LODs play an important part in optimizing a visual scene for drawing, because there is an effective limit on the number of polygons each real-time system can display.

Another point to be taken into account during this stage, is the texture mapping. Textures patterns are bitmapped images that are mapped onto polygons to give the Yuso database a photo-realistic appearance, without increasing the polygon count.

To get a good texture mapping within the Yuso database, the measure of the polygons, the texture size, the wrap method, the image resolution, the texture format, and the texture filters, have been studied. For example mapping a 2.5m x10m wall with an 32x32 pixel rgb file, is not the same as mapping the same wall with a 512x512 pixel rgb file. What is more, we have map the wall pondering the bricks size and the user's size in order to get a more real sense in our virtual world.

A very important point in this stage of the Yuso project was the lighting. The correct study of the lights and shades determined what kind of lights were the most accurate to recreate the monastery ambient. For example, instead of using local lights (Omnidirectional finite distance) or spot lights(Lobular finite distance) in the exterior, we preferred to use infinite lights (Omnidirectional infinite distance), whereas in the interior we used diffuse local lights.

Besides, the choice of the light source color and positioning light sources, were points studied too to get better results.

Finally, a good definition of the database hierarchy is needed when the real-time system is rendering, because the better "inheritance concept" is defined, the quicker the rendering can result. Besides, a not correct dependence relationship between nodes "father" and "children" can give place to wrong animations or movements.

Navigation Code
As it has been advanced in the introduction, the project of the virtual reconstruction of the Monastery of Yuso doesn't only have as objective obtaining a geometric modeling of the architectural complex, but programming and designing some algorithms of collisions detection, and navigation codes to fit our necessities. The study of the possibilities that VR offers to the engineering world, or education among others, is one of our priorities.

Several VR software tools were used to develop the Yuso application. It were the VRJuggler libraries (software library for VR application the Performer Libraries development, wholly developed in VRAC), the VEGA software, and obviously the Iris Performer Libraries.

Therefore, the most important details, obtained during the second stage of the Yuso Project, will be explained depending on the software used in each case.

VRJuggler libraries
With the purpose of using the model of the monastery of Yuso in the C2, the VRJuggler libraries were used.

The C2, or CAVE is a VR wholly immersing system where the virtual world is projected in a 4 sides room. The user has the opportunity to visit virtual environments that would not be possible in other case. By way of example, students of art at the Iowa State University have the opportunity to visit a sample of the Spanish Baroque art: the walls of San Millán. They can visit the chapels, or observe all the little details of the main altar, go for a walk for their cloister, or fly to watch the monastery from angles that in the real world would be completely impossible to reach.

The main characteristics of this VR library, and the Yuso application will be explain bellow.

VRJuggler is a library of classes of C++, entirely developed in VRAC, used as a frame for the development of Virtual Reality application. Its object-oriented system is designed to take advantage of the power and flexibility of C++. Most of the functionality of Juggler has been divided into a handful of distinct and well-defined components called Managers. Each manager encapsulates a specific set of system details. For example, there are several Managers to deal with different windowing systems, and others to handle aspects of drawing with different graphics systems. A single Manager oversees all input devices, while another control output devices. Other Managers control networking and system configuration. The Managers are united and controlled by a small kernel, which brokers all communication in the system.

Defining the Yuso application using VRJuggler, meant defining it like a class, allowing to the developer a bigger encapsulation degree, and control on the same one. The class included functions to draw the frames, to proceed on the necessary calculation for the PRE and the one POST drawing, and to manage the synchronization of the system. Using VRJuggler, each application is programmed like a class independently of the hardware devices that will be used. That is to say, VRJuggler provides common interfaces to different types of input devices, and the applications are not tied to any type of hardware device defined by default. What is more, it provides a dynamic reconfiguration without compiling and linking the code, in order to obtain different versions of an application.

Finally, the 19 of June 2000, the center VRAC inaugurated the C6, a three-dimensional virtual environment wholly immersing that consists of a room with four walls, floor and roof, where stereoscopic images are projected providing a total immersion to the participants. One of the DEMOs shows in the inauguration ceremony was that the Monastery of Yuso application.

Vega libraries

VEGA is a high-performance software environment and toolkit for real-time simulations and virtual reality applications. Among their modules, the APIs and the libraries have been used to get the objective that was sought. VEGA is constructed as a layer on top of Iris Performer.

VEGA is a library of C classes, and the Developer's Application is not a class but an executable.

Developing our own VEGA application meant defining the input file and programming the Yuso code, that is to say:

(A) On one hand, the input of a VEGA application is an ASCII file. It contains information about the geometrical model, texture mapping, the type of windows that are displayed, the number of observers and specifications of the illumination that is being used. But the main purpose of the input file is to initialize, define and configure the system.

(B) Once defined the input file, the properly VR application needs to be programmed. Here, the navigation code, the collision detection algorithms, and the definition of all the events such as sound reproductions or lighting are implemented.

The Yuso application, like many other VR applications, handles lots of information and needs an accurate memory resources management. Only the monastery model is an 8.2 megabyte database, accompanied by 63 megabytes of textures. So the Yuso application uses data structures to store the necessary information to define the user movement, or the impact volumes data.

Once the scene graph is loaded from the input file, and the objects declared, these are rendered into the windows. Each time the user uses the tracker (input device to define the user location into the virtual world), the Yuso application compute that information and calculate the following user's position according to the tracker movement. All those location information has to be stored and compute to get the newest positions.

In order to do that the Yuso application uses some functions to set and get the shared memory arena site.

At the same time the information about the objects the user can collide with, or see in each moment, need to be stored in order to be rendered or not, and to check that the user can not go through the walls, floor or ceilings.

A vital part in the computer graphics applications is the programming of the CALLBACKs, since they allow that the user interact with the virtual objects. In the application of the monastery they determine the kind of movement, the type of collision detection, as well as the actions and events that have to do with the keyboard, trackers, etc.

For example, to define a function that opens doors, and that causes the sound of the door hiss: this door is defined like DCS, a dynamic local coordinate system is fixed within this object, and the limits of the movement associated to its geometry, is defined too

Regarding the kind of navigation or movement , we are trying it with a couple of those ones in order to improve the speed, the quality and the "reality sense". By way of example, one of them is defined bellow:

1. Mouse Movement: Left-right | forward-backwards.

2. Mouse buttons: (Assuming that there are three programmable buttons in such as device) Up, Below and Stop respectively.

3. Keyboard: The yaw movement of the head.

Regarding the detection of collisions, several algorithms and different approaches are being used, depending on what is sought to get. That is to say, defining the objects you collide with, differs of defining those objects (places in the Yuso application), that have attached an event. In both cases some collision detection algorithms are used, but the way to use them and to use the ISECTORS and CALLBACKs are diverse. ISECTORES and VOLUMES are used to organize the structure of detection of collisions.

The VR application of the monastery of Yuso is configured to support the use of CrystalEyes, in order to let the user have the stereoscopic vision.

The stereoscopic vision is a technique that treats the images to obtain sensation of depth. An stereoscopic image arrives to the right eye and the left one from different points of view and different perspective, and the brain synthesizes them in an only image (exactly as the user you see it in the real world), but with the stereoscopic depth.

Finally, an exhaustive study about the sources of light and the materials to simulate the textures and the objects, have been taken into account within this stage of the Yuso Project, in order to get a more real world look.

FUTURE IMPLEMENTATIONS

One of the main reasons for using the Monastery of Yuso within this VR project was the historical importance of this place. The importance of the monasteries of San Millán de la Cogolla is not only the artistic value, but it's recognized as the birthplace of the CASTILIAN language. In fact, on the fourth of December 1997, the international organization of UNESCO decided to award the monasteries of SUSO and YUSO, incorporating them on the world heritage list.

Several initiatives and options to be undertaken in a future are going to be stated bellow:

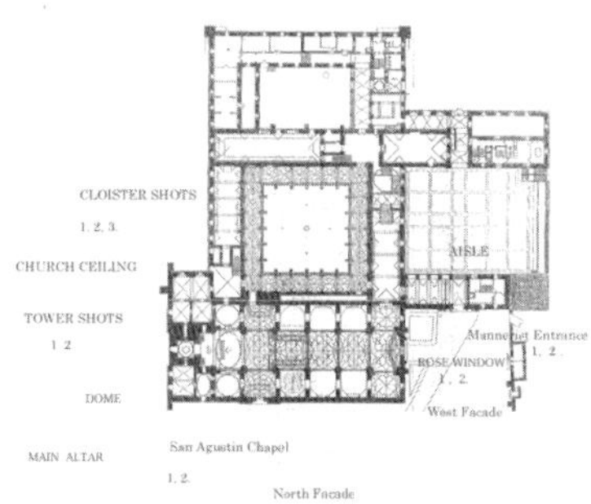

Figure 2: A monastery orthographic view.

1) The implementation of interactive devices to make easy the navigation in the virtual monastery. This device

consists on a location-mechanism that marks the user's position in the architectural complex. There are defined several channels; one of those shows us the perspective view, in which the user navigates. Another channel has associated a window using the orthographic view, that locates the user in the monastery.

2) One of the utilities that can make easier the user's navigation, is the possibility to implement a programmed navigation. For it, several tools can be used, but the most effective could be introducing a default path to be used during the simulation. The user is transported into the virtual world, but he can stop and interact with the objects during the trip. This way, the user can forget to handle with the joystick, tracker, or whatever input device that is being used, and enjoy the virtual trip.

3) Interactive device with informative goals. This device consists of several buttons that fit with some important topics related with San Millán. When anyone of the virtual buttons is pressed, a panel with the requested information pops up.

Figure 3: Interactive device to facilitate information.

4) Creation of a virtual repository: Because of the importance of the library of the monastery of Yuso, and with the purpose of letting the user interact with these virtual codices, this implementation is one of the most functional applications from the informative point of view.

This possibility would not only consist on the mere animation of the codices, but in creating a virtual repository and providing the user the possibility to handle virtually those codices. What is more, depending on the "professional degree" of the user, this one has access to some documents or has control to perform determined actions. This way all the researchers of the world would have access to that valuable information without moving to Spain, and asking for so many permissions.

5) Another important utility that can have the project of Yuso is not only the recreation of the monastery, but in general of some different historical buildings, with the purpose of showing the evolution that those ones have suffered along the time. This way, the different stadiums would show how the building has changed in a visual and interactive way.

6) Finally, and more focused on the construction world, the model can be used for evaluating the impact that possible restoration works could cause on the building. This way, the effect of a restoration work would be simulated, and some possible works could be evaluated.

CONCLUSIONS

The new technologies have acquired the maturity enough to be used in very different environments as for example history, medicine, education and of course the industry. Exploring and innovating the resources that these technologies offer us, in these and in other kind of fields, is in last term the purpose of our project.

The importance of the part of programming the codes, or the easy portability of the application, is the possibility to personalize the application as much as it is desired. The project is not limited to the projection of an elaborated geometric model, but the user has the possibility to choose one kind of navigation, and interact with the objects. For example, the user opens doors, he listens at any kind of sounds due to his presence, finds information about the historical importance of the monastery, or flies to the tower.

Depending on the expected results to obtain, different methods should be used. Even among the applications of Virtual Reality which goals are mainly informative, depending on the public those applications are meant for, is convenient to use one type of interfaces or another.

The study on new design of interfaces facilitates to personalize a "made-to-measure" application. This way, focusing the design on the functionality, and working with heterogeneous groups, many of the existent deficiencies in prototypes or previous versions, can improve.

The potent tools that the new technologies offer us, are in a position to let us innovate the world of the cultural diffusion. It is our work to use them.

ACKNOWLEDGMENTS

The authors gratefully acknowledge the provided information given by the "Fundación San Millán de la Cogolla" for this research work.

REFERENCES

1. Steve Aukstakalnis & David Blatner *"The Silicon Mirage: The art and Science of Virtual Reality"* PechPit Press 1992, ISBN0-938151-82-7.

ABOUT THE AUTHORS

María Aróstegui, studied the Licenciatura in Mathematics in the Basque Country and got an scholarship to finish her degree in Milan. She got two internships focused on VR technology to work in, LABEIN(Spain) and VRAC(USA), respectively. Among her research areas highlight Virtual Heritage, and computer graphics and visualization techniques for simulation purposes.
She has recently written some papers about Virtual Heritage for building magazines and the international meeting Culturtec2000.
E-mai: *arostegui@labein.es*

José Ignacio Barbero. was born in Legazpia (Spain) in 1965. He is Mechanical Engineer form the University of Navarra. He joined LABEIN in 1989 and currently is Manager of the Virtual Prototypes Technological Program. His main research interests are focused on Virtual Prototyping, Virtual Heritage, simulation and visualization techniques for engineering analysis and HPCN (High Performance Computing and Networking).
He has participated in multiple national and European research projects. E-mail: *jose@labein.es*

Carolina Cruz Neira is a Litton Assistant Professor at the Electrical and Computer Engineering Department and the Associate Director of the Virtual Reality Applications Center at Iowa State University.
In 2000 Dr. Cruz received the Iowa State Foundation Award for Early Achievement in Research.
Dr. Cruz completed her doctoral studies in Electrical Engineering and Computer Science(EECS) at the Electronic Visualization Laboratory(EVL) at the University of Illinois at Chicago in 1995.
E-mail: *carolina@vrac.iastate.edu*

A Vision-Driven Gesture-Based Pointer For Computer-Augmented Environments

Gabriele Baggiani (*), *Carlo Colombo* (*) *and Alberto Del Bimbo* (*)

(*) Dipartimento di Sistemi e Informatica, Via Santa Marta 3, I-50139 Firenze, ITALY
E-mail: {baggiani,colombo,delbimbo}@dsi.unifi.it

ABSTRACT
This paper describes a system for advanced man-machine interaction based on computer vision technology allowing users to manipulate information displayed on large wall panels by using their own hands as pointing devices. Graphic interface operation is the same as with standard computer mice; the main points of innovation concern naturality of interaction, low intrusiveness and a priori training, and low equipment cost. An experimental version of the system is currently being employed to provide museum visitors with advanced interaction services.

KEYWORDS: advanced human-computer interaction, computer vision, graphic computer interfaces.

INTRODUCTION
Computer interpretation of human action is the basis for advanced man-machine interface design. Mice and keyboards are gradually being complemented with new interaction devices that can monitor people as they act in everyday life [1, 2]. In particular, the possibility to capture the position and movements of user body parts in real-time and in a non intrusive way is of key importance for developing new applications in fields such as virtual and augmented reality, computer-supported cooperative work, telepresence and robotics. Current advanced interaction devices (e.g. 3D pointers or data helmets) do not appear to be immune to serious problems such as high intrusiveness, high cost, and low performance/cost ratio. Being intrinsically non intrusive and quite inexpensive, real-time computer vision is a technology that perfectly meets the basic requirements of next generation human-computer interaction applications [3]. Several vision-based interaction approaches have been presented so far [4]. In [5, 6, 7] a number of examples are presented of semantic interpretation of human hand gestures and facial expressions. Another side of applications is that exploting continuous user action to develop advanced pointing devices based on hand, head or eye motion [8, 9].

This paper illustrates the main features of a new vision-based interaction system capturing user actions in a three-dimensional environment. The system allows users to explore and select information displayed on large wall panels using their own hands as pointing devices. Graphic interface operation is the same as supported by standard interaction tools such as the computer mouse; the main points of innovation concern instead interaction style, and in particular naturality and usability issues. First of all, unlike both traditional and more advanced yet intrusive human-computer communication contexts, users do not have to wear or otherwise use any extra hardware device to carry out the interaction: they just have to point to an area of interest in the wall panel with their own hands. Secondly, while a traditional mouse must lay and be operated on a flat 2D surface such as a table, the system allows users to move freely in a 3D, real-life environment such as a room. Last, but not least, pointing in 3D is an everyday life operation, which therefore does not require a priori skills or training.

Figure 1: A computer-augmented museum.

A prototype of the system is being used to support interaction within an "augmented museum" room (Fig. 1). This is a museum environment provided with computer facilities, in which users can ask the system to display on the museum walls additional information about the paintings they are currently looking at by activating interface links associated to specific wall locations.

SYSTEM DESIGN

Overview

The main system components are shown in Fig. 2. The system gets its input from a pair of color cameras (placed so as to have the user in view); a personal computer performs image analysis and updates interaction parameters. Such parameters, which reflect current user status, are then transformed into graphic interface commands and output to the screen through a beamer.

Interface operation is based on both spatial and temporal characteristics of user action. On the spatial side, the screen location *P* currently pointed to by the user is continuously evaluated as the intersection of the pointing line *L* with the screen plane. To this end, user's hands and face are located and tracked from both images, and then used as the input of an affine stereo triangulation algorithm. On the temporal side, the system monitors pointing persistency: as point *P* remains inside a limited portion of the screen for an appropriate amount of time (i.e. about two seconds), a discrete event similar to a mouse click is generated for the interface. In conclusion, the overall interaction system behavior is that of a one-button mouse, whose "drags" and "clicks" reflect respectively changes and fixations of interest as communicated by the user through natural pointing actions.

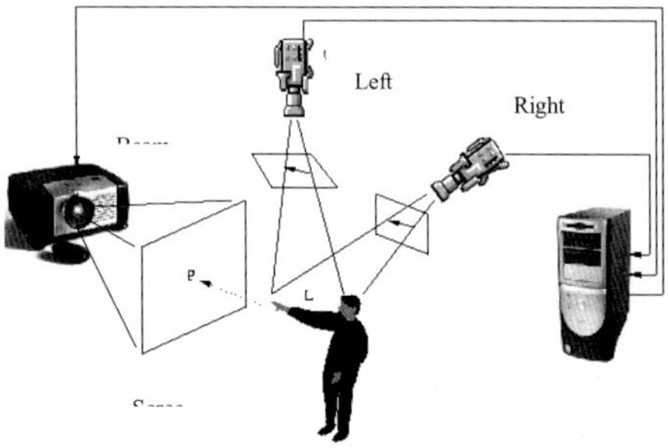

Figure 2: The main interaction components.

Interaction Geometry

A careful modeling of interaction geometry is essential to system design, since it heavily affects both localization accuracy and speed of operation. For an interaction scenario similar to the one depicted in Fig. 2, it is reasonable to assume that the cameras are enough far away from the user to neglect any nonlinear (perspective) imaging effect. The resulting affine camera model characterizes image projection as a linear mapping with six degrees of freedom $\{a_k\}$, $k = 1 \ldots 6$ [8].

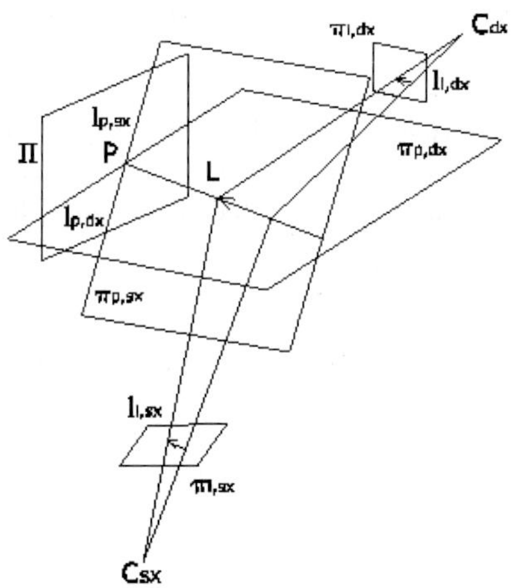

Figure 3: Interaction geometry. The subscripts 'sx' and 'dx' refer to entities associated respectively to the left and right camera.

The basic geometric problem to be solved is how to compute the location of interest P from image observations. Referring to Fig. 3, we can write $P = L \cap \Pi$, where Π is the screen plane. For each camera, an observable entity is the image line l_i, i.e. the projection of the pointing line L onto the image plane π_i. The projection plane π_p through L and l_i intersects Π in the screen line l_p; since P must belong to this line, a linear constraint of the type $P \in l_p$ can be associated to each image line l_i, whose parameters are continuously observed and updated. During normal operation (screen point remapping), two cameras are used to compute P from left and right screen line intersection. The affine stereo algorithm described here does not constrain the flat surface pointed to by the user to be visible by the cameras; as such, it can be thought of as an extension of the triangulation method proposed in [8].

In order to be used later for remapping, the affine projection parameters of each camera must be evaluated at system startup by a calibration procedure, which can be run simultaneously for the two cameras. Camera calibration requires a minimum of six known screen points, which are used together with image line observations in order to compute the projection coefficients $\{a_k\}$. In the current system implementation, nine points regularly sampled on the screen are used to calibrate, and the resulting overdetermined linear system is solved by singular value decomposition.

Image Analysis

Image analysis algorithms are used to monitor user action, and compute in real time image line parameters to be used during both calibration and remapping. Fig. 4 shows that image lines are computed as the lines through the head and pointing hand centroids.

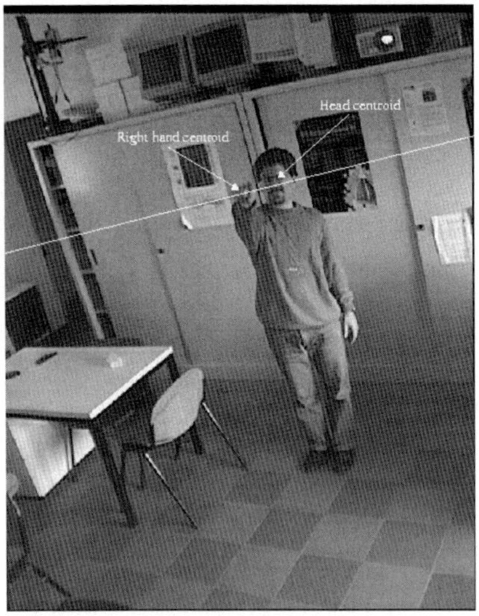

Figure 4: Run time extraction of an image line.

Before being tracked, the user is first located in the image by background subtraction. For this purpose, the background image is analyzed at a very low resolution (7×5 pixel for a 160×120 pixel image) in search of foreground information; once this is found, a recursive region growing algorithm starts, operating at different resolution levels. The multi-resolution nature of the algorithm allows a significant processing speed-up, as more than 90 percent of the raw image data are not analyzed at all. It is sufficient to grow the foreground region from a single point since the body is a connected shape. The growing algorithm expansion stops when the pixel encountered is very similar to the acquired background pixel in that position. Once a relevant foreground area is found, the user communicates his intention to start interacting with the system by setting himself in the reference body configuration shown in Fig. 5. A "locking" phase starts then, in which the anthopomorphic characteristics of the reference configuration are used by the system to capture a number of user features such as skin color, head and hand size, which are used later for user tracking.

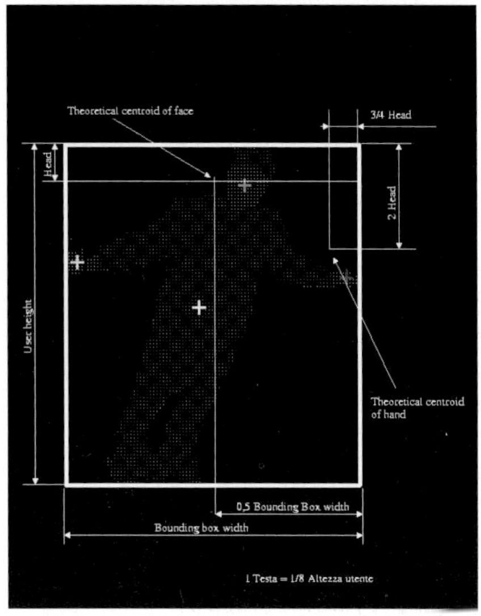

Figure 5: Anthropometric heuristics used during the locking phase.

Human body tracking at the image level is carried out by combining color and motion information. To check whether the pixel belongs to foreground or background, or whether it has or not a skin color, two different conditions are checked in two color spaces, *(Y,U,V)* and *(U/Y,V/Y)*, the latter being very useful in dark image portions. A region growing algorithm expanding from suitable foreground pixel "seeds" is used to evaluate the current position and shape of the hands and face regions; since these regions are very smooth and regular, skin pixel search only takes place in three directions (N, S-E, S-W) out of the 8-neighborhood canonical directions. Current region measurements are combined with previous ones using temporal low-pass filters, both to reduce noise and smoothen the tracking behavior; a constant velocity predictive filter is also used, whose effect is to produce growing seeds of better quality, thus speeding up the system response by significantly reducing the search range. Another important use of the prediction filter is the management of critical tracking situations, such as the detection of and the recovery from occlusions (hand-hand or hand-face).

EXPERIMENTAL RESULTS

The system has been implemented in C++ and runs in real time on a Pentium III PC with Linux operating system and GLUT/OpenGL graphic libraries. Two Sony EVI-D30 color cameras and a dual channel Linux Media Labs LML33 frame grabber board are used for image acquisition. As shown in Fig. 6, the system is structured into three different layers (physical: image acquisition and display, logical: image analysis and graphic synthesis, application: human-computer interfacing), communicating through a message-passing protocol. Videos showing examples of interaction with the system can be found at http://viplab.dsi.unifi.it/HCI.

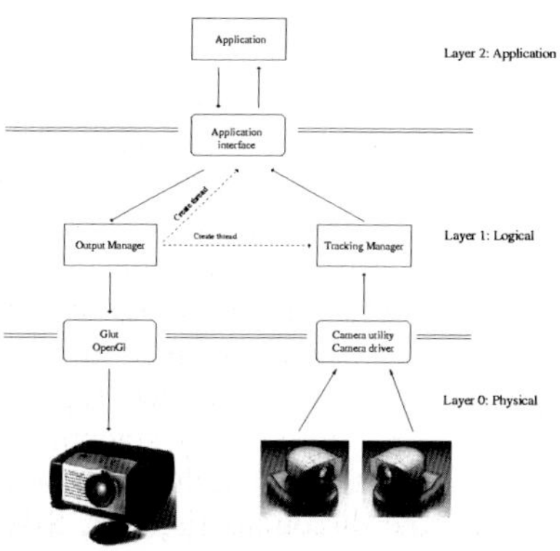

Figure 6: System architecture.

Several experiments were made to assess system performance both in terms of tracking performance and remapping accuracy. Concerning user tracking, the system has proved to be robust enough to support even long (i.e. several minutes) interaction sessions without missing the target. The main causes of tracking failure are occlusions; the system is very tolerant instead with respect to fast user movements. Concerning remapping error (i.e. the mismatch between the estimated and actual screen point locations), this depends on a number of factors, such as tracking and calibration accuracy, camera-user relative position and lighting conditions.

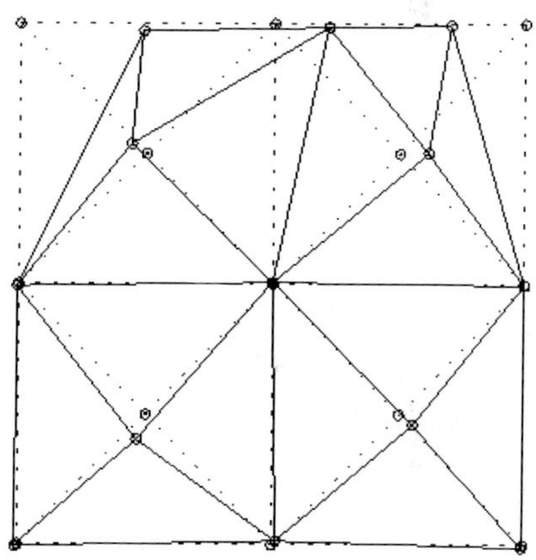

Figure 7: Remapping error: graphic data.

Results of remapping accuracy are given in Fig. 7. The user was asked to point in sequence at a number of reference locations arranged as a 13-point regular grid (dotted lines), while the corresponding locations remapped by the system were being recorded (solid-line grid). The mismatch between the two grids is generally very small, save for the three upper screen locations. Indeed, as for this experiment the two cameras were placed on the wall exactly at the same height and on either sides of these three points, the resulting image centroids of hand and head were so close to each other to cause ill-conditioning in image line measurement. This suggests that camera layout is a factor that dramatically affects system accuracy. More generally, remapping performance should increase by increasing the separation between image lines when pointing at nearby screen locations. However, such separation should be obtained with larger screens and not with wider camera angles, as in this last case the perspective effect would be so relevant to make the affine camera model unrealistic.

	$\Delta\varepsilon(X)$	$\Delta\varepsilon(Y)$	$\|\Delta\varepsilon\|$
P_1	0.241915	0.013934	0.242315
P_2	0.106528	0.006241	0.106711
P_3	-0.150202	0.002524	0.150223
P_4	-0.002958	0.004211	0.005146
P_5	-0.000634	-0.003825	0.003877
P_6	0.000961	0.000932	0.001339
P_7	-0.005624	0.002830	0.006296
P_8	0.007297	-0.007223	0.010267
P_9	-0.001437	0.005160	0.005356
P_{10}	-0.029769	-0.019459	0.035565
P_{11}	0.056718	-0.001845	0.056748
P_{12}	-0.017756	0.047626	0.050828
P_{13}	0.028386	0.019119	0.034224
			Average error: 0.05906

Table 1. Remapping error: numerical data (%).

Tab. 1 reports the numeric results (percentage of screen size) for the same experiment. Of the thirteen reference points, nine (P_1 through P_9, corresponding to the eight outer grid points, plus the grid center) had been used to calibrate. Apart from the bigger errors of the first line (P_1 through P_3), the error on the calibration points is very low (typically less than 1%). However, even far from calibration points, the error keeps limited to 5-10% of the screen size, thus making the system appropriate for use in a number of application interfaces and interaction scenarios.

REFERENCES

1. J. Nielsen Noncommand user interfaces. In *Communication of the ACM*, 36(4), 83-99.

2. R.J.K. Jacob Human-computer interaction: Input devices. In *ACM Computing Surveys*, 28(1), 177-179.

3. J.L. Crowley, J. Coutaz, F. Bérard Things that see. In *Communications of the ACM*, 43(3), 2000, 54-64.

4. R. Cipolla and A.P. Pentland, eds. Computer vision for human-machine interaction. In *Cambridge University Press*, 1998.

5. A.P. Pentland Smart rooms: Machine understanding of human behavior. In [4], chapter 1, pp. 3-21.

6. C. Maggioni, B. Kämmerer Gesture computer - history, design and applications. In [4], chapter 2, pp. 23-51.

7. K. Mase Human reader: A vision-based man-machine interface. In [4], chapter 3, pp. 53-81.

8. R. Cipolla, N.J. Hollinghurst A human-robot interface using pointing with uncalibrated stereo vision. In [4], chapter 5, pp. 97-110.

9. C. Colombo, A. Del Bimbo, S. De Magistris Interfacing through visual pointers. In [4], chapter 8, pp. 135-153.

ABOUT THE AUTHORS

Gabriele Baggiani graduated in 2001 in Computer Engineering at the University of Florence. His main research interests are in computer vision and advanced human-machine interaction.
E-mail: *baggiani@dsi.unifi.it*.

Carlo Colombo is an Assistant Professor at the Department of Sistemi e Informatica of the University of Florence. His research work is focussed on computer vision and its applications to semi-autonomous robotics, advanced human-machine interaction and multimedia systems technology.
E-mail: *www.dsi.unifi.it/~columbus*.

Alberto Del Bimbo is Full Professor of Computer Engineering at the Department of Sistemi e Informatica of the University of Florence, the Director of the Master in Multimedia, and Deputy Rector for Research and Innovation Transfer at the same University. His scientific interests address the subject of Image Technology and Multimedia, with particular reference to object recognition and image sequence analysis, content-based retrieval for image and video databases, visual languages and advanced man-machine interaction.
E-mail: *www.dsi.unifi.it/~delbimbo*.

Minerva: An Artificial Intelligent System for Composition of Museums

Francesco Amigoni[(*)], *Viola Schiaffonati*[(*)] *and Marco Somalvico*[(*)]

[(*)]Politecnico di Milano Artificial Intelligence and Robotics Project
Dipartimento di Elettronica e Informazione; Politecnico di Milano
Piazza Leonardo da Vinci, 32; I-20133 Milano (MI); Italy
E-mail: francesco.amigoni@elet.polimi.it

ABSTRACT
The employment of artificial intelligence techniques for supporting activities traditionally performed by humans and considered as creative has increasingly spread in recent years. In this paper, we present a novel artificial intelligence system, called *Minerva*, devoted to support one of such activities: the composition of museums.

KEYWORDS: Automatic composition of museums, multiagent systems.

INTRODUCTION

The adoption of machines for performing some of the intellectual and interactive activities of humans has recently spread. Initially, machines were only employed for carrying on somehow mechanical or "fabricative" human activities. Nowadays, they are also employed for supporting the execution of somehow artistic or "creative" human activities. In particular, we are referring to the class of machines, called *information machines*, which emulate some of the intelligent intellectual and interactive abilities of humans: computers and robots.

The purpose of this paper is to present a novel system of artificial intelligence, called *Minerva*, devoted to automatically compose museums in a virtual setting. The activity of composing museums, namely disposing a number of works of art in the spaces of a museum, is commonly considered as a creative activity traditionally performed by humans, since it involves a large number of complex artistic criteria.

In order to achieve the automation in the composition of museums, Minerva is based on the modern, powerful, and flexible technology of multiagent systems, developed and established in the field of artificial intelligence.

This paper is organized as follows. In the next section, we outline the general theoretical framework in which the significance of Minerva is captured. Then, we briefly illustrate the technical concepts of multiagent systems within artificial intelligence. At that point, we present both the architecture and the functioning of Minerva. Finally, we conclude the paper.

MAN AND MACHINES

We are observing nowadays computers and robots which are able to emulate

some intellectual and interactive human abilities. Within these activities we distinguish between the activities that can be described in rational terms (*fabricative* activities) and the activities that cannot be described in rational terms (*creative* activities) [2].
According to this framework, we stress that information machines, not only improve their skills to partially reproduce some of the fabricative activities of humans, but they are also designed and built in order to support humans in performing some of their creative activities. That is particularly interesting when we consider that the automatic composition of museums performed by Minerva is parallel and not substitutive to the traditional human composition. The composition of a museum resulting from the operation of Minerva is an activity not totally performed by a machine; humans still play a determinant role both in designing and in settling down the system by introducing their artistic criteria. The machine is not able to independently and autonomously carry on the whole creative activity of museum composition. However, we state that many processes in this composition can be automated and, therefore, delegated to information machines when it is considered useful.

MULTIAGENT SYSTEMS

In this section, we shortly introduce some technical issues about multiagent systems developed within distributed artificial intelligence (DAI, for short).
Multiagent systems [11] are systems of DAI that are composed of several entities, called *agents*, which are spatially distributed but connected together. In multiagent systems, the grain of the distributed entities is coarse; therefore, agents are loosely connected. There is no global agreement on the definition of agent [5]. For our purpose, we consider computational processes, computers, and robots as instances of agents. An agent is usually thought as exhibiting several properties [11], such as autonomy (the agent controls its actions and its internal state), social ability (the agent interacts with other agents), reactivity (the agent perceives and responds to changes in its environment), and pro-activeness (the agent exhibits goal-directed behavior).
A multiagent systems is, in general, composed of several interacting agents but, when these agents are cooperating, it is profitable to consider the whole system as a unitary machine called *agency* [3]. The adoption of the wording 'agency' emphasizes our conception of a multiagent system as a unitary machine, even if the composing agents have very complex natures. The concept of agency has been firstly introduced by Marvin Minsky under the metaphor of "The Society of Mind" [7] to set up a rich and precise description of the human intelligence. We consider an agency as a machine made up of intelligent agents interacting each other in order to offer or receive collaboration for achieving a global goal. By intelligent agents we mean that each component of the agency has inferential abilities [6]. Each agent is able to perform some functions, namely to carry on some high-level tasks, that require inference. In addition, each agent has the abilities for cooperating with other agents when it is necessary. Thus, an agency can be viewed as a machine of DAI able to perform distributed inferential reasoning. From the previous considerations it emerges that the agents have to be connected

together by a communication network, that represents the necessary infrastructure for cooperation and coordination among them.

Several proposals have been advanced for implementing the two essential activities of an agent involved in an agency: operation and cooperation. The operation activities of an agent can be based on architectural solutions that are usually classified as ranging from two extremes [10], namely the reactive architectures, which prescribe that the agent simply reacts to its sensory inputs, and the deliberative architectures, which conceive an agent as a system that integrates its sensory inputs with its past knowledge to determine its future actions. However, the most adopted solution is to develop agents with intermediate architectures between the two extreme ones. The cooperation activities of an agent can be based on different mechanisms such as, for example, negotiation [8] and auctions [10].

In our research, we have developed agencies that address several problems involving both computer and robotic agents (see [1] for a summary of these agencies). In this paper, we describe a new agency that addresses the problem of automatically composing a museum.

MINERVA SYSTEM

The experimental efforts we have taken up are inserted in a project, called *Minerva Project*, that aims to design, develop, and implement a system for the automatic composition of museums. As already stated, this is commonly considered an artistic activity since several subjective, creative, and imaginative abilities are involved in it. By adopting the term 'composition' we mean both the *preparation* and the *allocation* of a museum. The preparation is the arrangement of the given works of art to be displayed in ordered groups. The allocation is the placement of these works of art within a given geometrical space preserving their arrangement. The system resulting from Minerva Project is called *Minerva Agency* (we will refer to it as Minerva in the following). Hence, Minerva is an artificial intelligence system composed of several agents, where some agents are expert systems in particular domains and some other agents perform service activities, that cooperate together to compose museums.

The activity presents several difficulties that have to be overcome for obtaining a final result that exhibits a global coherent setting and that allows an effective (possibly virtual) visit of the museum. It has to take into account many different parameters: for instance, some works of art cannot be inserted in some rooms, some rooms can be scarcely lighted, some other rooms can be too hot or too cold to keep works of art in them. An interesting aspect is that the museums managed by Minerva may contain any kind of works of art, from archeological finds to paintings. All are virtually reproduced according to their three-dimensional models obtained either from the real objects or from other representations (e.g., pictures) of such objects. The allocation of the works of art is performed in a virtual environment that can be selected by the user and that can differ from the real environment in which the works of art are exposed. At the end of the operation of Minerva, it is presented to the user a subjective point of view in a virtual three-dimensional environment, which contains the works of art and that can be navigated by the user.

Minerva has multiple applications that show why it is interesting and useful adopting intelligent systems to compose museums. Simulation of proposed compositions for existing museums to study and evaluate them, simulation of existing compositions in existing museums for allowing their virtual visits, and simulation of compositions in imaginary museums for building personalized cultural paths are among the ways in which Minerva can be employed. Until now, Minerva has been tested on the collections of the Archeological Museum in Milan and of the Paolo Giovio Museum in Como.

The architecture and functioning of Minerva are shortly described in what follows (see Figure 1). In this paper we are not going to give a detailed presentation of all the technical issues of the Minerva; instead, we give an outlook to the main features of the system.

The user interacts with a *supervisor agent*, which presents different interfaces according to the user's level of knowledge. We have provided four levels and, thus, four typologies of users:

- the museum expert users, who are usually going to use Minerva for simulating proposed museums they have already conceived, in order to study and evaluate them;
- the artistic expert users, who are usually going to use Minerva by making requests on the works of art to be displayed, but who are not going to interfere in the composition of the museum;
- the inexperienced users, who are going to visit already composed museums;
- the technical users, who modify the system, for instance the database of collections of works of art and the database of environments.

In the following, in order to illustrate the functioning of Minerva, we will refer to the museum expert users who have the most complete control over the parameters of the system. Starting from the initial window (see Figure 2), these users select a collection (i.e., a set of works of art, such as paintings, sculptures, or archeological finds), that they would visit, and a real or imaginary environment (i.e., a set of rooms in a building), within which they would allocate and visit the collection. For example, a user can choose to visit the Leonardo da Vinci's works in the Versailles Residence. The collection and the environment are chosen from a set of collections and a set of environments, respectively, for which knowledge (descriptions, geometrical features, characteristics, …) has been inserted in Minerva.

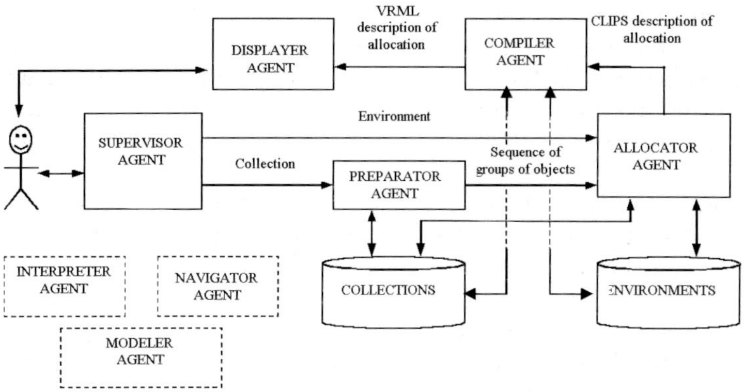

Figure 1 – The architecture of Minerva

Once the user has selected a collection and an environment, two further agents operate to find, respectively, a preparation and an allocation. More precisely, the first agent, called *preparator agent*, automatically determines the arrangement in which the objects of the collection will be displayed to the user. In order to achieve this goal the preparator agent exploits the expertise, inserted in its knowledge base, about museum criteria concerning the possible ways in which the pieces of a collection can be grouped. This expertise has been inserted in Minerva following a procedure of knowledge acquisition. By adopting the preparation criteria, the preparator agent divides the works of art of the collection into groups, each one composed of homogeneous objects. The criteria the preparator agent adopts to assemble similar objects in groups are based on both cultural (such as age, type, place of origin, and culture of origin) and physical (such as size, weight, shape, and material) characteristics of the objects. The user can tune the application of the criteria by acting on some parameters.

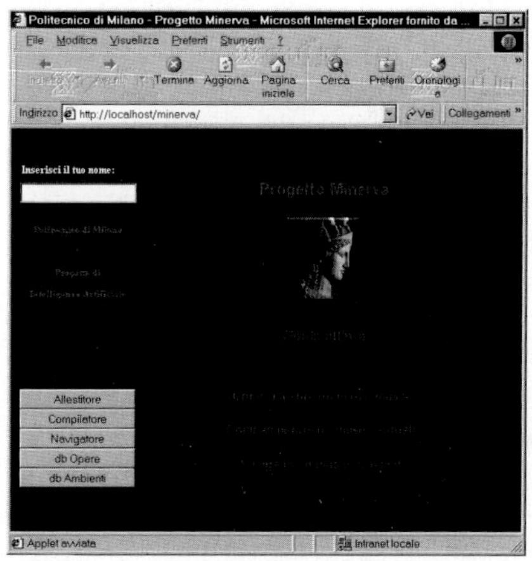

Figure 2 - The opening window of Minerva (this window, as the following ones, is in Italian since Minerva has been initially developed for Italian museums)

For example, the objects composing an hypothetical Leonardo da Vinci's collection can be grouped according to several different criteria (that can be selected by acting on the parameters): the chronological criterion, based on the ages of the pieces, the type criterion, based on the kind of the pieces (e.g., paintings, projects, drawings, and so on), or the subject criterion, based on the subject that pieces represent (e.g., about nature, about artifacts, about human body, and so on). These criteria can also be combined together to obtain, for example, a first division based on the subject, a second subdivision based on the type, and then a further subdivision based on the age. The preparator agent represents the obtained groups of works of art by a *taxonomic tree* (see Figure 3). Each leaf of the tree represents a group of artistic objects. The tree expresses the parental relations among groups and the criteria used for their formation. The groups are ordered according to their importance. The importance of a work of art is function of its historical value, aesthetic value, and condition. The importance of a group is function of the importance of the composing works of art. The taxonomic tree can be accessed by means of rules, called *access rules*, which act as methods for accessing and manipulating the taxonomic tree object. In conclusion, the preparator agent performs firstly a classification of the works of art in groups and then an ordering of these groups. The ordered groups of objects provided by the preparator agent and the environment selected by the user constitute the feed for the *allocator agent*.

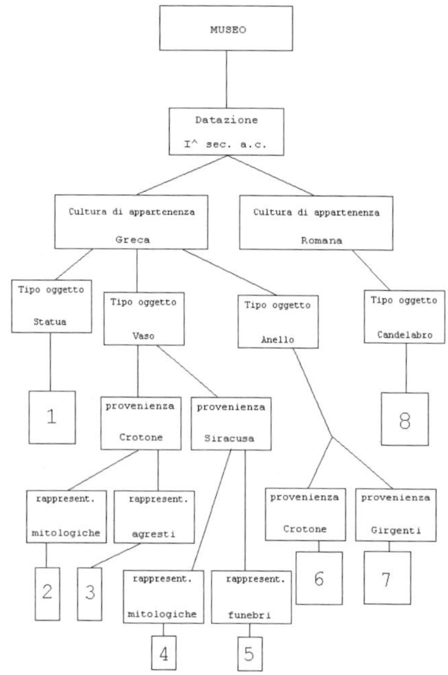

Figure 3 – The taxonomic tree of some archeological finds from I century BC

This agent automatically finds, within the given environment, the geometrical allocation for the objects of the groups in the taxonomic tree, preserving the ordering on groups imposed by the preparator agent. Again, in order to achieve this goal, the allocator agent exploits the knowledge, inserted in its knowledge base, about museum criteria concerning where to place the works of art in a given museum environment whose geometrical characteristics are known. These criteria concern, for example, the occupation of the space in a room, the occupation of the perimeter of the room, the collocation of the shrines according to the collocation of the doors, windows, and other shrines. The application of these criteria can be tuned by the user. The groups of works of art are allocated in the rooms in decreasing importance order (as established by the preparator agent). In particular, the allocator agent tries to allocate each group in a room; if it fails, the group is split over more rooms or more groups are allocated in the same room. In order to operate, the allocator agent accesses the taxonomic tree, generated by the preparator agent, by means of the access rules. The allocator agent discretizes each room of the environment to a matrix model (see Figure 4), in which it inserts the works of art. The result of the allocation is the geometrical description of the position of the works of art in the environment. The format of this description is a set of facts of CLIPS language [4], in which both the preparator and the allocator agents are coded. For example, the sequence of the Leonardo da Vinci's objects can be allocated in the Versailles Residence according to a criterion that places one object in each room, or to a criterion that places a group of related objects in each room, or to a criterion that considers the Residence as a single large space (namely, without considering the division of the whole space in different rooms) in which the objects can be placed. We explicitly note that both preparator and allocator agents are inferential, namely they perform their tasks exploiting inferential techniques of artificial intelligence [6].

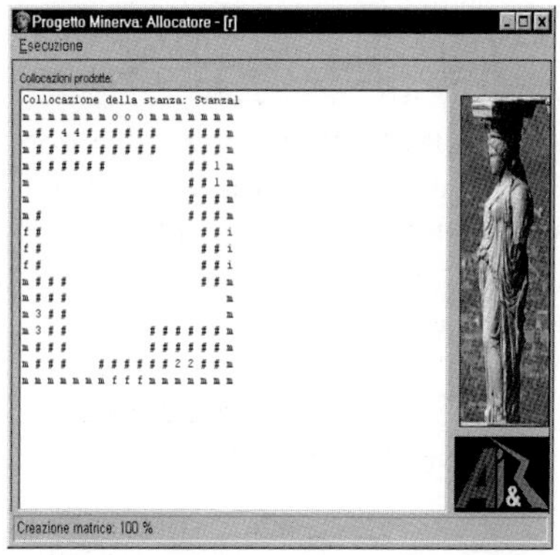

Figure 4 – The matrix representation of a room of the environment used by the allocator agent. Numbers represent the physical occupation of the artistic objects. # symbols represent the view space of the artistic objects. Letters m, f, i, o represent respectively walls, windows, in door, out door

Besides the fundamental preparator and allocator agents, Minerva contains other (simpler) agents devoted to carry on some technical tasks. The first one is the *compiler agent*. It translates the geometrical description of the allocation of the collection in the environment, which is provided by the allocator agent in its own format (set of facts in CLIPS language), to a format (VRML model [9]) suitable to be displayed to the user in a virtual reality setting. More precisely, the compiler agent reads the three-dimensional models of the works of art and inserts them in the three-dimensional model of the environment according to the geometrical result of the allocator.

Once the translation has been performed, the *displayer agent* displays the VRML model to the user that can navigate in it (see Figure 5). The displayer agent offers the possibility to follow a list of viewpoints through the museum and to access the information associated to the works of art by clicking on the works of art themselves (see Figure 6). The window that displays the information about a work of art allows also to move around the three-dimensional view of the object.

Figure 5 – The window of the displayer agent

The Minerva extremely flexible architecture allows for insertion of further agents. We are developing some new agents (depicted as dashed boxes in Figure 1) that will expand and enrich the system with new functionalities. The *navigator agent* will drive the user in the choose of the initial set of works of art and environments. Moreover, it will offer the opportunity of tailoring personal paths for users in already allocated museums. The *interpreter agent* will analyze the requests of the user, expressed in natural language, in order to select a particular set of interesting works of art. The *modeler agent* will allow to edit new environments, by means of the techniques of computer aided design, in order to enrich the sets of available environments.

From a technical point of view, Minerva exhibits many interesting features, such as the client-server division between agents and the possibility of using the system from a remote station which, however, we do not discuss in this paper.

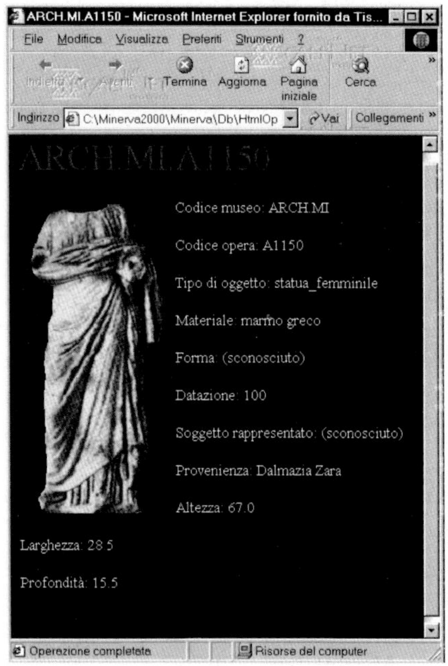

Figure 6 – A window that displays information about a work of art from the Archeological Museum of Milan

CONCLUSIONS

We have proposed an example of machine-supported artistic creativity in composition of museums, presenting the Minerva system for the automatic composition of museums. It is embedded in a theoretical framework in which the role of machines involved in some creative activities traditionally performed by humans is captured. Minerva is an agency composed of specialized agents that cooperate to the global goal of composing a museum. Minerva applications range from support the activity of museum experts to offer an environment for virtual visits to museums.

Future research work will address both the improvement of the technical apparatus of Minerva and the design of a system progressively more independent from specific applications. The goal is to obtain a system able to easily tackle the compositions of several different museums according to criteria that are specific for each museum. In this perspective, we will soon approach the automatic composition of the future Design Museum of Milan.

ACKNOWLEDGMENTS

The authors are glad to thank Riccardo Magini, Alberto Scipionato, Raffaele Regni, and the other students of Minerva Project.

REFERENCES

1. Amigoni, F. Dynamic Agency: A Methodology and Architecture for Multiagent Systems. *PhD Thesis*, Dipartimento di Elettronica e Informazione, Politecnico di Milano, Milan, Italy

2. Amigoni, F., Schiaffonati, V., Somalvico, M. Dynamic Agencies and Creative Scientific Discovery. In *Proceedings of the "AISB'99 Symposium on Artificial Intelligence and Scientific Creativity"*, Edinburgh, UK, April 6-9 1999, 72-81

3. Amigoni, F., Somalvico, M., Zanisi, D. A Theoretical Framework for the Conception of Agency. In *International Journal of Intelligent Systems*, John Wiley & Sons, 14(5), May 1999, 449-474

4. CLIPS, h*ttp://www.ghg.net/clips/CLIPS.html*

5. Franklin, S. and Graesser, A. Is It an Agent, or Just a Program?: A Taxonomy for Autonomous Agents. In Müller, J., Wooldridge, M., Jennings N., Eds. *Intelligent Agents III: Agent Theories, Architectures, and Languages*, Springer-Verlag, Berlin Heidelberg, Germany, 21-35

6. Genesereth, M. and Nilsson, N. *Logical Foundation of Artificial Intelligence*, Addison-Wesley, 1987

7. Minsky, M. *The Society of Mind*, Simon & Schuster, New York, USA, 1985

8. Rosenschein, J. and Zlotkin, G. Designing Conventions for Automated Negotiation. In *AI Magazine*, 15(3), Fall 1994, 29-46

9. Tittel, E., Scott, C., Wolfe, P., Sanders, C. *Building VRML Worlds*, Obsborne, 1996

10. Weiss, G. *Multiagent Systems: A Modern Approach to Distributed Artificial Intelligence*, MIT Press, 1999

11. Wooldridge, M. and Jennings, N. Intelligent Agents: Theory and Practice. In *Knowledge Engineering Review*, 10(2), 1995, 115-152

ABOUT THE AUTHORS

Francesco Amigoni is a postdoctoral research affiliate at the Dipartimento di Elettronica e Informazione of the Politecnico di Milano, Italy, where he is member of the Artificial Intelligence and Robotics Project. His research interests are in the areas of distributed artificial intelligence, distributed robotics, and epistemology of creativity.
E-mail: *amigoni@elet.polimi.it*

Viola Schiaffonati is a PhD student in Philosophy of Science at the Università degli Studi di Genova, Italy. She is carrying out part of her research at the Dipartimento di Elettronica e Informazione of the Politecnico di Milano, Italy, where she is member of the Artificial Intelligence and Robotics Project. Her research interests are in philosophy of artificial intelligence and virtual museums.
E-mail: *schiaffo@fusberta.elet.polimi.it*

Marco Somalvico is full professor of Artificial Intelligence at the Dipartimento di Elettronica e Informazione of the Politecnico di Milano, Italy, where he is the director of the Artificial Intelligence and Robotics Project, that he founded in 1971. His research interests are in distributed artificial intelligence, distributed robotics, virtual museums, intelligent systems for disabled people, epistemology of computer science, and epistemology of creativity.
E-mail: *somalvic@elet.polimi.it*

Case-studies

OMAR - An online database for oriental manuscripts

Matthias Brückner (*) *and Paul-Thomas Kandzia* (#)

(*) Oriental Seminar, Freiburg University, Germany
E-mail: m.brueckner@cyberfatwa.de
(#) Institute of Computer Science, Freiburg University, Germany
E-mail: kandzia@informatik.uni-freiburg.de

ABSTRACT
OMAR (**O**riental **Ma**nuscript **R**esource) [11] is a database which provides extensive search facilities for oriental manuscripts. It is currently based on the collections of the Oriental Seminar at the University of Freiburg which comprise about 2.500 manuscripts. Images of the manuscripts are stored in OMAR together with structured data containing information about bibliographic aspects, the author, the title etc.

KEYWORDS: database, orient, manuscripts

INTRODUCTION
Over the last decade information technology, especially databases and the internet, has become an efficient and indispensable tool for scientific work. Nowadays, in technology related disciplines, e.g. Computer Science, huge amounts of information exist (often purely) in digital form. However, in the Humanities the potential of information and communication technology is not exhausted by far. In the following we focus on the handling of texts and manuscripts in Islam Science. Here, the advantages of digital resources and access via the internet are obvious, since the community is dispersed all over the world, and copying and distributing their "working material" is not an easy task at all. Still, it is a quite usual practice that researchers travel far in order to examine a certain book. Furthermore, often appointments are necessary to get the permission to lay hands on a valuable worm-eaten loose manuscript.
Surprisingly, though several library catalogues are accessible online, there is no digital resource which provides extensive search facilities of the metadata (that is catalogue-like data about the documents in the library) as well as for the texts or manuscripts themselves. Our project is a first step in order to change that situation: We propose the prototypical core of a database which stores images of historical Arabic manuscripts together with the scientific information known about them. As a special feature there is a flexible and comfortable Web-interface. Furthermore, we point out some very specific problems, which differentiate the database modeling from well-known areas as (occidental) business databases.

ONLINE RESOURCES OF ISLAM SCIENCE
In the 1990s Turkey, Tunisia and Iran already had been connected to the Internet. The impact of the Internet on the Orient grew bigger just since the

mid-nineties. Now, with Saudi-Arabia and Syria getting connected, most of the Islamic Orient is online [10], and most of the scientists of Middle East Studies and Islam Sciences are using the Internet, too. So, how about the mass of Oriental manuscripts lying in the sand waiting for discovery? Digitalisation in this area is still exceptional, but of growing interest.

The American University in Cairo Library is providing a catalogue which could be displayed in Arabic with the help of Unicode and MS Arabic Support[1], however, it is not possible to get access to the texts. The database of one of the most famous fatwa[2]-online-services can only be searched by title, keyword, mufti, and topic [6]. The database of another important fatwa-service is just searchable by number, date and keyword [7]. The latter services belong to the main resources of modern authentic Islamic texts online [2, 3]. Unfortunately, those texts are rather short. The Qur'an and Hadithantologies[3] which can be searched by keyword are frequently found online, e.g. at *Musalman* [8]. The German *Orient-Institute* is providing a database with 1.380 texts [4], but these are only secondary texts about modern Oriental history. The largest recent book-collecting activity in Germany is the *Sondersammelgebiet Vorderer Orient/Nordafrika* (Near East/North Africa), which moved from Tübingen to Halle. Both collections can be accessed by a virtual library catalogue [13]. Again, there is no access to the texts. The same applies to the *Manuscripts' Catalogue* of the *British Library* [1]. Using a search machine with keyword "Arabic language", 204 resources were listed when searching, but none of them provides direct access to the texts. A small collection with some pictures of texts is accessible at the homepage of the *Islamic medical manuscripts at the* [US] *National Library of Medicine* [9].

In summary, one can say that surprisingly few Arabic texts are searchable and accessible through the Internet, and that the advantages of database technology seem to be not yet fully recognized by the community. The services which provide texts are limited in their scope, e.g. fatwa-online-services, and limited in their search facilities. For the time being, there is no database which allows for direct access to full texts through enhanced search-facilities and is open to a broad scope of Arabic texts.

THE OMAR PROJECT

OMAR, a joint project of the Oriental Seminary and the Institute of Computer Science, Freiburg University, has begun to close the gap described above. Starting point is the collection of the Oriental Seminary in Freiburg, consisting of about 240 original manuscripts of various origins and about 2.300 Mauritanian manuscripts [12] as microfilms. These comprise about 279.000 pages altogether. This material and knowledge about it is successively put into a database and made public over the Web. Already, a prototype is working, which we will describe in the following.

---

[1] MS Arabic Support is a package of arabic True Type Fonts, which is quite common in use for arabic websites.

[2] Islamic legal ruling, which consists of a question posed by a layman and an answer given by an expert.

[3] Hadiths are traditions of the prophet Muhammad and the second important source of law after the Qur'an.

A Data-model for Oriental Manuscript Research

While designing a data-model for the collection a lot of peculiarities occured. We are convinced that any database in this application domain will have to deal with similar problems. The most important are the following:

"Second Order" Information. Knowledge about historical facts is seldom fixed and secure. This leads to far-reaching consequences for the data model: In a business database the birth date and place will be unique attributes of a person. On the contrary, quite probably different statements may exist on the respective data of a Mauretanian author of the 14th century, dependent on the information source. It turns out that for OMAR often not facts are to be modelled, rather we want to store statements about the facts, originating from one or more sources (which also have to be stored). Such a situation is called second order by Logicians.

Consistency Checks. Modelling second order information has further implications. Whenever in a business database various birth dates of the same person occur, one has to assume an error. Hence, the uniqueness of birth dates will be used as a consistency check for the database. However, in the application domain of OMAR there are much less such conditions as we have seen above. In particular, there are no natural key attributes: In the modern western world a name together with the birth date, or a social security number etc., will determine the person with all its other data. There is no such key for historical oriental authors.

Arabic Names. Occidental names have a strict order of Christian name and surname. For Arabic names, categorization is quite more complex. The main categories for Arab names are Kunya, Ism, Nasab, Nisba and Laqab. The Kunya is a genealogic name, which is normally given according to the first born son. It consists of Abu (father of) or Umm (mother of) and the name of the son, e.g. Abu ′Ali (father of ′Ali). The Ism is what in the western world is called the Christian name, e.g. Muhammad, ′Ali etc. The Nasab is the Genealogy of the person consisting of many Ibn (son of) or Bint (daughter of). The Nisba is usually a geographic term which is denoting the origin, e.g. Tehrani, Baghdadi, Tunisi etc. The Laqab is like a nickname, e.g. the blind. Titles like Ayatollah, Shaykh and Pasha are being added either. Even worse, there is no general order of the components of an Arabic name, e.g. Ayatollah Khomeini (from the village Khomein in Iran) – the title stands in the first place, ′Ali Pasha – the title stands in the second place. As a consequence, ′Ali Pasha and Pasha ′Ali may be the names of two different persons, though both names consist exactly of the same components. Hence, for OMAR besides the different components of a name their placement is stored.

Dates. OMAR has to handle different calendars. At present, the Islamic calendar is used in the first place. An additional attribute contains the corresponding Christian date, this attribute may also be used for a fuzzy textual expression like "in the first half of the 13th century". A more advanced solution may be necessary in the future, which takes into account e.g. the Coptic calendar, too, or provides for automatic conversions.

Detailed Scientific Information. Since in Islam Science manuscripts reveal historical and social facts, they have been structured by scientists in many aspects. Marginal notes might tell historical stories. Important information can be concluded from the formal aspects of the manuscript, e.g. by knowing which paper or ink-colour has been manufactured and used in which time and place. Many such bibliographical aspects as well as certain relationships between persons, e.g. teacher-pupil or father-son, have been included in the database.

Altogether, the relational schema of OMAR is fairly complex, comprising 55 tables. Taking into account the growing user experience, extensions will follow in the future.

Technical Issues

The prototype has been realised with Oracle 8i on a Sun Ultra 5, using PHP 4 for the web interface [5]. Three types of users are distinguished: guests, normal users and administrators. A guest views the database as a black box, getting access via several search-facilities. It is also possible to display lists of the manuscripts ordered by title, author, keyword, signature or basmala[4]. Normal users may view schema information and even use arbitrary SQL-expressions for searching, however they have no writing permission. The latter is left to administrators.

Arabic Text Values. The values of some attributes (e.g. basmala) must be capable to contain Arabic text. However, a database has to support alternative modes of those values, as Original Arabic, transliterated Arabic (using an international alphabet, also available as True-Type font), and an ASCII-representation. At present, OMAR uses transliterated Arabic and the conversion of transliterated Arabic according to TUSTEP (Tübinger System von Textverarbeitungs-Programmen)[5].

Sorting. Of course, a database often has to output sorted lists. However, what is the meaning of sorted, if transliterated Arabic text values are concerned? For the time being, OMAR uses the alphabetical order of the internal ASCII-code. Of course, this is a somewhat arbitrary solution. On the other hand, there is no standard sorting order for transliterated Arabic, so this question is still open.

CONCLUSION AND OUTLOOK

At present, the metadata about the manuscripts in Freiburg have been imported into the database. The import of the images of the manuscripts will follow soon. Then, OMAR will start its work under real world conditions for German Islam scientists. The growing experience will be used to enhance the data model and search facilities. However, in order to be useful for an international community, a multilingual approach is essential. In particular, besides English, Arabic should be supported. Since the MS Arabic Support fonts seemingly are emerging as a standard, the Arabic extension will be based on it. During the work on OMAR we found quite few activities in this area. We hope that our project helps to raise a fruitful discussion on database and web technology for Islam Science.

---

4 Basmalas are innovative Arabic blessings at the beginning of a book, which allows to determine it. Often, there is no title written on Arabic manuscripts.

5 TUSTEP is a mighty word-processing-system based on FORTRAN.

REFERENCES

1. British Library *http://molcat.bl.uk/msscat/INDEX.ASP*

2. Brückner M. *Cyberfatwa, http://www.cyberfatwa.de*

3. Brückner M. Der Mufti im Netz. In: Lohlker R., et al. *Islam im Internet. Neue Formen der Religion im Cyberspace* Deutsches Orient-Institut, Hamburg, 2000, CD-ROM, Version 1.0

4. Deutsches Orient Institut *http://www.deutsches-orientinstitut-datenbank.de*

5. Fleig D. *Eine Datenbank zur Verwaltung orientalischer Handschriften, http://sansibar.informatik.uni-freiburg.de/fleig/diplom/html/index.html*

6. Islam Online *http://www.islam-online.net/fatwa/arabic/searchFatwa.asp*

7. Islam – Question & Answer *http://www.islam-qa.com/search.shtml*

8. Musalman *http://www.musalman.com/search*

9. National Library of Medicine *http://www.nlm.nih.gov/hmd/arabic/arabichome.html*

10. Network Startup Resource Center *http://www.nsrc.org/*

11. OMAR (Oriental Manuscript Resource) *http://www.uni-freiburg.de/islam/db/*

12. Rebstock U. *Sammlung arabischer Handschriften aus Mauretanien*, Otto Harrassowitz, Wiesbaden, 1989

13. Virtueller Katalog Sondersammelgebiet Vorderer Orient/Nordafrika, *http://www.ubka.uni-karlsruhe.de/hylib/vk_ssg_vo.html*

ABOUT THE AUTHORS

Matthias Brückner worked from 1996 to 2000 as system-administrator in the Oriental Seminar at the Freiburg University. Since 1999 he builds up an information website about fatwas (http://www.cyberfatwa.de). In 2000 he received the Master of Arts (M.A.) in Islam Science with a work about the distribution of fatwas about alcohol in the internet. Currently, he is working at the OMAR-database,
E-mail: *m.brueckner@cyberfatwa.de*

Dr. Paul-Thomas Kandzia has done research in the area of data models and object-oriented deductive databases. Currently, he is working as project manager of VIROR, the Virtual University of the Upper Rhine Valley (http://www.viror.de). VIROR is taking first steps to enhance the traditional lecture room scenario with more flexible computer supported settings,
E-mail:
kandzia@informatik.uni-freiburg.de

Guidelines for the Archaeological Map of Tuscany: some issues and experiences

Marinella Pasquinucci (*) *and Oreste Signore* (#)

(*) Università degli Studi di Pisa, Dipartimento di Scienze Storiche del Mondo Antico, Via Galvani 1, 56126 Pisa (Italy)
E-mail: pasquinucci@sta.unipi.it

(#) CNR - Istituto CNUCE - Area della Ricerca di Pisa - San Cataldo, Via G. Moruzzi, 1 – 56124 Pisa (Italy)
E-mail: oreste.signore@cnuce.cnr.it

ABSTRACT

Regione Toscana recently delivered guidelines for the archaeological map of Tuscany in the aim of fulfilling both administrative and research needs. Format proposed in 1995 by CIDOC has been slightly modified and enhanced, especially to represent chronological data in a more compact and formal way, and to add room for additional information. Wherever possible, normalisation of language is achieved by controlled dictionaries or thesauri. Main goals have been portability, scientific correctness, easy of use, coherence with national and international standards. Putting information in the right spatio-temporal context has been addressed as an important goal. In light of portability of data, and considering some limitations of traditional GIS, we decided to separate basic cartographic information from description of sites, that are georeferenced and have their own geometric properties stored as attributes. Comparison with national standards showed that there are many similarities, but some information is more adequately represented in our layout. Anyway, the two formats can generally be reconciled. Document is conceived as a structured document. To accommodate multiple or uncertain values, and to reduce atomization of information, in some cases syntactic rules drive writing of field content, so making the layout more flexible, readable and usable. The whole system is based upon well-established technologies and methods (DBMS, IRS, GIS), while Hypermedia approach and XML are used as unifying technologies. A test has been conducted upon data from Provincia di Livorno.

KEYWORDS: archaeological map, documentation form.

INTRODUCTION

Archeological maps have been widely addressed as a relevant tool for understanding and planning the territory. However, scholars' efforts and financial investments can be wasted as proprietary software solutions or differences in data organization often make impossible to effectively exchange, understand and use collected data. Hence the need of bringing together flexibility, scholars' and administrations' requirements, portability.

Methodological issues

Regione Toscana delivered guidelines for the archaeological map of Tuscany in the aim of fulfilling both administrative and research needs ([5]).

Archaeological maps are produced for study/research, conservation and administration purposes. These thematic maps are valid and predictive of archaeological sites when worked out on the ground of methodologically correct fieldwork, and detailed sites- and artefacts study. Data are to be collected in a diachronical perspective, using a multidisciplinary integrated approach, including:

- geologic, geomorphologic, paleoenvironmental and pedological research;
- non destructive techniques applied to landscape and maritime archaeology (remote sensing, geophysical and geochemical research);
- archaeological surveys and monitoring;
- stratigraphic excavations in at risk or scientifically relevant sites;
- archaeological and archaeometric study of finds;
- study of epigraphic and literary sources, documents and maps;
- bioarchaeologic and environmental analyses;
- study of toponyms.

Thanks to this research method sites and finds (including the underwater ones) are documented in their environmental and stratigraphic setting. Moreover, settlement patterns, territory exploitation (off site and intersite, agricultural and pastoral activities, woodland use), manufacturing activities and infrastructure works are diachronically identified and can be properly mapped. The importance of the geographic and geomorphologic setting of dated sites and finds has to be stressed, as prerequisite for the archaeological prediction. The archaeological map of Tuscany is also planned to include data about not georeferenced finds mentioned in documents and/or bibliography, in order to enhance its archaeological prediction potential.

TECHNOLOGICAL ISSUES

Main goal has been to implement a framework *easy to use*, coherent with international and national *standards*, *scientifically correct.* Software and hardware platform independence was an additional and relevant issue.
Many projects in the field are somehow similar, but there are differences that make results incompatible, so wasting scholars' efforts. Differences are sometimes in the technological platform, but the most relevant, difficult to overcome, arise from differences in *structuring* and *representing information.* It must be noted that in many cases there is an agreement on the semantics of single information fields, but there are differences in terminology and in document structure.
A relevant issue is the emphasis that many projects put on the cartography. Cartography is a basic and irrenonceable instrument for archaeological research, but its importance mainly results in the processing and tuning of models. Therefore, maps should be seen as one, among many others, interaction metaphors. In many cases, we can get similar results using low cost, platform independent tools. In essence, to produce an archaeological map, we need to correctly represent and describe object having *descriptive*, *geometrical* and *spatial* properties. To adequately represent descriptive information, we need an appropriate information structure and a clearly defined vocabulary, where every information item has its own semantics. GIS are adequate to represent spatial and geometrical properties, but lack in representing temporal and 3D data, while the spatio-temporal reasoning is

essential in the field ([4]).

To adequately support interchange of information among different project, at national or international level, *standards* play an essential role. Conformance to the standards also helps in saving of investments. However, we must stress that a *conceptual* standard (that is, an agreement on the semantics of information fields) is the real key point, and assures effective information exchange and unambiguous interpretation ([7]). In this project, we deeply considered both the standard issued by Istituto Centrale per il Catalogo e la Documentazione (ICCD) of Italian Ministry "Beni e Attività Culturali" ([3]) as well as the format ([1]) proposed by CIDOC (Comité international pour la documentation). A strong point is that the ICCD standard is obviously a must for any national project. The two resulted essentially similar in the organisation: the card is articulated in three levels of elements, each of them can be repeated several times, as needed. The main difference is in the high level of fragmentation in the ICCD schema, while CIDOC is a little more agile. Many information items are in common, but sometimes they are grouped in a different way.

As the project was aimed to administrative purposes, too, we remained a little more adherents to the CIDOC schema, certainly checking conformance to the ICCD schema. As final result we got a schema that is compatible with the ICCD one:

- some fields are absolutely equal;
- date fields are expressed in a much more simple and expressive way (for reference, see [11]);
- the more articulated fields in the ICCD format can be extracted from our schema with an appropriate post-processing;
- where we have real differences, maybe the ICCD schema should be amended (multiple occurrences of some fields, need for representing geometrical info).

We consider particularly significant some aspects:

- in the proposed schema, for each site we have the site geometry, too, specifying its shape (point, line, polygon) and recording the relevant coordinates (*x,y,z*). In this way, we have a sharp distinction between the basic cartographic information and the geometric properties of the archaeological evidence. The net result is high portability across different HW/SW platforms and openness to emerging new standards and technologies (e.g. SVG).
- Language normalisation is achieved making use of dictionaries and thesauri. Considering the difficulty in obtaining an effective communication when terms are too specific, requiring an in depth knowledge and, perhaps more important, a complete agreement on the conceptual organisation of information, we limited the thesauri depth to a maximum of three levels.

The card structure

Usually, sites are described using a "*card*", which contains the fields relevant to store the appropriate information. Local cultural and scholar standards normally dictates card layout. The card used in this project is organised as a *structured document*, ready to be marked according the new emerging XML standard. Some sections are mandatory, others are optional; in many cases, this property is dependent on the context, as a section is considered mandatory only if it is absolutely impossible

to omit the information. A mandatory field in an optional section just means that the field is required if the section is going to be filled. This approach gives the opportunity of rescuing data from previous databases, and possibly having an incremental filling of the card, as information will be available.
It is easy to understand that this choice has been done in light of having a flexible environment, avoiding the common problem of unfilled cards just because some detail was not available at the first compilation phase.
There are cases when it is impossible to fill in a single value, but we need two or more values from the relevant dictionary or thesaurus. In other cases, it can be necessary to express uncertainty between two values. In both cases, appropriate syntactic writing rules accommodate these needs.
The card is articulated in seven sections, which in turn are structured in subsections and fields:

1. *Site name and identification*. This section is mandatory, and contains site identification codes, reference to other cards, archaeological collections, objects, interventions.

2. *Localisation*. This section is mandatory, and is organised in several subsections. They will be filled or not, in different combinations, depending on the possibility of having an exact geographic localisation or an approximate one.

3. *Type and chronology*. Mandatory section, can be repeated in case of diachronic evolution of the site. Compiler can specify exact or generic (periods) dates. Site is interpreted according to its shape or function, using terms semantically arranged in a thesaurus including the expression "undetermined".

4. *Conservation status*. Optional, can be repeated, and is used to document the conservation status and date of information gathering.

5. *Juridical/administrative condition*. Optional, can be repeated, and is used to document juridical and administrative condition, its originator, type and dates.

6. *Toponym/Site name*. Optional, can be repeated, and is used to report ancient toponyms and their existence time range.

7. *Site description*. Section used to synthetically describe the site, using terms included in a controlled dictionary. Description will contain: geographic context, archaelogical evidence, interpretation, chronological clues and any additional info.

This last section will be object of new and hopefully interesting developments, as we plan to introduce in it a "*light markup*" to keep all the flexibility and richness of free text, clearly identifying, in the same time, relevant semantic items. These items could be used to facilitate the hypertextual association among documents, triggering appropriate interaction paradigms, like map, classification models, time ([[8], [9], [10]).

Technologies

There are some well-established and robust technologies that constitute the frame of reference for every application in the field. In short, they are DBMS, IRS and GIS.
DBMS is really powerful in representing complex relationships and structures, and is the hearth of any information system. However, this approach requires high formalisation and atom-

ised information items. This can produce forcing that result in semantic poorness, evident in a semantically rich field as the cultural heritage. IRS keep the semantic richness of natural language, but some lexical problems contribute to maintain a certain level of ambiguity, causing low precision and recall factors. Perhaps more important, they are unable to manage complex relationships among information items, as the IRS document model is poorly structured. GIS are the most suitable tools to manage geometric and spatial information, supporting queries on these properties. They also allow the processing of spatial models. Their counterpart is that they make use of proprietary models and software tools, and compatibility among them is scarce or lacking ([2], [6]).

Anyway, each one of these technologies exhibits strong and weak points: no system relying on only one of them can be completely satisfactory. On the other hand, there are some *unifying technologies* that can help in finding an appropriate solution, especially in the current technology trends, that aim to the implementation of totally web based information systems. We can briefly refer to them as hypermedia and XML technology family.

Hypermedia approach is the most natural user interface. It can be seen as an application interface towards underlying systems. In this way, it is possible to implement complex systems that make use of the basic technologies, and emphasise the associative aspect, so relevant in the specific field. We must therefore carefully consider the increasing diffusion of XML, the basic component of the next generation web. Markup of hypermedia documents can preserve the semantic richness of natural language, reducing, in the same time, some of its intrinsic ambiguities. In this way, it is possible to overcome some of the constraints typical of the DBMS approach, even if DBMS maintains its importance as the "glue" of information. It must be stressed that an appropriate markup (or DTD, in XML jargon) requires an in depth analysis of information items, and a clear understanding of their semantics and relationships.

The proposed schema is fully compatible with this technological trend.

A first test

In a first test phase, supported by the Provincia di Livorno, 445 complete cards concerning its territory, out of more than 600 available in not machine readable format, have been recorded, adding georeferencing data.

In Figure 1, 2 and 3 we report some examples that can help in understanding basic functionality and interaction style.

Figure 1 shows the usage of a dictionary in entering Section 1 data about a site that can't be checked again, because of the site has been cancelled by some human interventions ("*Sito asportato da eventi antropici*").

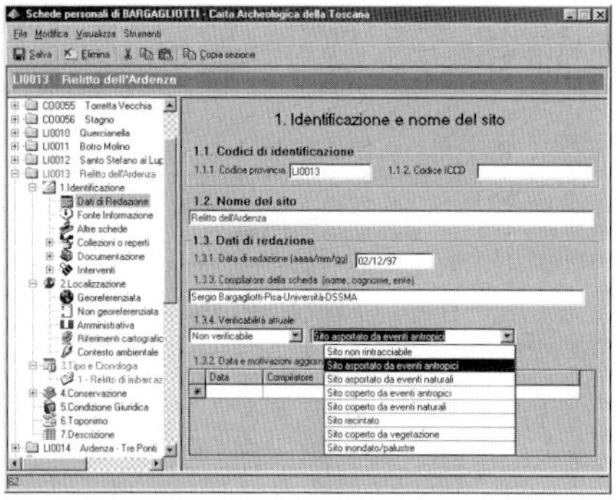

Figure 1: Entering data with help from a dictionary

Figure 2 shows as location data can be entered filling different combinations of subsections, and georeferenced data include the *z* coordinate (see fields 2.1.6, containing the geometry and 2.1.7 where co-ordinates can be entered).

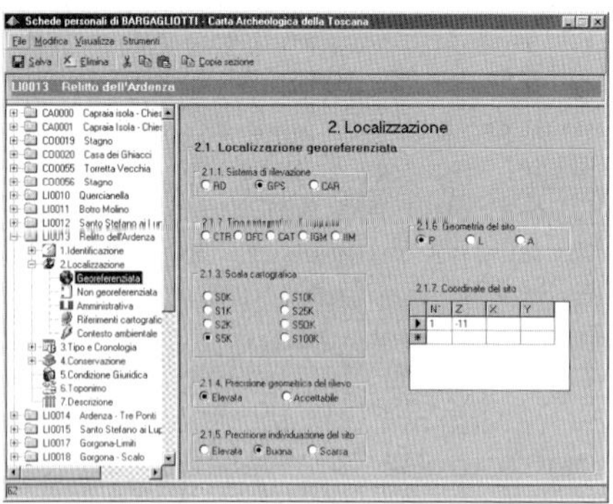

Figure 2: Coordinates include z.

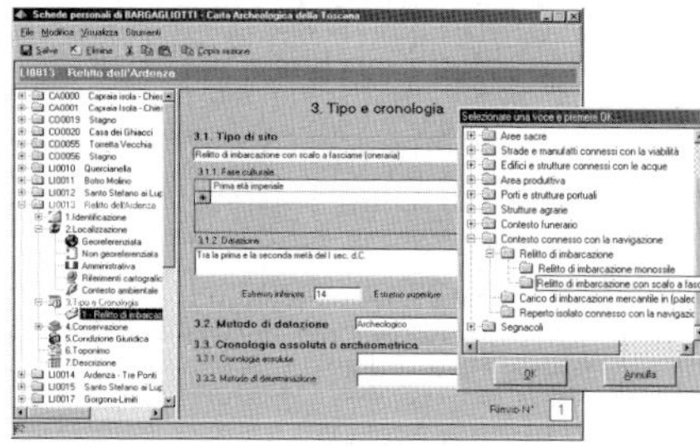

Figure 3: Browsing a thesaurus

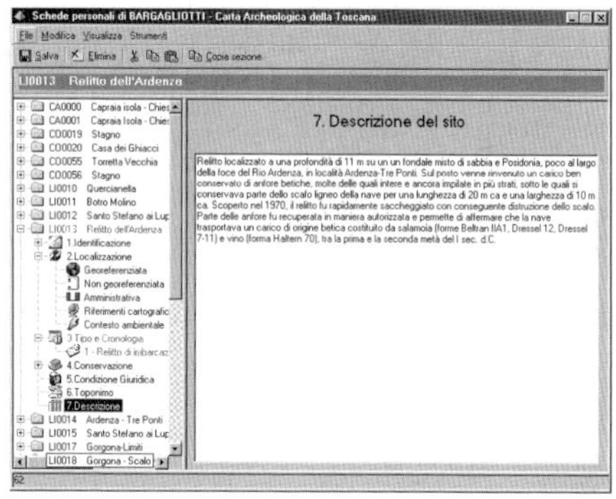

Figure 4: Free text section

In Figure 3 user can find the appropriate term browsing a thesaurus, looking for the exact name of a shipwreck type (*"Relitto di imbarcazione con scafo a fasciame (oneraria")*).

We stress that Section 7 of the card accommodates for a free text containing controlled terms, to semantically enrich the site description without constraining the user to fill a lot of fields (Figure 4).

During the experimental phase it was possible to verify the the model was adequate, obviously after some minor amendments of the first version. Therefore, during this first experimental phase it was possible to draw some indications about the adopted approach.

First of all, taking survey data and putting them into a common schema is useful to rethink previously collected data.

However, some effort is related to need of deeply entering into the system schema: this requires entering some tenth of "cards", to gather adequate experience.
In general, integration to the thesauri were very limited, as they were conceived with a large amount of terms, to accommodate for the foreseable scholars' needs. In essence, we had to insert some options to add to the term "*Other*" to enter information were data about the archaeological evidences do not allow to fully identify the appropriate term (e.g., sometimes the compiler is unable to state if an archaeological evidence is a "villa" or "necropoli", in such cases, (s)he can enter: "*villa/necropoli*").
Finally, the cultural dating (e.g. late republic, late empire, etc.) was used in conjunction with the absolute dating (expressed according [11]). However this task, mandatory for building appropriate historical maps, is really difficult, and is felt by the scholars as somehow too formal.

CONCLUSIONS

In defining the guidelines for the archaeological map of Tuscany, we considered both methodological and technological issues. We resulted in an approach that:

- Makes a sharp distinction among basic cartography and descriptive information, which includes all the elements needed to setting up position and geometry of the archaeological evidence.
- Emphasises role of dictionaries and thesauri, seen as the knowledge base shared between indexer and user.
- Defines a schema were number of fields is reasonable, scientific information can easily fit in, leaving room for insertion of semantically rich free (but somehow controlled) text.
- Is conformant to national and international standards.
- Is up to date with technology trends.

Is totally open to additional enhancements, and is portable in the Open Source Software environment.

Results from a first test phase were encouraging. We are going to check costs involved in migrating data gathered in other experiences to the new format, that can vary depending on the level of tightening to proprietary hardware and software environments.

ACKNOWLEDGEMENT

We warmly acknowledge members of the working group R. Francovich (Università di Siena), G. Nenci (Scuola Normale Superiore, Pisa), R. Palma di Cesnola and subsequently R. Gambassini (Università di Siena), M. Torelli (Università di Perugia), and staff members from ICCD, Soprintendenza Archeologica per la Toscana, Regione Toscana, who contributed to the work with their personal effort and were very effective in raising many methodological issues.
We also wish to thank the Regione Toscana, which promoted the initiative, and Provincia di Livorno, which supported the test phase.
S. Bargagliotti, A Del Rio and S. Menchelli where active partners in the whole development of the guidelines and their tuning after first tests.
Finally, G. Barletta, L. Cherubini, F. Cibecchini, M. Corsini, F. Fedeli, S. Meini and M. Vallebona contributed to record data.

REFERENCES

1. CIDOC (Comité international pour la documentation – Conseil

international des Musées, Groupe de travail sur les sites archéologiques (CIDOC-ICOM, Maison del l'Unesco) *Project de norme documentaire internationale pour les sites achéologiques*, 1995

2. Gillings M, Mattingly D., van Dalen J. *Geographical Information Systems & Landscape Archaeology*, Oxford Books, 1999

3. Papaldo S., Ruggeri Giove M., Gagliardi R., Matteucci D.R., Romano G.A., Signore O. *Proposta di strutturazione dei dati del catalogo: Beni mobili archeologici e storico-artistici.* (Edizione riveduta e aggiornata), Istituto Centrale per il Catalogo e la Documentazione (Roma) - Istituto CNUCE (Pisa), Multigrafica Editrice, Roma, 1988

4. Papaldo S., Signore O. *Un approccio metodologico per la realizzazione di una banca dati storico-geografica (A methodological approach to producing a historical/geographical databank)*, Multigrafica Editrice, Roma, 1989, ISBN 88-7597-105-6

5. Regione Toscana *Linee guida per la redazione della Carta Archeologica della Toscana*, (1998)

6. Sanjuán L. G., Wheathley D. W. The state of the Arc: differential rates of adoption of GIS for European heritage management. In *European Journal of Archaeology*, 2, (2), August 1999, 201-228

7. Signore O. From data structuring to data exchange: a simple path. In Marker, H. J. and Pagh, K. Eds. *Yesterday, Proceedings from the 6th International Conference Art History and Computing* (Odense, Denmark) August 28-30, 1991, Odense University press, 1994, 48-54, ISBN 87 7838 022 7

8. Signore O. Modelling Links in Hypertext/Hypermedia, in *Multimedia Computing and Museums, Selected papers from the Third International Conference on Hypermedia and Interactivity in Museums - (ICHIM'95 - MCN'95)* (San Diego, California, USA), October 9-13, 198-216, ISBN 1-88-5626-11-8

9. Signore O. Exploiting Navigation Capabilities in Hypertext/Hypermedia. In *HICSS-29 Annual Hawaii International Conference on System Science* (Maui, Hawaii) January 3-6, 1996, 165-175, ISBN 0-8186-7327-3, ISSN 1060-3425

10. Signore O., Bartoli R., Fresta G., Loffredo M. Implementing the Cognitive Layer of a Hypermedia. In Bearman, D. and Trant, J. Eds. *Museum Interactive Multimedia 1997: Cultural Heritage Systems Design and Interfaces – Selected papers from ICHIM 97 the Fourth International Conference on Hypermedia and Interactivity in Museums* (Paris, France) 3-5 September 1997, Archives & Museum Informatics 1997, 15-22, ISBN 1-885626-14-2

11. Signore O., Bartoli R., Fresta G., Marchetti A. Issues on historical geography. In Bearman, D. and Trant, J. Eds. *Museum Interactive Multimedia 1997: Cultural Heritage Systems Design and Interfaces – Selected papers from ICHIM 97 the Fourth International Conference on Hypermedia and Interactivity in Museums* (Paris, France) 3-5 September 1997, Archives & Museum Informatics 1997, 32-37, ISBN 1-885626-14-2

ABOUT THE AUTHORS

Marinella Pasquinucci is full professor of Ancient Topography, professor of Submarine Archaeology, professor of Ancient Topography in the PhD in Ancient History and in the PhD in Archaeology at the University of Pisa.
Her scientific interests are in landscape archaeology, ancient town planning, ports and calls, manufacturing activities and trade in Italy, from the hellenistic period to late Antiquity.
She published about 120 books and papers.
E-mail: *pasquinucci@sta.unipi.it*

Oreste Signore is full time staff member of CNUCE-CNR since 1973. His scientific interests are in the area of Information systems, covering hypermedia, user interface, web applications, information retrieval. He published more than 90 books and papers in national and international journals and conference proceedings. He had a relevant role in defining national standards for the cultural heritage. His present interests are in Semantic Web and adaptive systems.
E-mail: *oreste.signore@cnuce.cnr.it*

Private owners and public access; using the web to provide access to historic properties and their contents

Michael Hiley (*)

(*)Centre for Technology and the Arts, De Montfort University, Leicester, UK
E-mail: mrh@dmu.ac.uk

ABSTRACT
Museums and Galleries are the public repositories of national and regional collections of paintings, furniture, ceramics, silverware and other heritage items of great cultural significance. However in the United Kingdom there remain very extensive collections which remain largely intact located in more than a thousand historic houses which are privately owned. Many of these properties are open to the general public at least part of the year. The world wide web offers the opportunity to provide more extensive access, opening up collections for educational and academic purposes as well as encouraging tourist visits and thereby revenue generation. However changes in organisation and attitude are necessary to bring these rich resources together and to exploit to the full these new opportunities on–line.

KEYWORDS: web gateway, private collections, public access, historic houses, tourism

INTRODUCTION

The old church hymn refers to 'The rich man in his castle / The poor man at his gate.' This paper deals with historic houses and great estates in private ownership. In particular it looks at the way in which the development of web sites has begun to solve some of the problems of making public access to these private treasures possible.

From Jane Austen's *Mansfield Park* to Evelyn Waugh's *Brideshead* the great country houses – imagined and real – have had a central part to play in British cultural history. The contents of these historic properties are often of staggeringly high quality. The house used as *Brideshead* in the BBC television series – Castle Howard in Yorkshire – is reported to need several millions of pounds of work carrying out on the stonework of its mausoleum. However an album of drawings in the library has just been found to contain an early work by Michelangelo, worth perhaps eight million pounds. *The Guardian* newspaper reported:

> Castle Howard has a renowned art collection including several Holbeins and works by Rubens, Gainsborough and Reynolds. Simon Howard, who lives in one wing of the house, which featured in the television version of *Brideshead Revisited*, said the discovery came as a surprise – but meant the immediate loss of the drawing to the house. The cost of insuring it would be prohibitive, he said. "It

> should be on display in a national gallery or museum where everyone can enjoy it." James Millery, deputy chairman of Sotheby's, said: "It's like finding part of the Holy Grail. Everyone is extremely excited. It is the most significant Michelangelo work to be discovered in living memory." [1]

The drawing of a heavily draped woman is reported to be in superb condition, despite having been glued into the album almost 200 years ago.

How many public galleries in either the United Kingdom or in the United States have a work by Michelangelo in its collection? Very few. Yet it is still possible to see works by Michelangelo and Leonardo da Vinci on public display in the United Kingdom in *private* collections.

PUBLIC ACCESS VIA THE WEB

The growth of the World Wide Web has led to developments supporting public access to private collections which could not have been anticipated a decade ago. In 1844 the Duke of Devonshire allowed 'all persons whatsoever' to see Chatsworth every day in the year, Sundays not excepted, from 10 in the morning till 5 in the afternoon. In 2001 on–line visitors can see at least a part of some historic houses all day, every day of the year.

There are more than a thousand historic houses across the United Kingdom, many of them privately owned and lived in by families whose pedigree goes back for centuries. Unlike houses in some European countries they have not, for the most part, lost their treasures in the form of paintings, furniture, silverware and so on. Five years ago a consortium was formed called *Heritage on the Web* at an early stage in the development of the World Wide Web. This brought together myself as Project Manager, based at De Montfort University in Leicester, The Glass Page, a Leicester multimedia and web design company, and The Living Landscape Trust, a trust set up to provide access to Boughton House and its estate. Boughton is the Northamptonshire home of the Duke of Buccleuch, one of the largest landowners in Europe. His collection of art works includes paintings by Rembrandt, Canaletto, Van Dyck (forty of his paintings hang in one room alone at Boughton), El Greco, Reynolds and Leonardo da Vinci.

With the Duke's active support and encouragement Heritage on the Web was given a project award by the Department of Trade and Industry to develop a demonstrator web site for Boughton House. This went live in 1997 and is still under development. It is the largest and most advanced site for any historic property anywhere in the world. [2]

THE WEB SITE FOR BOUGHTON HOUSE

The project team built a site which uses a great variety of techniques to display hundreds of objects and items of information. It has been used by the British Government as a showcase of what is possible, demonstrating to others in the historic houses sector the benefits of going on–line. It has been to our great advantage that we have had the support of the Duke of Buccleuch and Queensberry, who is a very senior and respected member of the British aristocracy. This association has led to a growth in confidence amongst those who, without his example, would have been very reluctant to become involved with computer–based technology. As project manager I have had to act as the link and facilitator between the rapidly developing technology of the web and the world of the country house owner, which moves at quite a different pace and often with a very different perspective on the world.

The Boughton web site is designed to inform visitors how to find the house, what they will see when they visit, and much more. The nearest town is Kettering, which is unlikely to be known to many outside the English Midlands. Tourists tend to stick to the 'honey–pots' of British tourism such as London, Oxford, Stratford–upon–Avon, and so on. Boughton is in the middle of the countryside, off the main road and approached through villages. Maps giving accurate guidance and travel information are important for intending visitors, as are details of facilities such as tea rooms and children's playgrounds. It is significant that the Department of Trade and Industry and not the Department of Culture, Media and Sport set the project in motion. From the beginning one important objective was to establish a web site which would attract tourists and therefore business to the area. Coming from an academic background, the problem has been to reconcile these tourist concerns with the web presentation of the collections contained in the houses.

THE WEB AS AN AID TO SECURITY AND CONSERVATION

Boughton House is still used as a family home, but except for the private apartments, many rooms are open to visitors. The web site allows access to the Drawing Room, the Morning Room, the Rainbow Room, the Armoury and so on. Security considerations have meant that we have not provided a floor–plan of the house, nor have we gone out of our way to indicate which room leads to another. These decisions were made after consultation with specialist police advisers at New Scotland Yard in London, and with others familiar with security problems at stately homes.

Actual visitors to the house are kept back from valuable items by security barriers from items which are protected by alarms. Their extremely high value – a painting or piece of furniture might well be worth millions of pounds – has meant that visitors are not allowed to touch or handle items. To a limited extent this problem can be solved on the web, with virtual handling being provided in some instances. A more mundane problem which can be solved by web technology is to get the viewer closer to objects. A small painting on a far wall or an object in shadow on the other side of a room can be very difficult to see in detail at a distance. The web can help to overcome these practical difficulties.

Take, for example, the Drawing Room, where the rug on the floor is over four hundred years old and there are forty small paintings by Van Dyck on the walls. On the web site the on–line visitor can survey the room using an interactive QuickTime panorama taken from a point behind the safety barrier. The rug can be selected and examined, and each of the Van Dyke paintings can be chosen and viewed individually in close–up with accompanying details in a related frame of the web page. In other rooms we have experimented with animation of furniture and in one case a perspective effect in a ceiling painting. This is in the Third State Room, one of a sequence of rooms built in the late seventeenth century to entertain King William III. The State Rooms are only open occasionally to special parties of visitors and the light levels are always kept low for conservation reasons. The bed hangings, original upholstery on the chairs and the tapestries are very rare, valuable and vulnerable. But despite the real–world restrictions, all the State Rooms can be seen at any time on the Boughton web site. Handled with understanding and care, on–line access can provide new solutions to the concerns of private owners regarding security and conservation.

CHOICE OF WEB TECHNIQUES FOR DISPLAY

I will use the web pages covering the Great Hall at Boughton to describe some of the features of the techniques used on this site. The Great Hall page is accessed as part of the Tour of the House.

A variety of presentational techniques are used, including interactive image maps and links which provide both selected images and related information. The Hall was in the past lit by candles mounted on solid silver sconces attached to the walls. We have provided a QuickTime 3D object presentation of a candle sconce which can be 'handled' and turned on screen. This is a technique which could be used extensively in this type of situation. The two factors which restrict this approach are the time employed to take the large number of photographs required and to process them, and the fragility of objects which would have to be handled. It was on this second ground that we decided not to risk the beautiful and detailed porcelain swans at Boughton which had belonged to Madame de Pompadour. Instead they are shown as a still image.

It is difficult to feel confident when you are handling a fragile item of great value and in addition to this natural concern it is a fact that to photograph a VR object requires potentially hundreds of photographs taken under hot lights. Our experience has taught us that this process is very time consuming and full of issues relating to security and conservation. It is by no means easy to prepare images of the turning object against a neutral background at a consistent angle and distance from the camera lens. However, if these problems can be overcome, the results can be spectacular and justify the description of 'virtual handling' of objects which the public would not otherwise get near.

One of my favourite paintings at Boughton is the portrait of Elizabeth Vernon, the Countess of Southampton, which hangs next to that of her husband, the Earl. She was a great beauty of her day, a lady–in–waiting to Queen Elizabeth. The Countess is painted in the act of dressing, her fine clothes and her jewels laid out ready, in a work which both displays her beauty and wealth and at the same time points to human vanity in the Elizabethan tradition.

The detail in this painting is of particular interest; the jewels and the embroidery of her clothes can only be fully appreciated close–up. I had seen the ImagePump technology used by the Uffizi Gallery in Florence for some of their paintings and managed to arrange for a link to be made to the experimental display site of the Xippix company in California so that their zoom technology can be used to show the fine detail of the Countess portrait and of other objects at Boughton.

Many of these techniques are used on web pages for other rooms at Boughton and also for the grounds and estate. The initial support of the British Government and the backing of the Duke of Buccleuch have enabled us to make considerable advance in employing web display techniques to show priceless objects of great cultural importance.

PUBLIC ACCESS TO PRIVATE PROPERTY

We have given many demonstrations of the Boughton House web site to the owners of private houses, their curators and their archivists. In 1997 there had been almost no moves made to build web sites for privately owned historic houses in Britain. With no central funding available, a very patchwork, incomplete and often amateurish pattern of web sites has grown up on the web. Some sites branched off from local government web pages, others linked to hotel chains.

From the moment the Boughton web site went live it immediately began receiving visits on–line, and we received expressions of interest from all over the world. In the season following the setting up of the web site visitor numbers to the house were up by over ten per cent, and in particular American visitors we have spoken to in person describe how they have already visited Boughton in cyberspace, touring the house, opening cupboards and handling the silverware as the site allows, before they fly across the Atlantic.

Scholars visit the site. The portrait shown above of Elizabeth, Countess of Southampton hangs in the Great Hall, next to that of her husband, who was the patron of William Shakespeare. When recently it was claimed that Elizabeth was the Dark Lady of the Sonnets it was possible to refer queries to the relevant web page which provided details and also a colour image (at low resolution) to download. There were many queries both from academics and from the press worldwide.

UK TAX LAWS AND 'CONDITIONAL EXEMPTION'

A very important influence on patterns of opening historic houses has been the United Kingdom tax laws relating to what is known as conditional exemption. Under these arrangements works of art, buildings and land of outstanding historical interest can be exempted from Inheritance Tax. As the current rate of Inheritance Tax is forty per cent, and as it starts at just under a quarter of a million pounds, the impact of such a tax is enormous. The eight million pound Michelangelo could attract inheritance tax of over three million pounds. And that's just one drawing. The conditions placed on the tax exemption include arrangements for the maintenance and preservation of the heritage property and – of particular importance to anyone interested in communication over the web – a strong enforcement of making provision for public access.

All the works of art which over the years have been granted conditional exemption – paintings, furniture, ceramics, silverware and so on – are listed on the web by the Inland Revenue. [3] There are around twenty thousand tax–exempt items spread in private collections all over the country together worth an estimated total of one *billion* pounds on the list.

Unfortunately the web site only gives basic details and no pictures. Potentially these items could be presented on the web and provide a virtual national museum of immense importance and quality. So far my discussions with the Inland Revenue and the government have not yet led to the necessary changes in the tax regulations to make this possible, although my ideas have been discussed by a Parliamentary Select Committee. Eventually the creaking system of negotiating actual access to individual items will, I'm sure, give way to access on–line for everyone, and this is something which I think will be of immense benefit to the general public.

Heritage on the Web has to be very aware of the security fears of private owners – probably the most important reason for not putting more details of heritage assets on–line. The owner of a priceless piece of cultural heritage may well be a little old lady isolated in the middle of a deer park at considerable risk as she shares a huge house with millions of pounds worth of art treasures.
Many private owners are strongly opposed to having any information provided on-line about their property or their tax affairs. Our task is to gain the confidence of owners and to build web sites which deliver what they wish to present to the public. Our rôle is not to force anyone on–line.

HERITAGE ON THE WEB

The Boughton web site was the first stage in a much more extensive plan. Heritage on the Web established itself as a commercial company and set out to persuade historic properties to commission web sites for houses across the United Kingdom. This process has now been going on for over three years and over a dozen web sites are now live or under construction. No other company has built more than one web site for an historic house.

In more than one instance we have found that in order to 'get the house on line' the owner of an historic property, not sure how to proceed, turned to the bright offspring of one of his estate staff to build a simple web site. It's certainly true that people working on the estate know a great deal about these great houses and their grounds. But as the web continues to develop the basic sites they produce are revealed as elementary in terms of both design and function. Enthusiastic amateurs also produce their own unofficial sites for houses, sometimes linked to universities or to local tourist sites. Occasionally owners who, perhaps, have business connections in London commission web sites at great expense from new media studios working in the field of dot.com. These are the only historic house sites which even begin to display the sophistication to match the quality of the best museum and gallery web sites.

Many owners and their agents have little idea of what is possible over the web and have no more experience of presenting their properties and their contents than commissioning a local graphic design company to prepare and print a guide book. Even then they may be concerned that if they show too much of the property they may not be able to attract visitors in sufficient numbers. These are private houses and private collections, not public institutions with public funding and an obligation to provide public access. The result is often web sites of poor quality carrying little more than basic visitor information – individual, limited and isolated sites.

In order to locate web sites together and therefore increase traffic through them we have established a web gateway - Heritage on the Web. [4] This acts as a gateway leading to the sites which we have built for historic houses across the region. The gateway acts as an easy access point for visitors interested in historic houses and it also allows us to guarantee quality control and regularly updated information on sites reached through the gateway. In this section of Heritage on the Web are grouped houses including Boughton, Kelmscott Manor, Grimsthorpe Castle and so on. Each design is suited to the property which has commissioned the site. For example the house reconstructed at 78 Derngate in Northampton by the Scottish architect Charles Rennie Mackintosh reflects his design and style of lettering.

Each house has different demands. Renishaw Hall, near Sheffield, is particularly proud of the Italianate gardens facing to the South. It is the family home of the Sitwell family, famous as writers and poets. Often the cultural content of a house is potentially very extensive and important and I hope to be able to use money from academic research funds to enrich and develop the on-line resources to which houses are beginning to give us access.

We have developed a web site for Drumlanrig Castle, one of the homes in the Scottish Borders of the Duke of Buccleuch and Queensberry. It is reached through the Scottish section of the Heritage on the Web gateway, but it can, of course, also be accessed directly. Here the Leonardo painting *The Madonna of the Yarnwinder* is on display in the Staircase Hall. In the same room hangs a fine Rembrandt and a Holbein. At present these are not shown on the Drumlanrig web site. Part of the reason for this is security. Although the Duke recognises the importance of paintings such as these and the desirability of making them available them to the general public practical reasons mean that some items of his impressive collections are not on show. For example the collection of miniature paintings belonging to the Duke used to be on exhibition at Bowhill House, near Selkirk but concerns about security after a series of thefts in the Scottish Lowlands led to them being withdrawn from display.

At present we are receiving requests from commercial organisations that wish to draw on the traffic flowing through the web gateway in order to offer specialist services or goods for sale. Provided that developments of this kind are handled with care I see no objection to providing links between the commercial and the cultural worlds. Heritage on the Web has built a web site for Cottesbrooke Hall, one possible model for Jane Austen's *Mansfield Park*. In the garden there is a statue of Homer by Peter Scheemakers, which was originally in the Temple of Ancient Virtue at Stowe. Many visitors take as much pleasure from the gardens as the houses. Other estates have; for example, garden sculptures by Jan Van Nost the Elder. To me it seems only to add to the pleasure to know that it is possible to buy modern casts in lead taken from these originals and now cast by Bulbeck Foundry in Cambridgeshire. A planned section leading from the Heritage on the Web gateway will provide e–commerce services for historic houses.

Already many objects can be bought on-line, and we are developing this section of the web site to provide a service for houses and companies. The houses themselves have to employ teams of craftsmen and women to maintain the buildings and their contents. At Burghley House in Lincolnshire the courtyard buildings are full of small firms specialising in conservation and restoration. Perhaps it is in the very nature of these privately owned and maintained properties, requiring entrance fees from visitors to keep the place going, which makes an unforced linkage between commerce and culture.

THE FUTURE FOR HISTORIC HOUSES ON THE WEB

One disadvantage which historic houses face is that they are usually unable to draw on public funding to any great extent, despite the fact that they are repositories of enormously important cultural artefacts. Beyond their park gates the museums and galleries receive financial and technical support. Yet the private and public collections inhabit the same world and false divisions between them should not be made, particularly where this affects access to cultural heritage on–line. At present I am dealing with a request to make a web link directly from the on-line portraits of the Dukes of Buccleuch held by the National Portrait Gallery to the web site which we have built for Bowhill House, one of the Scottish homes of the Duke of Buccleuch.

I would like to go further and experiment with deep linking between sites so that the portraits which I mentioned which are in the Great Hall at Boughton of the Earl and Countess of Southampton can link directly to National Portrait Gallery paintings of them in later life. Perhaps it would also be possible to build a VR presentation of the Earl's ceremonial, engraved armour, which is in the Royal Armouries in Leeds in the North of England. Clearly the web can function very well to bring together items which are physically dispersed.

In many cases there are already close links between private collections of painting and furniture in the United Kingdom and public galleries such as the National Portrait Gallery and the Victoria and Albert Museum. It makes little sense to set up false divisions separating items in public and in private ownership - with the 'conditionally exempt' items holding the middle ground.

CONCLUSIONS

The web gateway which we have built is itself only a stage in the continuing evolution of the web and the way in which it can provide access to sites and objects of cultural interest. The way forward will involve interlinked databases and hopefully links between privately owned items and those in the great national museums and galleries, many of which originally came out of private houses.

One interesting development is the web site set up by the UEHHA (Union of European Historic Houses Association). [5] This has been sponsored by Pirelli and Compagnia di San Paolo. Its first phase covers privately–owned historic houses across Europe from the Château de Roquetaillade in France to La Casa de Los Balcones in Tenerife, and from Fraelemaborg in the Netherlands to the Palazzo Agliardo in Italy. Boughton House also features, though only basic details are given. International initiatives such as this and detailed national gateways such as Heritage on the Web point the way forward.

The fragmented sector of historic houses in the United Kingdom and across Europe will need guidance, organisation, help and funding to move forward and bring their web access up to new and changing standards. It is a great challenge, but the houses and their contents are of such international importance, as well as great tourist attractions, that the rewards to be gained if this can be achieved are themselves immense.

REFERENCES

1. Maev Kennedy, arts and heritage correspondent, '38m Michelangelo unearthed,' *The Guardian*, London, Wednesday October 11, 2000.

2. Boughton House, *http://www.boughtonhouse.org.uk*

3. Inland Revenue web site listing tax–exempt Heritage Assets, *http://www.cto.eds.co.uk*

4. Heritage on the Web, *http://www.heritageontheweb.co.uk*

5. Union of European Historic Houses Associations, *http://www.uehha.org*

ABOUT THE AUTHOR

Michael Hiley is Senior Lecturer in the Centre for Technology and the Arts at De Montfort University, Leicester, U.K. and is consultant for Heritage on the Web. He recently acted as national editor of the social history section of the U.K. universities' HELIX project, working with Getty Images, and is currently directing the development of a public access system for the collections of Burghley House, Stamford, U.K. E-mail: *mrh@dmu.ac.uk*

SIUSA Sistema Informativo Unificato per le Soprintendenze Archivistiche (Unified Information System for Record Office Superintendencies)

Daniela Bondielli [*], *Umberto Parrini* [*], *Maria Grazia Pastura* [#], *Giuseppe Alberto Romano* [*]

[*] Centro Ricerche Informatiche per i Beni Culturali,
Scuola Normale Superiore, Pisa, Italy
E-mail: bondielli@cribecu.sns.it, parrini@cribecu.sns.it, romano@cribecu.sns.it
[#] Ministero per i Beni e le Attività Culturali – Ufficio Centrale per i Beni Archivistici,
Divisione III – Vigilanza, Roma, Italy
E-mail: dirvigil@beniculturali.it

ABSTRACT
What we present here is a conceptual model to manage a complete information system related to the archival material. The model is divided into two sections: the first for the archivistic heritage descriptions, the second for the archivistic heritage management. The two sections are physically and logically separated but they are linked together so that the same records can be described in two different ways: according to the Archival Bureau requirements and according to the international standard description: ISAD (G) – ISAAR (CPF). The system has been designed to map only the high level structure of the archival documentation, but it will be also able to link the analytic descriptions located on different archival systems in Italy. The aim of the project is to set up a Archival Description Distributed System for the italian archival community and, at the same time, to store the old archival descriptions produced by the project *Anagrafe Informatizzata degli Archivi Italiani* during the last ten years.

KEYWORDS
Archival Information System, Archival Logic Structure, Distributed Information Access

INTRODUCTION
The proposed project has been developed by a team of the *Ufficio Centrale per i Beni Archivistici, Divisione III – Vigilanza* and aims at reengineering the national information system known as "Anagrafe informatizzata degli archivi italiani" for the part concerning the institutional functions of record office Superintendencies. The team, composed by members from the above mentioned office, as well as officials from both

Superintendencies and state Record Offices, is supported by members of the *Centro di Ricerche Informatiche per i Beni Culturali* (*CRiBeCu*) of the *Scuola Normale Superiore* in Pisa; the latter are responsible for analysing and developing the information system.
After a long and careful analysis, a conceptual model of SIUSA has been defined. The model is divided in two separate areas: an administrative managerial area and a descriptive area.
The two areas are physically and logically separate, though they can be linked together giving a twofold presentation of one and the same record, i.e. a brief presentation, as required by Superintendencies, or one based upon a standard model. The principle is to create a descriptive section to be accessed by external users, as well as an administrative managerial section for internal use of Superintendencies, i.e. to support them in quality of supervisors.
In defining the structure of the descriptive section we have followed international standards, ISAD (G) and ISAAR (CPF), to present information; as to the administrative managerial section, in particular concerning the definition of preservation tasks, the directions contained in a law issued on October 29th 1999 (decree n. 490) in matter of cultural and environmental heritage have been accounted for.

DESCRIPTIVE AREA

Three objects form the main data bases of the system:

- Archive
- Keeper
- Creator

For 'Archive' we mean a consistent and homogeneous set of records; indeed, in line with the aims of the project, we have considered only the 'higher' levels in the hierarchy of an archive such as fonds, subfonds, series, subseries. All descriptive information is presented according to ISAD (G) standard rules.
In the 'Keeper' file information about the preservation of records that might be of help for the user consulting them has been gathered.
Information regarding the 'Creator' is handled separately, in line with ISAAR (CPF) rules, and interrelated with the description of records.

Besides these major entities, the descriptive section also contains files that are stored into 'complementary' data bases, completing the information about the sets of records and/or their creators, i.e.

- Finding aids
- Location (places of residence and/or existence)
- Political or state context
- General institutional outline
- Bibliography
- Sources
- General documentary outline
- Compiler of the description

All these objects form independent data bases that can be linked up with the help of defined relations; these relations can be of the one to many (1:n) or of the many to many (n:m) kind. To establish the connections – associations or relations – between different objects the system assigns a code (ID) to each so as to identify it in univocal manner.

The relations that can be established between the major objects of the system and the remaining data bases are summed up in the diagram below:

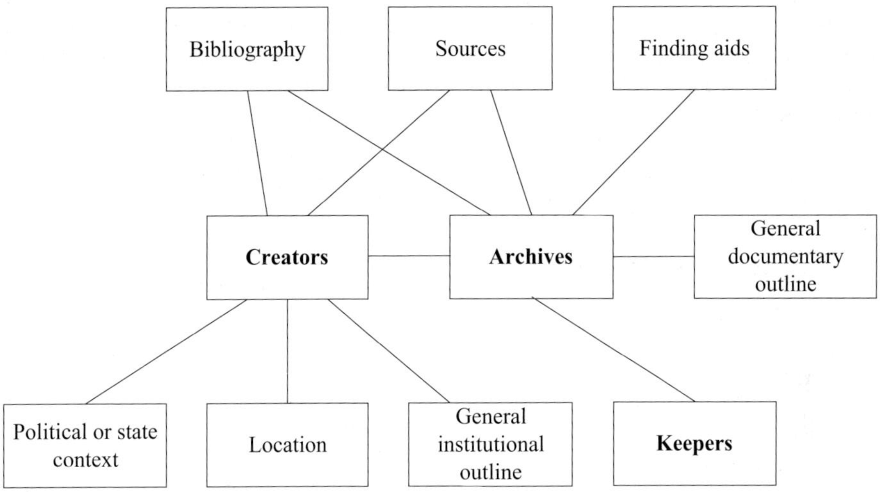

ADMINISTRATIVE AREA

The administrative part of the system is divided into three main "complexes":

- Archive controlled
- Owner
- Place of conservation

For "Archive controlled" we intend a set of records, identified and described in all its parts or only very briefly, subject to operations aiming at its preservation.

For "Owner" we intend people, families, public or private institutions who own, possess or keep records, with which Superintendencies interact in their activity of conservation and supervision.

For "Place of conservation" we intend the place were records are kept; in particular, the address as well as specific information about the exact "Location" are given, since several archives may be kept in different rooms on one and the same premises: it is, therefore, essential to provide precise information on the place where records are actually stored. The place must fulfil certain technical requirements to be declared suitable for the preservation of records.

Besides these main objects, in the administrative area we find files forming "complementary" data bases that complete the information about Archive controlled, Place of conservation and Owner of the records. The system also makes it possible to provide information on the "Administrative persons", i.e. staff or people who interact with the Superintendency on various grounds.

Moreover, the Owner file can be linked to the "Responsible", identified on the basis of his/her personal data, who in some cases, particularly in large archives, may represent the liaison between Owner and Superintendency

and be directly in charge of the archive. The Archive controlled, on the other hand, may be linked up with two different data bases: the first containing a brief description of "Supports other than paper" of which it might be composed, the second describing the "Finding aids" deposited at the Superintendency.
Both Owner and Archive controlled, moreover, can be linked up to the descriptive area of the system: the Owner will be linked, when appropriate, with the corresponding Keeper; the Archive controlled will be linked with the Archive, usually at the highest level of hierarchy (fond): the description of records according to standard rules can therefore be seen also from the administrative managerial area.
The described objects form independent data bases that can be jointed using defined relations. We have already mentioned the kind of links that may be established between different components of the system; here we simply sketch the possible links between objects of the administrative managerial area:

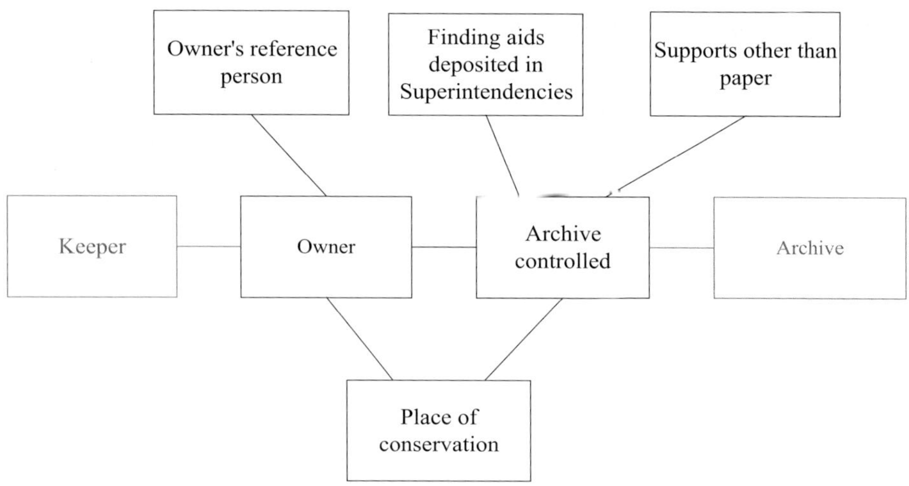

Besides all these entities, in the administrative area we may also register information that have been identified as fundamental for the tasks of survey and preservation performed by Superintendencies. We therefore call *Functions* the acts (of permission, assent or declaration) and the tasks (inspections, interventions, controls) that derive or are part of the supervisory function according to current regulations and, in particular, to the directions issued by law 490/1999. The identified functions are suitably plotted and, when appropriate, linked up with other objects mentioned above,

according to whether the function is performed on Archive controlled, Place of conservation or Owner; each Function can be further linked to other Functions deriving from it.
The architecture of the system as a whole permits the organisation of a data base whose information is handled locally and that can also be consulted, entirely or only in part, via http; some of the recorded data, moreover, can be transmitted to the headquarters. For example: the information stored in the administrative area will help to manage the task of supervision within each single Superintendency and will, therefore, hardly be visible to the end user, though some of the data will necessarily form part of the periodical reports that Superintendencies have to produce to their Ministry.
As to the descriptive area, most of the information contained in it will have to be visible to the end user; this may be implemented either with systems spread over the territory or with a single centralised data base that receives only part of the data and links up with local systems. In the latter case we will have to determine kind and amount of centralised information: for this purpose suitable fields have been identified in the descriptive area providing standard denominations and creating headings to access information.
Finally, we must point out that the SIUSA system allows only the description of higher level records (fonds or parts of these); other data bases (i.e. catalogues) have to be set up using different software tools. For this reason files contained in the archive should have a URL reference enabling the access to analytic descriptions of records, where there exist.

REFERENCES

1. International Council on Archives, *ISAD (G): General International Standard Archival Description*, Second Edition, Adopted by the Committee on Descriptive Standards, Stockholm, Sweden, 19-22 September 1999.

2. International Council on Archives, *ISAAR (CPF): International Standard Archival Authority Record for Corporate Bodies, Persons and Families*, Prepared by the Ad Hoc Commission on Descriptive Standards, Paris, France, 15-20 November 1995.

Oriental Brightness into an European City – Bucharest of '30 years

Laura Ciocoiu (*) *and Dan Falcan* (#)
(*) Research Institute for Informatics, Romania
E-mail: ciocoiu@ici.ro
(#) The Museum of Bucharest, Romania

ABSTRACT
This paper is concerning with the cultural heritage presented Web musems used for the reconstructing human memory of the interethnical influences into the culture.
The museums represent the cultural heritage and patrimony, the Web musems contain both the digital documents and multimedia techniques.So, the paper describes the instruments used to reveal the people the oriental influnces into Romanian culture (mentality, arhitecture etc) and to recontruct, by simulating them, the demolished cities during nineteenth and twentyth century, like the old Bucharest.
The Web museums has the role to disseminate knowledge about the Romanian culture as an interference cultural zone (life and interests of our ancestors) and to present the demolished cities such as:
Bucharest of the end of nineteenth century and the beginning twentyth century as "The little Paris", as it was known in that period, characterized by its renaissance buildings, monuments and its café life, most of the demolished in 30' years and during communism period
The Web museum reconstructs the memory from patrimony, because the memory has an important role of remembering and learning; so, it helps the young people to know and understand the culture the history and the evolution of the town and life in our country.
Our paper will address the specific technologies and pedagogical approaches, such as art-and-science Web sites.

KEYWORDS: cultural heritage, reconstructing human memory, digital documents, multimedia, hypertext database, metadata schemas

INTRODUCTION
Situated at a crossroads of civilizations, graced with a harmoniously distributed relief, Romania brought an original touch in the concert of universal history and culture, undoubtedly, part of the European one.
Latin as origin and christians of Orthodox rite as faith, the Romanians lived from the Middle Ages to the modern times on the location of influences of the Byzantium Empire, Ottoman Empire, Czarist Russia, the Habsburg Empire (in Transilvania) and French and Italian culture; they managed to preserve their state entity, faith and civilization.
Beginning with the XVIIIth century the modernization represented the major goals of the Romanian national movement. Capitalist elements of develop, in the Bucharestan economy. Commerce and manufactures are

flourishing. In the late Middle Ages they built both in the italian and oriental style; in modern times, Bucharest got a new look – besides the existing buildings with oriental influence they are built in french renaissance style. Unfotunately, in **1930's Bucharest suffered the sistematization work** and a lot of superbe buildings of the patrimoine in just downtown were demolished

But, the **huge damage was made by the communism regime.** All the country's elite destroyed, the mechanisms of social life were paralysed, all the country's resources was used for building mammoth projects, to satisfy the dictators' primitive megalomania, brought about a decline in the population's living standard down to the bearable limit and the country's isolation from the rest of Europe. The total misery. The huge part patrimoine on entire country, especially in Bucharest, was destoyed and transformed into a desert.

So, there are very important reason to present, to the world and to the young romanian people, the culture as an interference cultural zone (life and interests of our ancestors) especially the demolished cities such as: Bucharest of the end of nineteenth century and the beginning twentyth century as "The little Paris", as it was known in that period, characterized by its renaissance buildings, monuments and its café life,

PROJECT GOAL AND OBJECTIVES

The Web Romanian cities museums with sites as Bucharest,"the little Paris" and countryside offers a virtual tour into the romanian cities and villages of the end of nineteenth and the beginning twentyth century, organized as a collection of web pages which offer a sort of guide both into the museums and in time in those cities (*http://risc.ici.ro/mbucur/*). The similar sites for countryside are under construction

The goal of the virtual Web museums project is to make the museum content more accessible to visitors especially the young romaninan. Another purpose is an educational one and it consists of reconstructing memory.

Starting with simple virtual Web museum, this project creates specific **agents** for a reading and understanding the hypertext and construct a metadata record for each branch of hypertree. Then, according to the metaschemas the information are regrouped and catalogued by thematics as they are shown in the **results.**

The resources in virtual Web Museums exist in various but known servers and sites, the use of metaschemas of agents permit the search of information specific for a thematics or a tree of thematics. The metadata contains cultural information and they can be used to improve resources also has the potential to enhance Internet search accuracy, improve resource discovery capabilities, and facilitate cross-domain interoperability.

The objectives for the project are:

- define agents for metadata schemas to catalogue the information by thematics.
- create an algorithm to search into the site hypergraph
- develop a system of of pointers to metadata.
- design an interface to access indexed to the tree of hypertext

Methods, Models, Methodology

The Web Romanian cities museums are based on hypermedia design; the information is divided pages by pages corresponding to the context; the layouts of the screen (color, fonts, and other visual elements) and the relations are creating according to the the context of that period. The navigation among the information permits a tour, in time and space, in that disapeared cities and villages. Small portions of the image can be viewed in detail (bitmap mask), together with the explanations and details about personalities involved; we've used scripts or JavaScript. The images are divided into dozens or hundreds of invisible regions. When the visitor points their cursor at a region, a description is displayed next to the cursor. Thus, a mass of data is navigated. The visitor only sees the image and one description at a time. As they point their cursor at a sample location, the corresponding section and description is displayed. Only the painting and one section are viewed. Different places and their images in time so the visitor cam make comparison; The visitor navigates a virtual reality scene, looking up, down and around. explain the historical context.

Technology

The information is organized hierarchically; the nodes relate to one another within a conceptual architecture,

Relationships between the different components of a Web site is used to create the hypergraph and finally the hypermedia applications. Reading and understanding the Web virtual museum needs technical knowledge and cultural design; knowledge about cultural design and its context needs to be contained as rules and facts into base. This little known vector represents process of imagining Web page design.

Constraints consist of :

- defining the hypergraph with complete localization of interesting parts of the site
- graphics with localized labels, captions, and legends, and access the content.
- alternative ways to navigate a site, including search by thematic.

link labels that make actions predictable for all users.

Results

In the next paragraph, we present the design of the project that contains the thematic : oriental influence and brightness face to the renaissance french buildings into an romanian city – Bucharest, as they are organized from the hierarchy of sites about Bucharest (*http://risc.ici.ro/mbucur/*) and another sites about romanian towns and village, all at the end of nineteenth and the beginning twentyth century (not on the server yet).

- Buildings in **oriental style** - in the eighteenth century

Stavropoleos monastery and curch (**bellow**), built by a rich greek monk named Stavropoleos, in 17th c. In 18th c., the monastery was used as a inn. The curch has a superb decoration of colored faianta (blue, yellow and green). They represent a jewellery as arhitecture and style.

The "Manuc" Inn (17th c.), built by a very rich owner of land and shops named Manuc (armenian as origin). This inn was famous by its beauty and used for diplomats and commercants.

Beginning with 18th c, the architecture of the country was strongly influenced by in **italian and french renaissace style.**

Places in downtown in the nineteenth century for representing the interference cultural zone, as shown bellow:

- An example of interference of oriental style and brightness (the ordoxe curch) face to the renaissance french influnces (the buildings in the background and the old clock)

The **italian and french renaissace style** - at the beginning of nineteenth century –

One of the **typical commercial place -** the **italian and french** influences are observed on city squares with radial streets and buildings:

- The flower market

At the **end of of nineteenth century**, Bucharest received strong inluences from occident (high schools were founded, first stome paved streets, railroad stations, tramway line (with horses), electric illumination system, the Royal Palace, National Theatre, the Public Gardens,
electric tramway line, frst cinema.
The café-life in "the little Paris" (The Otetelesanu Garden famous for the artists ho used to come here)

A walk in time, on the streets and boulevards of this beautiful town, by the end of the nineteenth century - under the vaults of the boarding trees, so pleasantly shady during the summer heat - will lead us through a town that is no more and which fully deserved its name, "The little Paris".

- The main street "Podul Mogosoaiei", famous for its theaters, café, hotels and restaurants.

We shall observe the French influence on some city squares with radial streets or we shall reach streets flooded with vegetation. We shall admire monuments, and we shall try to make revive the atmosphere of the last century; we shall notice its changes and evolution...

<<Bucharest is a creation of our own nation, it expresses our originality, it is what were able to do... By its own vitality, Bucharest has managed to impose itself as a unique Capital, that had received all the attentive care of some great personalities we have not fully understood, to show then the gratitude they deserved.>> - Nicolae Iorga*) - 1939

*) a famous romanian historiographer

Further Work

Further work will consist in both the developing the Web sites for reconstructing the towns and villages of demolished or destryed during communism period, outstanding the the interference of the french, italian, oriental influence and brightness with romanian original architecture and the simulation of parts of these villages and towns as they looked before the communism period using multimedia technologies.

The developing of these sites permits to present the culture of this part of the world both in educational purpose for the young people and for the foreigners.

CONCLUSIONS

Web sites.are very important for the confluence of education and art and creates a vital, dynamic medium between schools and museums. These sites represent a medium developed in a multicultural environment to facilitate diverse audience needs. As the old Bucharest was a multicultural place where the european influences, especially freanch and italian ones, interference with the oriental decoration style; users received useful information supportive of the content for reconstructing memory or an old experience.

Our point is to suggest that the dissemination of cultural heritage in the age of digital networks deals with the transmission of memory as experience and that Web Museums have a specific role to play as far as this reconstructing memory approach to memory is concerned. Consequently particular attention will be given to those artworks for which the net is not only a tool for technical reproduction but the very texture of the artwork and the exhibition place for its appearance or disappearance.

The goal of the virtual Web museums project is to make accessible to visitors especially the young romaninan the museum content, thus an significant part of a beautiful city that exists no more. Another purpose is an educational one and it consists of reconstructing memory or an old experience in a multicultural environment.

ACKNOWLEDGMENTS

The authors wish to acknowledge the outstanding research performed by Dan Donciulescu, Cornelia Lepadatu and Anamaria Borozan. My paper uses the technology developed by them. We also wish to thank Cosmin Boca, Dora Coardos and Ana Mitescu for their skill and generous work and Sfetcu Paul, Elena Jitaru for their advice.

REFERENCES

1. Fernandes, T. (1995) - Global interface design. Boston: Academic Press
2. Hall, E.T. (1989) - Beyond culture. New York: Anchor Books
3. Nielsen, J. (1999) - The top ten new mistakes of Web
4. design., The Alertbox: Current Issues in Web Usability
   i. consulted February 14, 2000
5. Schriver, K. (1997) - Dynamics in document design. New York, Wiley
6. *** *http://www.ici.ro/romania/*
7. *** *http://nt1.ici.ro/history/*

ABOUT THE AUTHORS

Laura Ciocoiu is a I am Senior Research and Professor Associate. I just submitt my PhD thesis in “Advanced Methods for Higher Education“. My main research work is on **Intelligent methods** for network simulation **in teaching process - hierarhical problem- solving using intelligent agents** andon Object-Oriented-Design **on** Intranet and Extranet systems and Web site design. I‘ve had training grants USA and France.
E-mail: *ciocoiu@ici.ro*

Dan Falcan is an historian and journalist, specialist of Bucharest **history;** member of several organizations for **historical monuments protections;** member of editorial board of **“Corint” magazine (Grecia);** Address : Bucharest Museum, 2 Bd Bratianu

The Joachim DeBrum Photograph Project; Digital Restoration and Archiving in Micronesia

Eric Lindborg(*), Cristina Lindborg(*) and Sue Rosoff(#)
Marshallese Cultural Society, Kwajalein, RMI
PO Box 1593 APO, AP 96555 USA
(*) Email: lindborg@tcsn.net
(#) Email foto_zelda@yahoo.com or sooz@sunset.net

ABSTRACT
The Marshallese Cultural Society is undertaking digitization, retouch/restoration, printing and electronic storage and archiving of over 2400 photographic glass plate negatives created between 1890-1930 in the Marshall Islands. This poster paper presents a summary of the technical, organizational, cultural processes that have brought the project to fruition.

KEYWORDS: glass plate digitization, Micronesian history, Joaquim DeBrum

DEBRUM PLATE E-200 (DEBRUM COLLECTION)

PROJECT DETAILS

The Marshallese Cultural Society, a nonprofit organization located in the Marshall Islands, is now undertaking the digitization of the glass plates with the goal of capturing the images before further deterioration. The plates, ranging in size from 2"x3" to 8"x10", are being scanned and archived as 1600 dpi TIF images. The images will then undergo a restoration effort as needed. Restoration includes adjusting the scan's tonal range to maximize photographic detail as well as retouching areas of damage (cracks in the glass, cracks in the emulsion, and deterioration of the emulsion due to time, mold, salt and a variety of unknown elements). For ease in identification, a re-sized (falling within 7.5" x 10") 200 dpi image is made for printing purposes.

There are almost 2500 glass plate negatives that survive today and are being housed currently in a facility that generally has controlled air quality. For years, however, the plates were housed on an outer island in elements not conducive to long life of emulsion. That we have the plates at all with images intact is probably a miracle. They were saved from WW2 bombings by being removed to a small island and buried for a time.

The project has complemented technical efforts with a focus on gathering background information on the plates. Commentary on the photographs has been solicited and recorded from Joaquim's surviving son Leonard DeBrum, who at present time is 85 years old. Comments are documented in writing or recorded on videotapes and audiotapes. Comments are then transcribed as background information for each photograph whenever possible. An electronic searchable database of photographs and comments has been established.
The material end product of efforts has included the following:

- High resolution digital files of the entire original DeBrum collection
- Digital files of retouched and restored photographs stored on CD's
- Archival quality digital prints of the restored photographs
- An electronic data base for the photographs

THE PROJECT PROCESS – TECHNICAL ASPECTS

The glass plates look to be a combination of collodian or wet plates and dry plates. They were lettered and numbered with what appears to be India ink at some point, but it isn't clear whether this was done by Joaquim or someone else, or if it was done at the time the images were taken, or much later. The glass plates were organized by a couple from Illinois working on a grant sometime in the late 1970's, and at some point they were catalogued, but the records we have of this are incomplete.
In a traditional darkroom situation, the glass plate negatives would be contact-printed plate emulsion to paper emulsion, rendering the positive on the light-sensitive surface with processing. To accomplish this on a scanner, a transparency adapter is used to supply the light that passes through the negative. To avoid the problem of Newton rings which occur when glass touches glass we have developed cardboard shims that hold the plates just off the glass surface of the scanner, while also taking into account the depth of focus capabilities of the scanner.

When there is surface damage to one side of the plate, it has been found that more detail is acquired by scanning the plate with the unscarred side towards the scanner and then flipping the image digitally once it is in Photoshop.

DeBrum plate E-55 detail scanned with damaged glass towards scanner. (DeBrum Collection)

DeBrum Plate E-55 detail scanned with damaged glass towards the transparency adapter and flipped in Photoshop. (DeBrum Collection)

The scanning software is utilized to get the most accurate exposure for each plate. Since the exposures of the plates vary immensely, so do the scanning exposures. The final scan is done at 1600 dpi, maximum optical resolution for this scanner. Once the image is scanned and is opened in Photoshop then a variety of tools can be utilized for cropping to the image and adjusting tonal qualities. For now, these primary scanned images are stored on CD's to be moved to whatever stable storage media is next. From these images come the 1600 dpi images being restored digitally. This takes place on a separate layer in Photoshop. These images will eventually become reunited with their primary scan once sufficient work is done.

TECHNICAL HARDWARE AND SOFTWARE

2 Macintosh G4's with 512 MB RAM
Adobe Photoshop 6.0
Epson 1600 scanner with transparency adapter
Yamaha CD/RW burner
Adaptec Toast CD burning software
Epson Stylus Photo 2000P Printer

THE PROJECT PROCESS: EXPANDING SKILLS AND ACCESS

The digitization project has not only been a process for restoring and archiving lost photographs, it has been a learning opportunity for Marshall Islanders associated with the project. In a general sense it has brought a new awareness of the kinds of things that digital imaging technology can accomplish. In more concrete terms it has resulted in a core of individuals who are becoming proficient in scanning and manipulation of digital images. Individuals who began with minimal computer skills are now comfortable and confident with fairly complex computer tasks. Engaging indigenous people with an exploration of their cultural roots has yielded expertise that several are already applying in other areas of work and personal interest. We are hopeful that the finesse and expertise gained during the project will translate into discriminating use of appropriate new technologies in this part of Micronesia.

THE FUTURE

With the integrity of the images on the glass plate negatives ensured in the new digital format there is a tremendous range of opportunities for future uses. The project members envision three significant subsequent projects. One is to convert the images and accompanying commentary into an html format to allow widespread access to low-resolution versions of the materials via the web. A second goal is a book of high quality DeBrum photos with accompanying notes and historical background. A third goal would be to produce multimedia cd's incorporating the images, video, audio, etc. that could be disseminated to outer islands where they may have computers that could read CD's but not Internet access. Any future undertakings will of course depend on support from the DeBrum family. We are hopeful that we will be able to identify a confluence of interests wherein there are benefits to the DeBrum family while the Marshallese and the greater Pacific communities gain from easy access to the photograph collection.

BACKGROUND

The Marshall Islands are located in eastern Micronesia just above the Equator. The islands are renowned as the home of some of the best indigenous navigators in the Pacific. The islands are also famous for Bikini Atoll—site of United States nuclear testing in the 1950's and source of inspiration for naming the bikini swimsuit.

Much less known and incompletely documented is the cultural history of the islands. The scattered atolls were neglected backwaters of the Spanish Empire from the time of Magellan's circumnavigation until the islands were formally ceded to Germany in 1886. American missionaries first arrived in 1857.
Traders followed and by the 1870's the islands had emerged as a center for the lucrative Pacific copra trade with a strong German trade presence. The islands remained a German colony until the League of Nations awarded the islands to Japan in 1918. The German colonial period was a time of major transition for the Marshalls: the culture transformed from an isolated and traditional semi-nomadic island culture to a trade economy familiar with Western ways and artifacts. Documentation of the social and material changes experienced by the Marshall Islanders during this time is limited.

JOAQUIM DEBRUM

Joaquim DeBrum (1860-1937) compiled the most complete visual record of life in the Marshall Islands during the German colonial period as a series of extraordinary glass plate negative photographs. Joaquim's father was a Portuguese seaman who was one of the first European traders in the Marshalls. His mother was the daughter of Marshallese chief. Thus, not only does DeBrum's life coincide with a major transformation of Marshallese society, he has a unique and intimate bicultural perspective on the process.

DeBrum Plate J826 – Joachim DeBrum (DeBrum Collection)

THE DEBRUM PHOTOGRAPH COLLECTION

During his lifetime Joaquim DeBrum took almost 2500 photographs of people and places in Micronesia. These photographs represent an invaluable record of Micronesian life and times. An occasional photograph has accompanied past historical and ethnographic publications. The overwhelming majority of the photos have not been preserved or collected for public and professional scrutiny. DeBrum's remarkable work survives, however, in the form of his original glass negative plates. Some of these plates are over 100 years old and all are suffering the ravages of time in varying degrees.

DeBrum Plate E-44 (DeBrum Collection)

THE PROJECT PROCESS: THE TECHNOLOGICAL LINKING OF PAST, PRESENT, AND FUTURE

When Joachim DeBrum first brought his camera equipment to his island home in the Marshalls, he astounded the islanders with his near-magical ability to capture freeze his colorful living subjects into thin gray scale pictures. In his choice of subjects and photographic composition there is little doubt that DeBrum was doing more than creating photographs for albums; he clearly was aware of his role as a photographic ethnographer. The glass plate technology was a means of preserving images of a fast-changing island culture for posterity. Some of his photographs were, in fact, used contemporaneously in various German publications to convey a visual sense of the islands and its people. The passing of years has demonstrated unavoidable limits to the glass plate and photograph media. The original photographic prints themselves have been lost, filed away, given to various subjects or families, or have lost detail and quality. The glass plates have proved to be a durable medium so far but printing of new photographs entails risks to the integrity of the plates. The plates themselves demonstrate a clear and inevitable degradation.

DeBrum Plate D-69 Black areas on upper and lower parts of the plate indicate the loss of emulsion. (DeBrum Collection)

The technology used in this project is neither cutting edge nor radical. Nonetheless it provides striking advantages over the traditional photographic approach. Handling the fragile glass plates is minimized. The more flexible and forgiving process of digital scanning and software image processing replace the challenging art of traditional dark room developing of prints. From the original high resolution scanned image multiple future images can be produced in the future without degeneration since the digital image is utilized as the negative. The electronic medium allows unprecedented portability and transmissibility as well as eventual archival retrieval for both the public and researchers. Last, with proper care and backup the images--both original scans and their incarnations as retouched photos--will endure as long as an archivist maintains the basic hardware/software/storage to retrieve the pictures…a kind of archival immortality. The technology allows generations of future scholars and island residents to study the first photographic documentation of Marshallese people and places.

DeBrum Plate E-36 (DeBrum Collection)

THE PROJECT PROCESS: ARCHIVAL RIPPLE EFFECT

An unexpected byproduct of the project was what we have come to call the "archival ripple effect". As we began our efforts we showed how photographs as well as negatives could be scanned and how fading or damaged photographs could be reproduced and revitalized through digital reproduction. It became quickly known within several of the island communities that old photographs could be brought to life (and small images could be enlarged for viewing as well as ease of identification). As a consequence we have received a steady stream of old photographs to the project office for reproduction and enlargement. Photographs that were undergoing a slow degradation in the perpetually warm, humid and saline climate came to us for restoration. Photographs that were hidden away as family treasures came to us for reproduction. In return for processing these old photographs the Cultural Center has been allowed to keep digital versions and prints of the photographs to add to archive files. Success has bred success. A growing central file of pictures from dating from the 1890's to the 1950's has accompanied the development of the DeBrum photograph digital project.

CHALLENGES AND SHORTCOMINGS

The most obvious challenge in putting together a project of this sort in a remote atoll (we are 2400 miles west of Hawaii—tiny specks of land in the vast Pacific Ocean midway between Hawaii and Guam) is logistical. After funding was obtained in the form of a grant from AusAid, it took almost 3 months to select, purchase, and transport needed equipment, software, and supplies here, (No overnight delivery service in the islands!) In these first few months we have encountered no technical or equipment problems but troubleshooting and repair of failing equipment could cripple comparable projects in similarly remote areas.

The most critical challenge was not technical but socio-cultural. The DeBrum family has recognized for many years that the glass plate collection is exceptionally valuable. (There is a family story about Joaquim's sons taking the plates to an uninhabited island and burying them during the latter part of World War II for burial to protect them from air strikes that never materialized.) The family has been quite cautious in allowing reproductions of the photographs. There has been a proud sense of ownership accompanied by apprehension that others may copy the images for their own profit diminishing the value to the DeBrum family. One of the unfortunate results is that good copies of prints have been difficult or impossible to obtain in the past. Another result is that many of the plates have no readily accessible hard copy in recent years. In order to proceed with the project we needed to gain the trust and support of the 85-year-old son of Joaquim who has been the custodian of the films. Success in gaining this support evolved slowly. We first explained the digital process, producing some examples of digitally restored photographs that the project photographer had produced in the past (as well as some digitally restored prints of images that Leonard DeBrum had given us of family members in our first contact). We then successfully conveyed the news that the technology offered the opportunity to preserve and restore the images that would be ultimately lost in the glass plate medium. Finally and most importantly we were able to gain the trust of the DeBrum family. We believe this emerged from our enthusiasm for the pictures, our respect and administration for the Marshallese culture, and an unhurried deferential approach to hammering out a final understanding about the scope and goals of the project. It is interesting to note that although one of the attributes of the digital medium is the opportunity to distribute images widely, the family has been very explicit about limiting copies to those under control of the two Marshall Islands cultural museums.

A CONCLUDING OVERVIEW

Through digitization the DeBrum Photograph Project has reclaimed for perpetuity this collection of unique images--for the Marshallese people as well as ethnographers. The original glass plate negatives were a unique encounter of Micronesian culture and an early photographic technology put to use by a talented islander with intelligence and vision. The new digital technology results in unprecedented access, organization, and flexibility in archiving and reproducing images providing key insights to Micronesian history and culture. Through restoration and retouching, the photographs have gained an otherwise impossible clarity and completeness. Through digital storage and archiving the images have the capability for future widespread dissemination for public viewing and education as well as Micronesian cultural/historical study. Through involvement of indigenous Marshallese the project has provided local individuals and institutions training and insights in the use of digital tools to preserve, archive and disseminate images. In the final analysis the current success and future development of projects are dependent on interpersonal relationships—a working partnership of technical experts, institutional leaders, and those who possess the artifacts and knowledge of the past.

This project was made possible by a grant from the Australian Agency for International Development

ABOUT THE AUTHORS

Eric Lindborg is the President of the Marshallese Cultural Society and a 20-year resident of the Marshall Islands

Cristina Lindborg is the Artistic Director and Exhibit Designer for the Marshallese Cultural Society and a 20-year resident of the Marshall Islands.

Sue Rosoff is a photographer/digital imager for the Kwajalein Missile Range as well as for this project. She received an MA in Photography from UC Berkeley where she was the 1983 Dorothea Lange Fellow. She taught photography at colleges in Northern California and was a photographer/digital imager in Special Collections at California State University at Chico. She has resided in the Marshall Islands for almost 2 years

Images of the early 20th century. The Photographic Fund of the Istituto di Arti Grafiche of Bergamo for "Emporium" magazine

Federico Brunetti (*), Ilaria Alberici (*), Giuliana Ricci (#)

(*) Politecnico di Milano, III Faculty of Architecture- Design, Italia

(#) Politecnico di Milano, Department of Conservation of Architecture, Italia

E-mail: federico.brunetti@polimi.it

E-mail: giuliana.ricci@polimi.it

E-mail: ilaria.alberici@libero.it

ABSTRACT

The Institute of Graphic Arts (I.A.G.B. Istituto di Arti Grafiche) in Bergamo was the first Italian publisher to equip itself with typographic apparatus for the printing of photographs on the same page as graphics and texts.

Financially supported by the Italian Ministry of Education (at that time also responsible for cultural activities), it was particularly notorious for the literary contributions of eminent members of the cultural society of the early 20th century, who through this publisher found new ways to communicate their own work.

Particularly, the monthly magazine "Emporium" became an extremely advanced place for cultural discussion and divulgation from art, architecture, geography and sciences, to the most varied and eclectic cultural news typical of that period.

Part of this archive consists in albums of photographs, which had been previously microfilmed at the Società Alinari in Florence. A second part is conserved in around 5000 glass negative plates.

The fragility of this part of the archive had till now allowed only a few consultations and an approximate indexing.

In the context of a research study carried out for a graduation thesis at the Faculty of Architecture of the Polytechnic in Milan, a research stage was stipulated, thanks to which it was possible to study these materials more thoroughly and digitally copy the fund of plates that had architectural themes. In this way it is possible to examine the entire fund, digitally made positive, allowing a safe, clear and extensive consultation.

This new digital recording technique allowed the extensive study on the section of the fund regarding architectural photography (about 10,000 images) in a relatively short time and permitted the highlighting of the most significant case studies, with the aim of defining certain materials as "guiding elements" to the whole archive.

The whole digitalised fund, mastered to CD-ROM, will soon be compiled in reference to the information concerning the images, according to the parameters

designated by SCHEDA F.
This fund is a very clear example of a early editorial photographic archive, where the accumulation of iconographic materials was not yet completely organized, (we could not fin any general catalogue, many negative plates were not printed); but now digital technologies allows us to explore these delicate photographic memories with acceptable qualities for the aim to understand, in our present researches, the reason that moved these interesting reportages.

KEYWORDS: Twentieth century architecture, digital and analog photographic archives, classification methods, history of architecture, Editorial private funds

INTRODUCTION

This investigation originates from the arising opportunity to deepen our knowledge of a historic photographic background characterized by marked interest in architectural and landscape themes, which was traced during a recent Doctorate of Research[1]. With this information it has been possible to explore, catalogue and analyze the photographic "fund" of the Institute of Graphic Arts in Bergamo.
proceedings a single, high quality appearance, ICHIM'01 requires final submissions of accepted papers to be in "camera-ready" format. Both for short and long papers, authors must follow some simple guidelines described in the rest of this document. In essence, we ask you *to make your paper look exactly like this document*. The easiest way to do this is to simply remove the "instructional" content in one of the templates and add your own text.
The object of this research shows us a consolidated experience in the use of photography and, particularly for the first time in Italy, photography intended for reproduction in typography and acquired by large editorial organisations.
The media event that these materials signify is a new possibility, representing works and architectonic objects through photography, exemplifying and distributing the possibilities of visual acquaintance through printed reproductions in the pages of a widely circulated magazine.
This reproductive typographic technology gave rise to new potential in social fruition and the divulgation of knowledge, in the methodologies already defined in the second half of the 19th century and based on indications from the most expert cultural consultants of the State. These had begun to construct a catalogue of the artistic-monumental heritage.

HISTORICAL CONTEXT

During the second half of the 19th century a clearly positivistic initiative developed in this way, aimed at building a catalogue of news and images of objects and situations considered to be of historic value.
In Italy, at the beginning of the 20th century, following the first phase mentioned, a debate opened and an attitude for research developed in the fields of historic and geographic studies. This shows a new attention for the value of cultural material: an approach that is not only linked to the archetypical exceptionality of certain classic works,

[1] Federico Brunetti, *"L'archivio delle immagini nell'insegnamento dell'architettura. La fototeca storica dell'accademia di Brera"*, Genoa, 1997. Doctorate Thesis in "Problemi di metodo della Progettazione Architettonica". Faculty of Architecture, University of Genoa

or to the epochs that were stylistically canonised by the academies, but which is also aimed at the study of all the anthropic levels that shape the territorial, urban and architectural forms.

These are read as signs of sedimentation and are interpreted by a new sensitivity promoted by the studies of the forms of cultural material.

In this way, within the findings from the past, the quality of the minor objects is studied – the cultural traces present in the usual forms in everyday life, the fragments spread around environments modified by man and even the smallest social organisations: those aspects which were normally ignored during the reading of historic phenomena tied to the great periodic divisions of the typical history books.

It is in this context of studies on cultural forms (expressed as material sedimentations in locations marked by anthropic activity) that the minor centres are brought to light, better understood and given more value, as depositing grounds for the socialised forms of knowledge.

Through the theoretic discussions conducted by personalities such as Corrado Ricci and subsequently Gustavo Giovannoni, the knowledge of historic-artistic resources and the organisation of monuments on the territory become part of a synergy in which monuments and works of art are not any longer studied as objects of single and absolute significance, but on the other hand are studied in the context of the entire environment that contains them and forms their background.

In this period at the start of the 20th century, the State proceeds with the systematic and extensive construction of the entities and cultural institutions born with the foundation of the National State (not only infrastructures, railways and bridges, but also schools, museums, hospitals and theatres).

The homologation within the newly unified National State originates not only from the building of a national language (imposed on local dialects) but also from the construction and standardisation of divulgative systems that are increasingly accessible and common to an ever-growing section of the people.

The experience of "Emporium" is situated in this era of multiplication and reproducibility, not so much for the divulgation and reproduction of places and objects present in historic-artistic literature, but more for the presentation of the characteristics of minor centres of the Italian regions and other places that were the object of the last colonial explorations of the world. A kind of cross between the exposure and exclusiveness of artistic heritage (that scientific divulgation was able to disseminate), and the common man's curious desire to know the world and be an iconographic participant.

Most of the materials in the I.I.A.G.B. fund, images which were created for an editorial archive and whose importance was related to publishing, are actually unpublished: they have a specific value because they were never issued. They therefore are of more current interest than the images that were selected, published and distributed in their time.

The Institute of Graphic Arts (Istituto di Arti Grafiche) in Bergamo was the first Italian publisher to equip itself with typographic apparatus for the printing of photographs on the same page as graphics and texts.

Financially supported by the Italian Ministry of Education (at that time also responsible for cultural activities), it was particularly notorious for the literary contributions of eminent members of the cultural society of the

early 20th century, who through this publisher found new ways to communicate their own work.
Particularly, the monthly magazine "Emporium" became an extremely advanced place for cultural discussion and divulgation from art, architecture, geography and sciences, to the most varied and eclectic cultural news typical of that period.
The contextual presence of photographs in a page enhanced the visual impact of the message, allowing an immediate interaction between the graphic text and the photographic image.
This new digital recording technique allowed the extensive study on the section of the fund regarding architectural photography (about 10,000 images) in a relatively short time and permitted the highlighting of the most significant case studies, with the aim of defining certain materials as "guiding elements" to the whole archive.
The whole digitalised fund, mastered to CD-ROM, will soon be compiled in reference to the information concerning the images, according to the parameters designated by SCHEDA F.

CONCLUSIONS: IMAGES FOR MEMORY

This fund is a very clear example of a early editorial photographic archive, where the accumulation of iconographic materials was not yet completely organized, (we could not fin any general catalogue, many negative plates were not printed); but now digital technologies allows us to explore these delicate photographic memories with acceptable qualities for the aim to understand, in our present researches, the reason that moved these interesting reportages.
The existence and survival of these materials, rejects of topicality, make us ask ourselves about the way the Present transforms itself into discards and rejects of topicality. On the other hand they make us aware of the importance of representing a time that so rapidly became the past, with the fast speed of utilization that the media has imposed on every object and method of memory storage.
We can therefore affirm that these are images, memories, and glances of objects, works of art, cities and landscapes. They are treated as documents of the objects but can also be reread as the introduction of one epoch to another.

PAESAGGI DEI CENTRI MINORI

FIG.16 Via Claudia Bieda
Neg. Su Lastra
21X27 cm.; 0772

APPENDIX:
THE NEW "SCHEDA F" OF THE CENTRAL SYSTEM FOR NATIONAL CATALOGUING

(with technical support of Ilaria Alberici)

The scheda F (Form F) surfaces in the context of the search for methods linked to the cataloguing of photographs, which are considered as a heritage of historic and artistic value, therefore being endowed with an autonomous, expressive and creative language whose cultural importance cannot only be measured by the subjects they represent.

The acceptance of semantic values connected to the photographic image has led to the creation of a cataloguing process that accounts for all typologies of photographs, preserved and collected in different contexts.
The essential elements for the description of photographic works and the transmission of their relevant data have been defined in a mode that is independent from the cataloguing traditions of the various Institutions in which these pieces are kept.
The cataloguing task was carried out with the participation of many expert researchers and Institutions such as the *I.C.C.D.* (Istituto Centrale del Catalogo for the Sole Catalogue of Italian libraries and bibliographic data), the *Istituto Nazionale per la Grafica* (National Graphic Institute) and the *Archivio Centrale dello Stato* (Central Archive of the State).
Parallel to Scheda F, in order to allow the exchange of data between official boards and various institutions, the plan for the transmission of data in UNIMARC format was laid out:
The UNIMARC codes, internationally stipulated for the structuring of bibliographic records (a specific result of the ISO 2709 rule), allowed the creation of an unequivocal format for the transmission of catalogue data concerning photographic material.

Identification criteria in photographic images

The identification of a photographic image requires the fine-tuning of a language that allows it to be read correctly and as objectively as possible, beyond the purely connotative aspects that may be attributed to it.
Lists of GENRES and families of CONTENTS were elaborated as references for the initial codifying of photographic documents.
The genres defined for photography (since its beginnings) are traditionally similar to those of painting, which is an analogous art form.
Several other types of photographic representation follow, such as the landscape, intended not only as recording of monuments, but also as a document of exploratory journeys to known and less-known lands; photographs of architecture; the picture story as a testimony of real facts and events; photography in fiction, related to specially created scenes; photography of Art; artistic photography, formulating and studying its own languages; scientific photography, with fields of application mostly regarding biology, astronomy and medicine; industrial photography, born at the end of the 19th century with the intent to document all the aspects of production; and finally reproductive photography, intended as a clear and precise recording of other images and documents.

Structuring of data in the catalogue files

Scheda F is divided into paragraphs composed of simple fields and/or fields organised in sub-fields, according to the scheme already used for the collection of data during the cataloguing of other historic-artistic movables (OA-D-N, S-MI, etc.).
As listed the following reference table, a different graphical figure was used to indicate paragraphs, simple or structured fields and sub-fields corresponding to the structural data scheme of the catalogue file:

AA – PARAGRAPH
BBB – STRUCTURED FIELD
BBBC – Sub-field
DDD – Simple field

The presence of an asterisk next to the paragraphs, fields or sub-fields shows the need for their completion, so that the file may be considered as valid. An asterisk in brackets indicates the compulsoriness "in context", i.e. the particular cases in which under certain conditions it is necessary to compile a field, or one or the other of the sub-fields making up a specific field.
Also indicated is the possible presence of a closed or open vocabulary, i.e. a list of terms used to complete certain fields and sub-fields.
Also specified is that cataloguing Boards can make use of this vocabulary (in keeping with the ICCD rules concerning the transfer of data 1998).
The symbol # is used to indicate data which, through the Unimarc format, can possibly be used in information exchange between different catalogue formats.

Method of acquisition and digitalising of the images

Digital images are acquired using particular instruments known as *scanners*.
The type of scanner to be used varies according to the format and quality desired in the digitalising process.
For photographic plates and prints flat scanners are recommended, while film scanners are suitable for negatives (35mm).
Commercial software such as Paint Shop Pro and Photoshop are available for the successive treatment of the images.

Parameters for the digitalisation of images

Are recommended a minimum of 1536X1024 pixels for an original measuring 24X36mm: this format creates a 4.5-megabyte file if in colour or a 1.5-megabyte file if in black and white. For most images, if a stable original photograph or the original subject is still available, this resolution can be considered as sufficient.
For valued images where one does not have access to a stable original (for example colour prints or slides) we recommend a resolution of 3072X2048 pixels for negatives or slides measuring 24X36mm: this creates an 18-megabyte file.
For extremely valuable images, for which it may be necessary to create a computerised matrix suitable for professional printing, we advise a resolution of 4096X6144 pixels, giving rise to a 72-megabyte file.

Digital file formats

According to the quality of the digitalised image there are three categories: Category A (High resolution), Category B (Medium resolution) and Category C (Low resolution). Each category can include various types (extensions) of file.
For Category A .PNG, .TIF, .PCD KODAK (base 16) files are recommended. For Category B .PNG, .TIF, .PCD KODAK (base 4), .JPG files are recommended. For Category C the .JPG format is suitable.
It is to be pointed out that the formats used in high resolutions maintain the image information thanks to the use of memory and therefore do not destructively alter the acquired data. The .JPG format, used at low resolutions for network transmission, is

a compressed format: this guarantees a lower use of memory but the compression process eliminates some of the original data. The resulting information loss causes the lowering of image quality, especially during the printing phase.

ACKNOWLEDGMENTS
Marina Miraglia, Giuseppina Cattaneo, Leonardo Devoti Laura Ubodi (T.C.I.), Cristina de Vecchi, Cristina Gatelli, Silvia Bergna (Library of I Faculty of Architecture Politecnico di Milano)

REFERENCES

1. J. LE GOFF, *Documento-monumento,* Enciclopedia Einaudi, vol V, Torino, 1978, pp 567-670.
2. M. MIRAGLIA, *La costruzione iconografica del monumento architettonico,* in Rassegna n° 20 " Fotografie di Architettura, Milano, 1984
3. R. CASSANELLI, F. BRUNETTI, G. AGOSTI, *Camillo Boito. La raccolta fotografica. Una selezione,* Milano, 1996
4. G. BENASSATI, *La fotografia, manuale di catalogazione*, Bologna, 1990
5. G.MIRANDOLA, *"Emporium" e l'Istituto d'Arti Grafiche 1895-1915,* Bergamo, 1985

ABOUT THE AUTHORS

Federico Brunetti, *Architect. PHD,* Doctor of Research: Faculty of Architecture in Genoa: *The Image Archive in the Teaching of Architecture.* Research about the origin of the photographic representation of architecture in unedited institutional funds of the 19th century. Since 1996 Lecturer of Design and coordinator of the *Laboratory of Design* in the III Faculty of Architecture-Design, Department of Technology Politecnico di Milano, where has been assistant from 1986.
E-mail: *federico.brunetti@polimi.it* or *fedbrune@tin.it*
Dipartimento Di.Tec; Via Durando 38/a 20133 Milano (ITALY) Fax. +39.02 2399. 5977/ 7282

Ilaria Alberici Graduated in Architecture, Politecnico di Milano, Milano, Italy. Her first studies, starting from Graduation Thesis mentioned in XI International Competition ISAL/Lions Club Cenacolo, regard methods for indexing, cataloguing and digital recording tecnique for photographic funds. She followed several courses about preservation of historic photographt. Actually she is working for Associazione Jacqueline Vodot e Bruno Danese di Milano
E-mail: *ilaria.alberici@libero .it*

Giuliana Ricci Politecnico di Milano, Departement of Conservation of Architecture; Ordinary Professor in History of Architecture. Her first studies regard theatre design and construction and stage designs, especially for sets based on the 18th century.
Much of her work concerns the analysis of the architectonic culture from the second half of the 18th century till the first decades of the 20th, dealing with institutional and theoretical aspects and the creation of buildings and urban locations of that period.
E-mail: *giuliana.ricci@polimi.it*

The New Norwegian Rock Art Database

Wenche Helliksen (*), *Inger Marie Holm-Olsen* (*)
and Olaug Hana Nesheim (*),
Bjørg Pettersen and Randi Sønderland
(*) Norwegian Institute for Cultural Heritage Research, NIKU, Oslo, Norway
E-mail: wenche.helliksen@nikuosl.ninaniku.no

ABSTRACT
The Rock Art Project was set up by the Directorate for Cultural Heritage on behalf of the Norwegian Ministry of the Environment in 1996. The database was initiated within the EU-supported INTERREG IIA, a joint Norwegian-Swedish project called Rock Art in the Borderlands 1998-99. The Norwegian Rock Art Database combines photographs, drawings and digital maps to provide a useful multimedia source. Any feature described in the text part of the database can be linked to one or more photographs. The photos can also be linked to an overview of the selected site, which is represented either by a tracing or by another form of small-scale map. An important function of the database is to store the complete documented history of each site. By selecting a specific site from the database, the user is presented with a list of all previous scientific evaluations and records for that particular example, including dates of earlier field studies, the names of the project leaders, and types of documents available. The application is designed for the data input and interrogation using an Oracle database. The software used to develop the application is Visual Basic and Map Objects. The application executes 100 percent on a Citrix application server and only mouse clicks and screen updates are transmitted over the network between server and client. This allows the use of inexpensive computers and enables high application performance even over limited bandwidth connections. The regional institutions and the regional museums are responsible for updating the database.

KEYWORDS: Rock art, database, Citrix client-server

INTRODUCTION

The Rock Art Project [1] was set up by the Directorate for Cultural Heritage (Riksantikvaren) [2] on behalf of the Norwegian Ministry of the Environment in 1996. Alarming reports had already been coming in for several years before that date. The rock art is in generally poor condition, due to weathering, vandalism and other human impact. Around 95% of Norway's rock art is damaged in some way, and the speed of deterioration is steadily accelerating. Around 1100 sites representing more than 2000 individual examples have been documented to date, the majority of which are rock carvings. The rock art mostly dates to the Mesolithic, Neolithic and Bronze Age periods, but also extends a few hundred years into the early Iron Age. A current priority is to create standardized methods of recording and maintaining 300 targeted sites. In order to do this, the Norwegian

Institute for Cultural Heritage Research (NIKU) [3] has developed a rock art database to manage the protected sites. The database was initiated within the EU-supported INTERREG IIA, a joint Norwegian-Swedish project called Rock Art in the Borderlands 1998-99. The database is available on the Internet [4] for the purposes of research and local and national government administration.

THE NORWEGIAN ROCK ART DATABASE

The Norwegian Rock Art Database combines photographs, drawings and digital maps to provide a useful multimedia source. Any feature described in the text part of the database can be linked to one or more photographs. The photos can also be linked to an overview of the selected site, which is represented either by a tracing or by another form of small-scale map (figure 1).

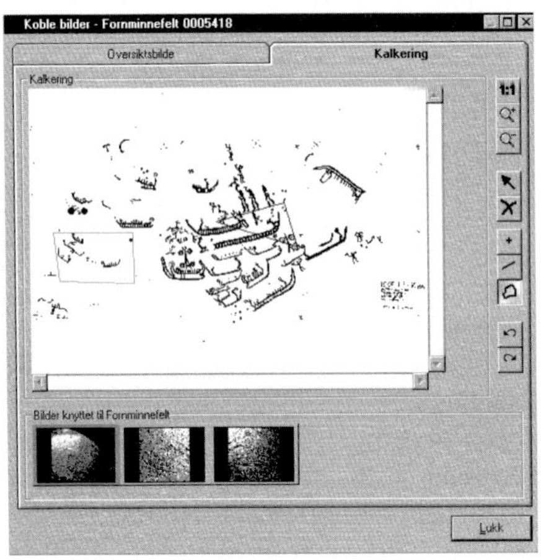

Figure 1: Tracing and photos.

The database also includes an integrated GIS-function, allowing any selection of sites from the database to be given their own signatures and to be presented on a digital map.

Most data fields are supplied with a dropdown list with options giving the legal terms or values for that field, thus ensuring content consistency.

Information relating to the landowners on whose property rock art sites are found can be automatically retrieved from the national property register via the database.

An important function of the Norwegian Rock Art Database is to store the complete documented history of each site. By selecting a specific site from the database, the user is presented with a list of all previous scientific evaluations and records for that particular example, including dates of earlier field studies, the names of the project leaders, and types of documents available.

TECHNICAL SOLUTION: THE ROCK ART DATABASE

The application is designed for the data input and interrogation using an Oracle database. The software used to develop the application is Visual Basic and Map Objects.

We are using a "one server" solution, placing both the database server (Oracle) and the application server software (Citrix) on the same physical machine.

Additional software includes: Windows NT Terminal Server 4.0, Citrix Metaframe Application Server v. 1.8, ODBC Merant Client Access, MapObjects and Oracle 8i standard edition.

The software and application is run on a 800 MHZ, Pentium III server with dual processors, 1 GB RAM and 2 * 9 GB (os mirrored) and 3 * 9 GB (striped) RAID discs. The quality and size of the server is critical for this kind of solution

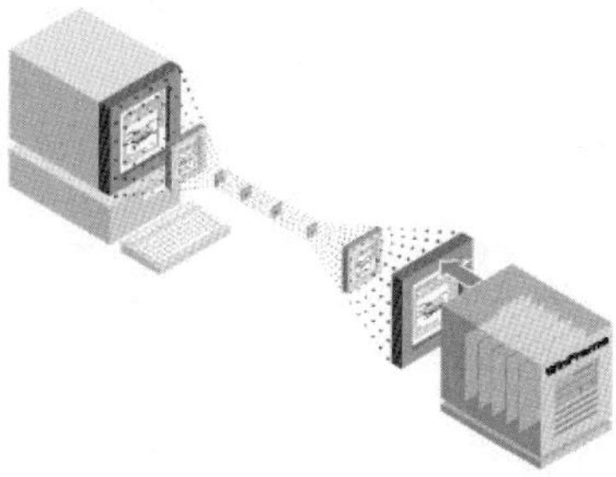

Figure 2: Thin clients

The application executes 100 percent on the server and only mouse clicks and screen updates are transmitted over the network between server and client. This allows the use of inexpensive computers, which act as "thin clients" (figure 2). The only software needed on the user's side, is a very small Citrix file for win95/nt/2000 (ica32.exe), which is downloaded from our website. The technique enables high application performance even over limited bandwidth connections such is ISDN (64 K) and modems (28.800 baud).

USERS AND DISTRIBUTION

The most important users are the regional government administrations and regional museums, about 30 institutions in all. The Rock Art Database is intended for use in regional and local planning, as well as in the management of cultural heritage at all levels. It is also designed as a tool for research.

The regional government administrations and the regional museums are responsible for updating the Rock Art Database. The database and the application are located on a central server at the National Institute for Cultural Heritage (NIKU), and are distributed to the users by means of a thin client (Citrix Metaframe). This includes all data-types: text, maps, drawings and photos. The regional institutions, which are responsible for updating the database, are given read-write access regulated by security passwords.

NIKU offers introductory courses for all new users. To date four such courses have been held and the plan is to run several more each year throughout the country. We believe these courses are necessary to ensure that a high standard of data is maintained in the database. In addition to this there are established protocols to verify the data. This means that NIKU monitors database updates. During this process the data is "locked" so that no changes can be made by the users without special permission.

User support is critical for all distributed data-systems. It is important that new users in particular can get rapid solutions to any problems. NIKU offers user support by e-mail and telephone. There is a website with information about user support (figure 3).

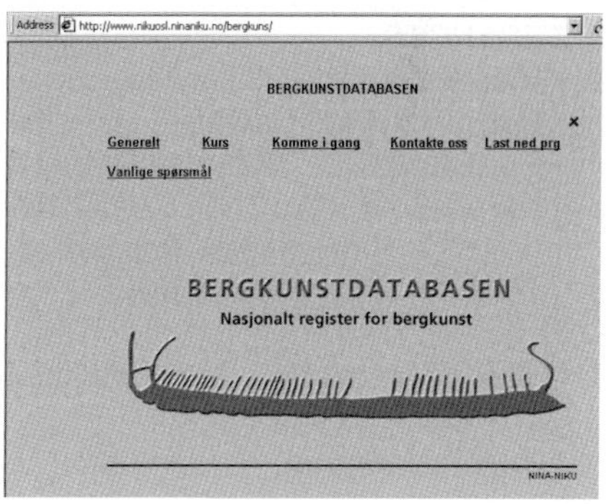

Figure 3: Website for support

We have also planned to expand this site with an additional section covering items such as: "frequently asked questions", information on changes in the applications, and any new procedures.

CONCLUSION

The first version of the Rock Art Database is now available. The aim of the database is to centralize all relevant governmental information about rock art. This includes data from archaeologists, geologists, botanists and curators, and other relevant disciplines. This provides a cross-discipline information resource, which we hope will help to reduce the further deterioration of Norwegian rock art.

The use of "thin client" technology has made it possible to access the Rock Art Database through analogous telephone lines and mobiles. This enables the use of the database during fieldwork.

There will, of course, be a constant need for improvement and adaptation as user-demand dictates. New suggestions will be recorded and acted upon to ensure that updated versions of the Rock Art Database are continuously developed.

REFERENCES

1. Karl Kallhovd and Jan Magnusson (Ed.): Rock Carvings in the Borderlands. Bohuslän/Dalsland and Østfold - An INTERREG IIA project - Final Report Rapport nr: 2000:56. ISSN 1403 - 168X.

2. The Directorate for Cultural Heritage,Riksantikvaren. *http://www.ra.no*

3. Norwegian Institute for Cultural Heritage Research, NIKU. Foundation for Nature Research and Cultural Heritage Research, NINA - NIKU. English website: *http://ninaniku.no/Estart.htm.* Norwegian website *http://www.ninaniku.no/*

4. *http://www.bergkonst.org*

ABOUT THE AUTHORS

Wenche Helliksen works as a research scientist and archaeologist at NIKU. She is the leader for the Rock Art database project, and responsible for user support and management of the database. She contributed to the development of the documentation standard, upon which the development of the Rock Art Database is based.
E-mail:
wenche.helliksen@nikuosl.ninaniku.no

Inger Marie Holm-Olsen is a research scientist and archaeologist at NIKU. She is responsible for user support and management of the database. She contributed to the development of the documentation standard, upon which the development of the Rock Art Database is based.
E-mail: *Inger.m.Holm-Olsen@nikutos.ninaniku.no*

Olaug Hana Nesheim (PhD) is a research scientist and information scientist who formerly worked with landscape analysis at NIKU and as a member of the Rock Art database project team.
E-mail:
olaug.nesheim@statskonsult.dep.no

Bjørg Pettersen is a system developer at NINA-NIKU. She is the database administrator for the Rock Art Database.
E-mail:
bjorg.pettersen@admosl.ninaniku.no

Randi Sønderland is a system developer at NINA-NIKU working with the Rock Art Database project.
E-mail:
randi.sonderland@admtrd.ninaniku.no

Getting a multimedia message across in Palermo Information technology and Artistic Heritage in an area with a low rate of development

Loredana Benincasa (*) *and Cinzia Ferrara* (*)

(*) Institute of Industrial Design the Faculty of Architecture
of the University of Palermo, Italy
E-mail: lorebeni@inwind.it
E-mail: acyxfe@tin.it

ABSTRACT
For the cultural development an effective communication energy. Palermo's image abroad can be promoted by taking advantage of its exceptional cultural heritage. Giacomo Serpotta, Palermo's great sculptor, could be promoted on an international level.

KEYWORDS: an area with a low rate of economic development, communication, cultural heritage, technology, preservation.

INTRODUCTION

Cultural activities all over the world and the public's interest in them have increased enormously over the last few years while on the other hand one of the most prevalent social activities in a globalized world will be tourism. As such, introducing profiles of our artistic heritage, places of interest and beauty spots into the worldwide information arena could offer a contribution to economic growth in an area with a low rate of development. It is clear that to enable this to happen, an effictive communication strategy is indispensable.

COMMUNICATING INFORMATION IN PALERMO

Today the Mediterranean Basin is the one area in Europe which really calls for this type of approach. Palermo, for example, has a rich artistic heritage, but lacks a plan to use of means communication to make it accessible to potential visitors.

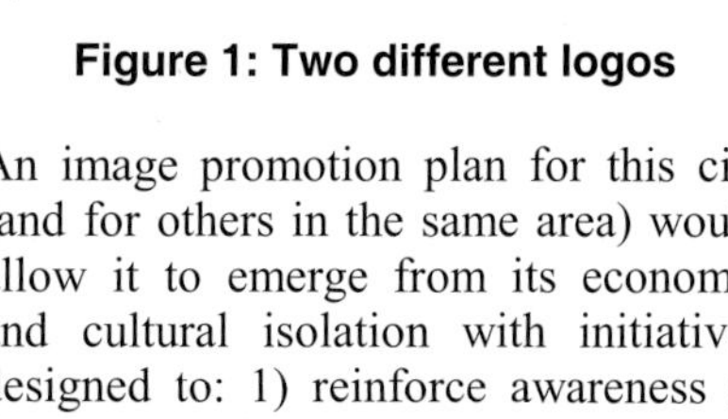

Figure 1: Two different logos

An image promotion plan for this city (and for others in the same area) would allow it to emerge from its economic and cultural isolation with initiatives designed to: 1) reinforce awareness in the community of the value of our artistic heritage as a main factor in the make-up of the local identity and as a bridge-builder with other cultures; 2) underline the importance of restoring and preserving our artistic heritage which has been neglected for too long;

3) give incentives to the development of social and economic activities deriving from tourism and related industries and to the increase the production of locally-produced saleable items, usually arts and crafts and this includes paying close attention to the quality of packaging and other selling aspects of the high quality traditional food and wine products from the area.

Figure 2: Cover of a calendar

One of the institutions from this area which has taken into consideration these factors is the Institute of Industrial Design of the Faculty of Architecture of the University of Palermo. This Insitute has been very active in promoting both Sicily's cultural heritage and local products. It has been doing this by means of tools like product design, visual design and image promotion. These are all things which can help to promote the socio-economic development of the island. Recently its attention has been drawn to one of Palermo's most highly appreciated artists. The 17th century sculptor Giacomo Serpotta's works of art are to be found in 12 churches, 6 oratories and 2 museums, in the very heart of the city of Palermo. Unfortunately, not much attention is paid to them by the administrative bodies responsible for the preservation of the city's artistic wealth, even though this artist has been defined as the most important Italian sculptor of his age. During the academic year of 1999-2000, the Institute of Industrial Design organized a project in image promotion on the subject of "Giacomo Serpotta in Palermo". A workshop was set up and students from both the degree course and the School for Professional Qualifications from the Institute were involved. Qualified professionals and tutors were brought in to run the course, namely Prof. V. Pasca (Course Director), F. Achilli (Graphics Supervisor), the architects V. Trapani, M. Ferrara, S. Giacomarra (course tutors) and M. A. Malleo and S. Scalia (multimedia communication experts). The course was divided into three stages. During the first stage visits were made to the sites of the artworks and photographs were taken. The aim here was to get to know the history behind the works and plan areas of research which might follow the outlines of an image promotion strategy.

Figure 3: A poster

During the second phase, practical decisions like poster design, choise of logo an so on were taken. During the third phase, an analysis was made of the local budget available for financing the use of information technology. The overall aim of the project is to bring Giacomo Serpotta's works of art to a wider audience and allow the city to better define its own identity by retrieving an otherwise forgotten patrimony. An initial attempt was made to get the Artistic Heritage Department, the Town Council and Diocese involved but, as often happens here in Southern Italy, while showing interest, they only offered moral support. Steaming ahead regardless, the project will promote a celebratory year dedicated to " Serpotta in Palermo" in 2001, the 270th anniversary of the artist.

Figure 4: A folder

The idea is to get the whole city involved at every level. An introductory conference on Serpotta where scholars of international repute take part would pave the way for a week long celebration dedicated to visiting the sites of the artworks and other cultural activities. As a preliminary, much effort would be put into the graphic design project and into publicising the event by all available means. In order to give the event drawing-power, package tours would be arranged through travel agents. As well as the necessary arrangements for the reception of the incoming flux of tourists, all sorts of information would be made available to them through maps and "Serpotta Itineraries", specific street signs and infopoints. At the same time a web site would be launched and continually updated with information about main events and related activities, including realtime images of the venues. A year-long calendar of events would be compiled to include concerts of music from the period and shows involving other forms of related artistic expression, aswell as guided tours and further conferences. Furthermore a merchandising activity would also be set up. The challenge will be to bring the works of Giacomo Serpotta out into the forefront, give them national and international resonance and brush off the layers of cultural dust under which they have lain for so long. The example of what Glasgow has done for C. R. Mackintosh and how the city has taken advantage of its illustrious son to renew its own cultural identity comes to mind here. The whole operation has a dual intent: to set the ball rolling for Serpotta and prod the local institutions into greater awareness of his value. As Cesare Brandi puts it: "I hope a careful review [of his work] gets the authorities on their feet...Serpotta's oratories, their altars deconsecrated, are probably being used as storehouses and are going to rack and ruin...what a wonderful sight the supreme jewel of Serpotta's stuccoes might otherwise be for the visitor to Palermo".

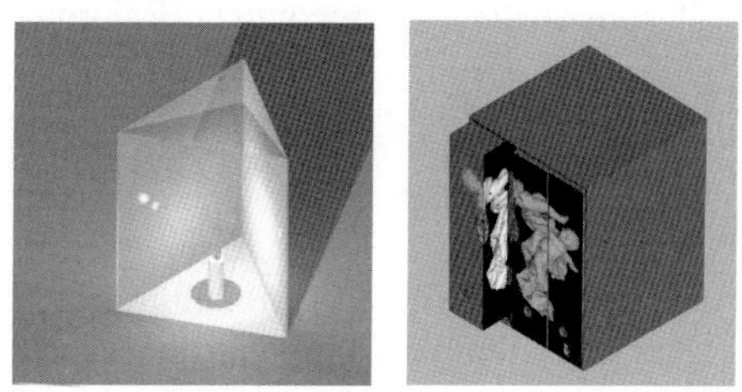

Figure 5: Some products

REFERENCES

1. VV. AA., *Giacomo Serpotta. Architettura e apparati decorativi settecenteschi a Palermo*, Flaccovio, Palermo, 1996

2. VV. AA., *I colori del bianco*, Ed. di Salvare Palermo, Palermo, 1995

3. VV. AA., *I beni culturali una risorsa economica e sociale*, Luxograph, Palermo, 1995

4. Antinucci F., Se i musei sono immagini puoi vederli e capirli di più, is in *Télema*, n.6, autumn 1996

5. Aurigemma M. G., *Oratori del Serpotta a Palermo*, Palombi, Roma, 1989

6. Bagdali S., *Il museo come azienda*, Etaslibri, Milano, 1997

7. Bellafiore G., *Palermo, guida della città e dintorni*, Palermo, 1986

8. Campana M., In mostra, is in *Linea Grafica*, n.292, July 1994

9. Carandente G., *Giacomo Serpotta*, Eri, Torino, 1967

10. Chirco C., *Palermo, la città ritrovata*, Flaccovio, Palermo, 1997

11. Consolo V., *Retablo*, Mondadori, Milano, 1992

12. De Seta C., I beni culturali: un patrimonio non da "sfruttare" ma da godere, is in *Télema*, n.22/23, summer/autumn 2000

13. Fernandez D., *La zattera della Gorgone*, Sellerio, Palermo, 1992

14. Garstang D., *Giacomo Serpotta e gli stuccatori di Palermo*, Sellerio, Palermo, 1990

15. Grossi R., Debbia S., *Cantiere* cultura. Beni culturali e turismo come risorsa di sviluppo locale: progetti, strumenti, esperienze, Il Sole 24 Ore, Milano, 1998

16. Isman F., Un'opera restaurata rivive e l'elettronica può salvarla, is in *Télema*, n.6, autunno 1996

17. Meli F., *Giacomo Serpotta, vita ed opere*, Palermo, 1934

18. Paolini M. G., *Giacomo Serpotta*, Novecento, Palermo, 1983

19. Pronti S., Boccenti L., *Il museo tra istituzione e azienda*, Il Sole 24 Ore Pirola, Milano, 1996

20. Università del Progetto, Bologna dei musei: corporate image per l'Europa, is in *Ottagono*, n.134, September/October 1999

ABOUT THE AUTHORS

Loredana Benincasa (Palermo 1972) graduated in 1998 from the Faculty of Architecture of the University of Palermo; she took a professional qualification about Industrial Design (Palermo 2001), and started teaching and doing research work as an assistant in "History of Industrial Design" at Faculty of Architecture of Palermo, under Prof. V. Pasca. She than took a further qualification as a "Tourist Information Provider". She took part in the "Workshop" conference, in Palermo, submitting "Autoproduzione nell'editoria" and now carries professional activities in Palermo.
E-mail: *lorebeni@inwind.it*

Cinzia Ferrara (Palermo 1964) graduated in 1992 from the Faculty of Architecture of the University of Palermo; she took a professional qualification about Industrial Design (Palermo 2001), and started teaching and doing research work as an assistant in "History of Industrial Design" at

Faculty of Architecture of Palermo, under Prof. V. Pasca. She took part in the "Workshop" and "Salone del Mobile" conferences (Palermo 2000); wrote "Pensieri" on Aedo-ba, n. 02, Milano 2000, and now carries professional activities in Palermo.
E-mail: *acyxfe@tin.it*

The Virtual Museum of Architecture

Eleonora Cardellini(*)

(*) Università degli Studi di Roma La Sapienza, Facoltà di Architettura,
LaMA (Laboratorio Multimediale di Architettura) –
DPAU (Dipartimento di Progettazione Architettonica e Urbana), Italy
E-mail: dpau@.uniroma1.it

ABSTRACT
This paper proposes a virtual space project for the web which is capable of 'containing' virtual buildings or representations of real buildings (through computer rendering, video or photographs). Its theme requires a detailed study of the museum (from the 'meta-project' to the architectural development of the project itself) in order to demonstrate how the realization of a virtual project goes through those same criteria and methods acquired with traditional architectural studies and how it integrates new tools of representation and new disciplines like web-design.

KEYWORDS: virtual architetcure, web museum

INTRODUCTION
The work reflects upon which 'forms' are able to contain a virtual museum but more generally it traces the new role of architecture within Information Society.Therefore, the aim is to give life to a virtual space which then stimulates the introduction of 'digital design' into an educational context collecting the precious work carried by professors and students in shared databases.
Moreover an architecture museum can be conceived for a youth market who may know little about architecture. It would include a 'permanent' display of important buildings, thematic exhibits on particular aspects and clear explanations on the processes tied to the execution of architectural works. Other spaces such as Forums, Chat lines or on-line libraries and workshops, can offer opportunities for immediate exchange of ideas and the chance to discover more detailed information on a particular subject using.
The idea is to make 'architetcure' easy to understand for everyone, thanks to a metaphorical structuring of the museum space and a reasoned application of the tools of representation offered by computers. We are convinced that this museum can feed a new impulse to the further development and spread of architectural themes.

VIRTUAL ARCHITECTURE ON THE WEB
From the analysis undertaken, and allowing for some exceptions, we can affirm that even today most museum sites follow a simple two-dimensional layout, while still exploiting the dynamic nature of hypertext, and that the realisation of websites remains in the realm of graphic design and apes all the cliches of the printed page.
Despite the use of attractive graphics and layout, the results are often disappointing and standardised. On the other hand, at least for now and with rare exceptions, the on-line museum does not present itself as an alternative

to the real museum but acts as a means to promote it, postponing the three-dimensional experience to an actual museum visit. But if the works exhibited are digital or virtual in themselves, then the virtual museum becomes a space which has no substitute and which requires accurate definition before it can be visited.
In situations such as these it becomes necessary for the designer to use three dimensions and here lies the architect's role, despite any difficulties of on-line publishing.

THE VIRTUAL MUSEUM OF ARCHITECTURE

As previously explained, the aim of virtual museums which contain representations of existing buildings/architectural works or which have a corresponding museum in the real world is not to substitute the physical experience of the works exhibited but to entice it. To convince a person, who may even live many kilometres away, to visit the museum itself. It is obvious that this is not the case for virtual museums which are accessible only on computer or via Internet and exhibit only digital works.
However the virtual architecture museum, as it is generically intended, comprises both these situations and is therefore a specific exception. This can be attributed to the unique nature of architectural disciplines which are the 'art of defining space' [1] and which include not only those works exisitng in physical reality but those expressed in a 'parallel' world, whether it be digital or whether it be defined by other forms of representation (cinema, painting, drawing…).
In addition, due to the vary nature of architectural works, the virtual architecture museum can only contain representations of architecture which has been realised with the aim of inciting people to visit the physical buildings. It is interesting to note that the same occurs for a real museum which deals with the same subject. Such a museum, in fact, cannot contain the actual works being exhibited because of obvious factors such as physical size and context: it exhibits its representations.
The virtual museum differs from the real one because it allows access to digital spaces acting as a place of transit. Hence we see both the uniqueness and the necessity of a virtual architecture museum and that it should not be considered inferior to its real-life alter-ego.
It is also clear that a physical museum carries out an irreplaceable role in the conservation of forms of architectural representation (original projects, sketches, models…) but in terms of 'capacity' the virtual museum offers unlimited space for exhibition and storage with remarkable and innovative results.
We can thus affirm that the architecture museum, be it digital or real, is first of all a 'vestibule' before travelling elsewhere, a presentation of spaces, of building materials and processes. It is therefore essential that it be considered as an open organism, as a 'vestibule' for a logical sequence of paths and openings. These considerations are not new to museum studies but in the digital age they become enriched with new meanings.

DESIGN PATHS

Three points of reflection have been singled out and analysed and they apply to any museum or work of architecture in general.

1-The container
2-The contained
3-Services
The container is the form or layout which defines the museum spatially, the contained is the database of architectural works and finally services are defined as all those support functions required by the museum and its contents.

The container
It has been decided 'a priori' not to depart from an existing form or combination of forms but to allow the structure to emerge through an analysis of the relationships between that which is contained. The transformation of thought, of its flow in architectural space is one of the most interesting aspects of virtual space design.
What kind of architectural script must we use to understand the meaning of the information revolution? The answer does not lie simply in atoms and their capacity for change but it must delve into their dynamic mode of interconnection. Just as digital civilisation has developed in a completely different fashion to the mechanical one, so has a new architecture developed which has been profoundly touched by electronic script, an architecture of nerves and not of body.[2]
Such a choice is also confirmed by today's ways of building. The envelope is becoming less and less important, making way for technical elements, services, bio-climatic control as if the nervous system were becoming more important than muscle or bone.
To use a 'dynamic' methodology for the design of virtual museums is therefore spontaneous since it is the 'neural' network of the relationship between the parts which will define its form.
Another aspect which features in current architectural debate is the capacity of an architectural work to be reversible and light (dismountable structures, recycling, light architecture..).
In the realm of virtual design, the concept of reversibility becomes that of non-consequentiality, that is the possibility to go or not to go into an virtual place, to activate and to experience it or to go back, to deactivate it and move into another space. This is tied to the dynamics of the inter-active and of hyper-text which allows us to choose from a number of different, non-sequential paths.
The concept of lightness is not to be associated only with the material aspect. It should not be intended as a mere reduction of weight but as a possibility to embrace the complexities of the world with levity and clarity. A wall surface acts as a barrier to information and blocks all that which tries to penetrate it. The sensible envelope, light and changeable, is like a skin which communicates with the outside, a filter which receives and transmits information. Transparency is, therefore, a unifying factor, a bridge with nature which implies mobility and communication.
The dichotomy which has always been seen to exist between architecture and nature, because of the former being 'fixed' and 'inorganic', the latter 'dynamic' and 'organic' now offers new themes for debate. Today's architectural research passes through fluid spaces, biomorphic forms, transparent filters,

through environments with soft transitions and light, changing thickness and hybrid materials whose transformations can be simulated and experienced through the computer.
The integration between architecture can only occur through architecture's capacity for change.
"...The butterfly, who projects the image of the universe in its organic movement, is the product of a series of creative phases: of an organism which goes from larva to chrysalis and to flying insect. How can architecture make this process visible? By acquiring a spiritual dimension and by borrowing its future forms from nature."[2] It is inevitable that such a thing should happen: thanks to increased communication a new consciousness has spread. It leads towards a respect for nature, to co-penetration, to the formation of consciousness and, because there is a relationship between organising and thinking about architectural form, this forms a prelude to new scenarios and environments.
As we know, the form of nature's organisms is not something 'a priori' but is the result of evolution, the explication of a balance which has developed through the years between each of its parts. Form this point of view, a biomorphic architectural form should not be a stylistic choice on the part of the architect but 'a consequence' of the relationship between its parts and between these parts and the environment.
Frei Otto points out that being close to nature "does not mean to design buildings which appear similar to nature without being at all natural".
This appears even more pertinent when we discuss virtual architecture. Since restrictions of space, time and context, which influence form, no longer exist, virtual architecture can emerge naturally from flows and relationships therefore closely resembling an actual organism. It is also a unique characteristic of organisms to be able to communicate at a high level, directly, evocatively and metaphorically despite the fact that internally they are very complex. This capacity is influenced by the means of communication and derives from a simplification of language and is undoubtedly fruit of a 'formal transposition' which the computer forces us to make.
J.W.Mitchell says 'the web is an environment which is in no place in particular but at the same time everywhere' [3]. The necessity to overcome that sense of 'estrangement' produced by that which is digital because of its lack of canonical reference points and its 'anti-spatiality', has brought about the use of spatial metaphors such as site, address, navigation, web, desk, folder, mail...) which have become particularly effective because they are based on concepts that exist within our imagination and force us to make constant excursions into our subconscious, to an intimate self-exploration and a subjective form of interpretation. It is therefore inevitable that metaphor should become particularly effective when describing organic forms because it allows us to unite our fascination for creativity and the human intellect to that for creation and the works of nature.
When we refer to Cyberspace (in its more extensive range of meanings as 'space of information', or 'conceptual environment without any physical existence') or more simply to Internet, we allude to an entity which has no

structure which corresponds to that of any real-life model. On the Internet it makes no sense to say that a piece of information is 'above', 'below', 'to the left' or 'to the right'. If so it is only through a metaphorical use of terms which refer to an environment which has no spatial points of reference.
But it is possible to transform, abstract or represent information so as to allow someone to become aware of it in a way which is analogous to the experience they have with space and day to day objects in the real world. It is here that virtual spaces are structured in a similar way to physical space. The level of interaction that the user has with the digital allows to move through the space, changing perspective and point of view thus making the design of virtual space much closer to architectural design.
The virtual museum as an 'information object' in its entirety (and metaphorically conceived in its form) has no directionality or dimensions in respect to the web, it can be above or below, very small or very large. Its representation therefore has no scale and no points of reference. But in terms of subjective experience and interaction between space and user it becomes necessary to provide dimensions and reference points to the field of vision. Those metaphorical elements which are part of the museum and which define the internal spaces (ramps, lifts, columns...) will have plausible dimensions that the eye will read as such. In an absolute sense, these elements have no real dimensions because in reference to the information web they can be of any size and have any orientation. However, in relation to the necessity of their being perceived as defined elements such as ramps, lifts and columns it is essential that they be described in terms of relative dimensions.
This does not refer to a banal metric definition of internal environments but more to a spatial definition. Such a necessity comes from a desire to give 'information' on the content through a metaphor which allows room for misunderstandings.

The contained

The works of architecture have been subdivided into the following categories:
1. Digital architecture
2. Hybrids
3. Non-digital architecture :
1. Digital architecture comprises those architectural works which come directly out of the computer and are expressed inside the visual space where they are visited.
2. Hybrids are works which are conceived with the help of a computer, they are very closely tied to it but are carried out in the real world. Therefore they are spaces which can be visited both through the computer and physically in real space.
3. Non-digital architecture comprises:
a) Architecture of reality.
All those works existing in the real world and conceived entirely through a 'traditional' use of design tools.
b) Architecture of representation.
Architecture which is represented in cinema, photographs, drawings or in written works where they find their spatial definition within their own representation. These are experienced in a space which is neither real nor digital. These representations can be either of existing architecture, of hybrids, of digital architecture.

Services

The Services are: library, photographic library, picture gallery, cinema, web-cam, simulator, museum shop, bookshop, offices, workshops, web-mastering and discussion forum.

As previously outlined, it is a prerequisite that each service be reached easily and from any point within the virtual museum. This characteristic has taken on a graphic form through a reticular system extended throughout the museum.

From a more detailed analysis it emerges that there are close ties between some service typologies and the categories of architectural works which can influence their placement within the museum.The web-cam, for example, is tied to the filming of real works in specific geographical positions within the real world.

THE MEDUSA

Before finally reaching a convincing form which could express completely the internal relationships of the museum, a number of schemes and diagrams have been developed which place realtionships in evidence.

From an initial analysis, the necessity emerged to build an organism which could 'contain' the non-digital works while at the same time 'remaining open' to the digital ones. In addition it needed to provide a hybrid zone where interior and exterior remain undefined and a axis along which elements can be placed in relation to each other vertically.

Keeping in mind those considerations made in relation to the 'informative capacity of transparency', it becomes obvious that the object must present this characteristic while remaining symbolically light and allowing for introspection.

These considerations have led to the development of an organism similar to a medusa, a metaphor of navigation and mobility, of strength and lightness.

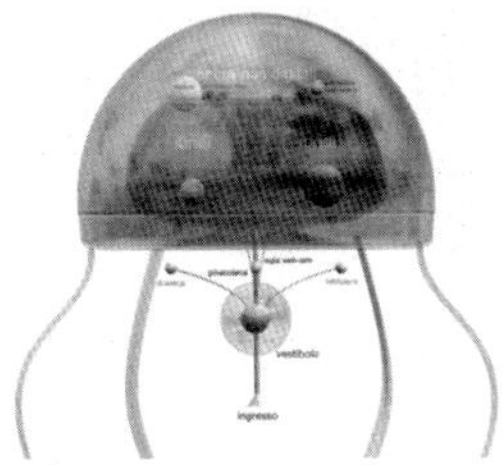

Figure 1: The Medusa

The initial form is intentionally and explicitly biomorphic but it has undergone a number of changes.

The tentacles have been eliminated to become a single,
central axis of communication which connects the entry to the three basic experiential levels: digital works (first level), hybrids (second level), non-digital works (third level).

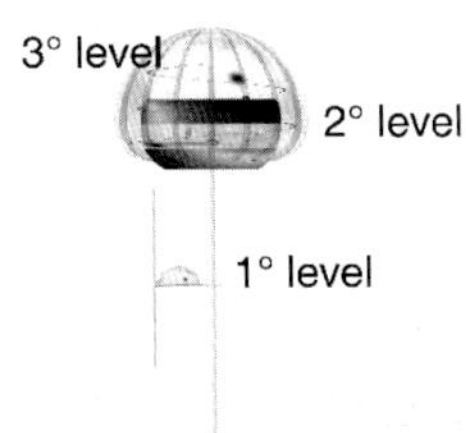

Figure 2: The three levels

A mobile disc-shaped element represents the museum's main navigation tool and provides symbolic access to several services and database.

The lowest level contains the digital works area and is completely immersed in a metaphorical sea (cybersapace).

Here a large ring generates a symbolic aquatic vortex inside which the link to the digital works can occur.

Figure 3: The first level

The second level regards the hybrids visualized by light spherical bodies flowing nomadically through water inside an open container.

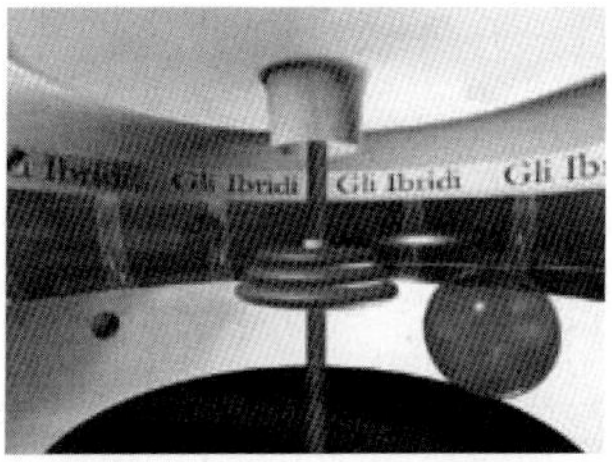

Figure 4: The second level

The third level is made from the concave back of the container filled to the brim by the water coming from the level underneath. In the middle of this body of water we find the 'islands' of real architecture.

Figure 5: The third level

THE DATABASE

The database is the information heart of the museum and it contains all the works exhibited.

They are divided into the following categories:

1.permanent exhibition

2.thematic exhibitions

The museum's Database is a constantly evolving archive connected to other institutes of culture.

The works which make up the permanent exhibition can be listed or searched by name, by author, by date, by place through a specially tailored search engine.

In the thematic exhibitions the works are placed in relation to each other according to a particular theme (which can also be suggested by the user) making it possible for the user to go on a kind of guided tour. When a search is made on the database and a particular work or the title of an exhibition is selected the museum will immediately be activated ion the relevant sector and the visit can begin.

THE ITINERANT MUSEUM

Upon completion of the Virtual architecture museum, it is deemed appropriate that a series of mobile elements be designed where terminals may be placed allowing access to the virtual museum and the web.

These elements will be envisioned as the virtual museum's physical offspring, elements which recall an object existing in another reality.

The mobile elements will have to be designed with stylistic elements which recall the virtual museum.

These will be dismountable (and therefore reversible) and can be placed in town squares, in front of real museums and in particular in front of architecture schools.

Figure 6: The itinerant museum

METHODS OF REPRESENTATION

Non-digital architecture can be filmed or simulated (real architecture) and then reproduced (architecture of representation).

Hybrids may be visited virtually through a link and represented in their reality through film or simulation.

Digital architecture can only be visited through a transition or link.

By simulation we intend 3D renderings which can be explored, by reproduction we intend moving images (films, documentaries) and static images (books, photos, paintings, drawings) in digital form and finally by transition we intend the possibility to access other spaces or sites through transitions or links.

The virtual museum is an environment 'window and screen' which is capable of representing and reproducing a space, a 'simulation container', 'a vestibule' for transitions and finally a 'grid' which interconnects each part.

METHOD FOR USING MUSEUM SPACE

It is believed that the visit to the virtual museum should happen through subjectivity, through a linked series of screens which are controlled by the user. The user moves from one image to another through simple transitions, by clicking on the left side of the screen the user moves to the left and so on.

The inherent simplicity of this system allows us to concentrate on a high quality graphic definition of the spaces combined with an interesting and captivating soundtrack.

In addition, fragments of animation are used to obtain a more realistic movement and they are also tended to offer the user pleasant surprises.

By using the mouse, the user is also able to carry out a number of operations (such as opening and closing doors, pushing buttons and turning handles).

CONCLUSIONS

We really believe the Information Society can be built also by architects but in the meanwhile we have to accept that a new type of architecture is emerging from the Information Society itself . In regards to architects and digital space arch. A. Eloueini states provocatively: "It is no longer the case that we operate in the real world, architecture today extends into the virtual. Architecture of the digital era must be able to operate both in the real and the virtual worlds. To build in the twentieth century does not mean building and intervening in today's metropoli, but it means building virtual cities which, like the first cities, were not built by architects."[4]

REFERENCES

1. Gillo Dorfles, *Il divenire delle arti*, Editore Bompiani ,Milano,1996

2. Luigi Prestinenza Puglisi, *HyperArchitettura*, Editore Testo e Immagine, Collegno (To), 1998.

3. William J.Mitchell, *City of Bits*, Mit Press Edition,Cambridge, Massachusetts,1996

4. Ammar Eloueini , *Architecture e macchine astratte*, Introduction by Ammar Eloueini for the conference held by him in Florence 11/12/1998 at
the PMPE department of the Architecture Faculty

ABOUT THE AUTHOR
Eleonora Cardellini is architect and webdesigner.
Graduated with the experimental thesis "The Virtual Museum of Architecture", she has recently joined the Multimedia Laboratory of Architecture in La Sapienza (Roma). She participated to the conference "Architetcure and Multimedia" (La Sapienza, Rome, 2000), with a 'concept web-site project' called 'The swan' and won the contest "New Ideas for Rome Post-Jubilee 2000" with the 'web-cam installation' titled "Eyes on the city".
E-mail: *leo.n@libero.it*

Museum of Natural Heritage: an Interactive Hypermedia Presentation

Mario Bochicchio (*), *Roberto Paiano* (*),
E. Andreassi (*), *Tiziana Montanaro* (*)

(*) SETLAB-Software Engineering and Telemedia Laboratory, Department of Engineering, Lecce, Italy
E-mail: {bomal,rpaiano}@ingle01.unile.it

ABSTRACT
A Museum of Natural Heritage is quite different from a traditional Cultural Heritage museum. In fact, most part of the contents of such kind of Museum are in a natural environment unlike the traditional museum that exhibits the contents present in their building. This kind of museum, managed by Universities, is also devoted to a larger non-expert public and to University students and researchers, therefore it requires different ways of communication. This requirement associated to the need to present the museum goals and activity, when it is still under construction, has driven us to develop an intermediate application (waiting for developing the final hypermedia application when the museum building will be finished) using a tool, named LEZI, usually adopted to develop interactive lectures, that requires minimal computer expertise in order to develop, quickly, educational multimedia applications based on interactive videos and e-documents.

KEYWORDS: Lezi, HDM, natural heritage, educational hypermedia application.

INTRODUCTION

The Department of Natural Sciences of the University of Lecce is devoted to research and to spreading the natural heritage of the Salento Region, in the South East of Italy.

Their researches concern both the sea and the earth environment and they manage the Sea Park of Porto Cesareo (a sea resort in the Salento Region). Their goals are both to understand the natural life and its evolution during time and to contribute to the preservation of the natural habitat.

They also reached an important result, among others, with an international visibility about the reproduction cycle of a particular jellyfish, *Turritopsis nutricula* that can reverse this cycle becoming an octopus.

The Department is realizing also a Botanical Garden to preserve and to spread the more typical species often unknown to non-expert people.

In this scenario, the University of Lecce has decided to build a Museum of Natural Heritage in order to enhance the effort to disseminate the natural environment knowledge among the new generations joined together with actions to preserve it supplying a new high education instrument to students.

This project is partially founded by EU

and is currently in progress and it will be finished at the end of this year.

In the Museum will be several kinds of fossil, photos and videos to be used directly or by means of special information kiosks.

In the meantime the Department asked us to develop a multimedia application, which could be used both on the Web and on CD-Roms, conceived as a way to present the contents of the future Museum and to allow in depth study of some arguments. The real hypermedia applications that will present the Museum on the Web and will run also on the information kiosks are to be developed later when the museum is finished.

In the University of Lecce, at the Department of Innovation Engineering, the Telemedia[1] Laboratory is devoted to the development of Multimedia applications, some of them concerning educational subjects. "Messapia"[2], for example, is a recent application illustrating the history and the culture of the "Messapi", a pre-Roman population living in the southeast corner of Italy.

Within this activity several professors, especially from the Humanities, have been "exposed to the activity of the laboratory". Many of them asked for developing their own "cultural" application: it was clear that they did not realize how much effort was needed, and that they may not have the budget or the human resources[3] to carry out the job. On the other hand, the staff of the Lab realized that there were a few solid assets, to start with:

- the teachers knew very well their subjects, and could deliver wonderful lectures on them;
- they had good bibliographies at hand, from where students could get detailed information;
- some of the references were already in a digital format;
- an "interactive lecture", rather than a full size hypertext,was the basic need;
- low costs and little time span for the development were absolute requirements.

To satisfy these requirements Telemedia Laboratory developed a tool named LEZI (from the Italian "Lezione" or lecture) mainly based on video and text, reaching the goal to build lectures in a short time. We used this tool, described in the next section, to develop a very special "lecture" concerning the Museum of Natural Heritage.

THE LEZI TOOL

Let us briefly recall the main requirements for the tool:

- the tool should allow the development of an interactive lecture;
- It must be possible to introduce video fragments of the lecture;
- It must be possible to use the tool to access specific topics (very focused) within the lecture, or broad topics, involving most of the lecture, if not all of it;
- It must be possible to get short explanations and also deep

[1] Now the Telemedia Lab has become part of SET, Software Engineering and Telemedia Laboratory.

[2] Project partially funded by the European Commission in collaboration with the local Museum

[3] The Telemedia Lab could provide the technical support; it could not provide expertise on content development.

details, at level suitable for University students;
- The cost for development should be very contained;
- The resulting application should be delivered as an off-line CD-ROM or an on-line website.

With the above requirements in mind, a prototype was developed at HOC, the Hypermedia Laboratory of the Politecnico di Milano; after an initial trial and experimentation, a new version was developed by the University of Lecce, in the context of a large effort[4] to develop innovative, culture oriented, applications.

The tool, named LEZI [1], is a development environment that does not require a strong computer expertise in order to develop in a quick manner educational multimedia application.

The basic functions of the tool are the following:

- The "core" of the application consists of digitized video. Starting from a video VHS taken from a lecture, in half a day a digitized version can be obtained. The tool can support several formats; in particular it supports Real-Video[5], an efficient format that can be played also over Internet.
- The digitized video can be cut into small pieces, which can be assembled in larger pieces, hierarchically. The smallest fragments should last no less than 2 minutes and non more than 5 minutes: this is not a technical requirement, but it is suggested by empirical evidence and Human Computer Interaction researches.
- The process above described can take (starting from a lecture on a 2 hours videotape) between 1 and 3 days, according with the complexity of the subject. The result is a tree-shaped index of topics, each one consisting in a number of subtopics. It should be noted that starting from a lecture of 2 hours, it is difficult to obtain more than 45-50 minutes of good digitized video. The rest can be discarded, having "little content" associated to it.
- The next step is to provide a few lines of comment for each topic: between 5 and 15 lines is the proper measure. Depending upon the complexity of the application, it may take again from 1 to 3 days of work.
- The next step is to add, for each topic, the proper bibliographic references; if the electronic format of some material (e.g. papers) is available, they can be inserted in the application. A competent teacher can complete this task half day or 1 day.
- A relational Data Base, invisible to the teacher, who uses simple commands to insert, modify or delete content, manages all the material of the application.
- As a last step a "generation" (taking 1 hour of work) can produce the wished application, i.e. an offline application or a WWW site.

Therefore, overall, in about 10 days, and with little cost, an interactive multimedia lecture (assuming that good traditional content is already there) can be delivered, to the students.

These are the actions allowed to the

[4] The project, partially sponsored by the European Commission, involved the University of Lecce and Catania.

[5] Real-Video is trademark of RealNetworks, Inc.

user (a student, in general):

- To interactively play the video (with the usual commands for interactive video), choosing the topics from the tree: s/he can choose a very narrow topic or the full lecture;
- To look at the short comments to decide whether a topic is interesting or not;
- To get bibliographic references;
- To get possibly additional contribution in digital format.

Our experience shows that students (and also teachers) of humanities can easily learn how to use similar applications, with very little effort. The applications themselves can be very satisfactory or not, dependently upon the quality of the content (as always it is the case).

THE MUSEUM

The application is conceived to satisfy two different aspects of education: it allows to present the future Museum and its contents and to supply relevant scientific arguments devoted to specialists.

While the tool was conceived to digitalize videos from a lecture or an interview with the attached text and for an in-depth study of the concepts associated with documents in PDF format, the contents of this application has various different formats. There are underwater video, several photos, and also video of lectures and interviews.

To satisfy both the structure of LEZI and the needs of usability, we transformed all contributions in "digitalized videos" adding audio comments to the underwater videos and editing the photos in a slide show with audio comments.

The structure of the application, shown in the left frame of Fig. 1, associates the presentation of the Museum and its goal - realized by interviews with the professors involved in the specific areas and by a collection of video about the most important contributions - with an area explaining the researches on specific arguments. In Fig.1 we can see the video related to the Policheti in the sea habitat while in Fig. 2 there is an important research contribution about the desertification of the coasts due to the illegal poaching of seafood. The user can see the associated videos or the pdf documents by using the buttons in the bottom area of the right frame after selecting the argument in the left frame.

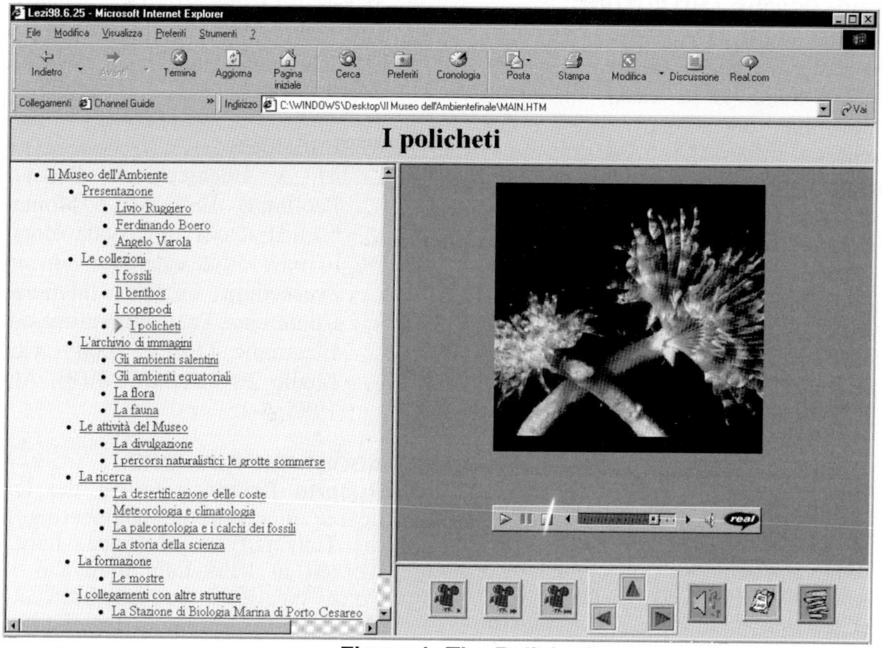

Figure 1: The Policheti

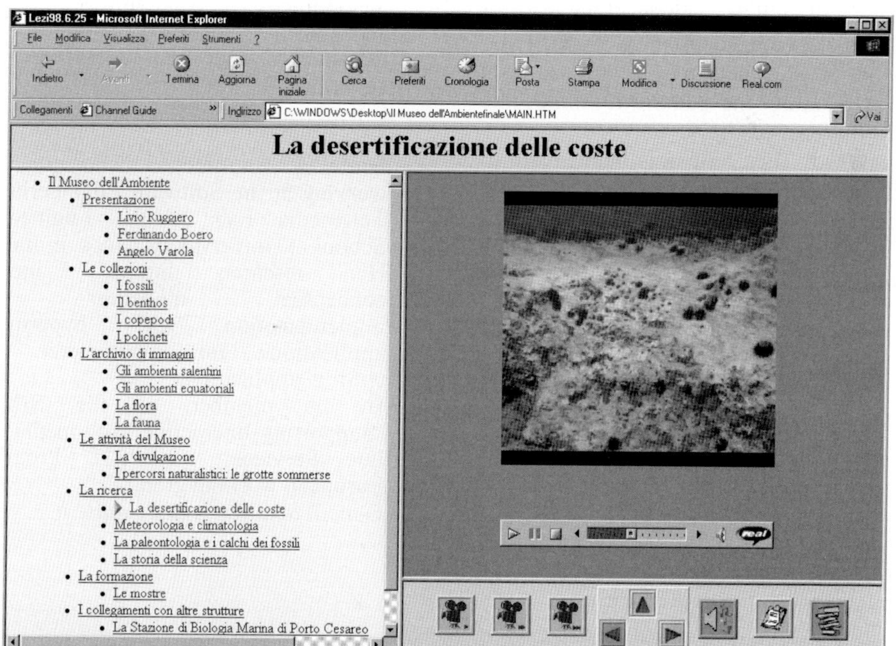

Figure 2: The desertification of coasts

CONCLUSIONS AND FUTURE WORKS

The idea presented in this paper is quite simple: there is a need for good multimedia educational titles, that teachers can develop by themselves, with little time, little effort and little budget.

The University of Lecce has developed 3 applications on cultural and educational subjects. After all these developments and with data about the students' reaction, a final version of the tool, LEZI, is planned by the end of year 2001.

We are convinced that several excellent teachers, mostly in non-technical fields, by using LEZI (or similar tools), could develop, with little effort, a large body of high quality multimedia educational materials that could greatly enhance the education process, and also improve the cultural exchange among different institutions.

The tool will be enhanced to support complete courses oriented to different kind of students reusing the contents and to allow the authors to choice the Web layout easily.

Although this tool was conceived to develop interactive lecture, it's demonstrating its flexibility to be used in other applicative context like museum.

The traditional museum hypermedia application is strictly related to both the contents and its disposition into the museum structure. This kind of application are conceived to realize the paradigm of the visit to museum.

The Lezi tool could be used to realize an overview of the museum, such as a brochure, driven by the interviews of experts people in order to give an intelligent summary to the user.

ACKNOWLEDGEMENTS

The authors wish to thank Prof.Boero, Prof. Ruggiero and Prof. Varola for their lectures.

REFERENCES

1. M. A. Bochicchio,R. Paiano, P. Paolini, E. Andreassi, T. Montanaro " LEZI: a tool for easy development of interactive video for education" Proceedings on CD International Conference Telecommunication and Electronic TELEC 2000 - Cuba - Luglio 2000 *ACM*, 38 (8), August 1995, 8-15

ABOUT THE AUTHORS

Roberto Paiano received the DrEng degree in electronic engineering from the University of Bologna, Italy. He worked in IBM Italy until the 1996. Since the 1997 he has been at the Department of Engineering, University of Lecce, where he's assistant professor. His research interests include the methodology of Web application design and distance learning tools. He is member of the IEEE and the IEEE Computer Society.
E-mail: *roberto.paiano@unile.it*

Mario A. Bochicchio is a researcher working at the Software Engineering & Telemedia Lab of the Engineering School at University of Lecce, in Italy.
His research interests include conceptual modeling and implementation of hypermedia applications for e-commerce and cultural applications.
He is member of the SIMOS (Supporting Interactive Multimedia On-line Services - ESPRIT Program) European working group.
E-mail: *bomal@ingle01.unile.it*

IBC WEB SITE: a multi-purpose site developed from W2000 model

Maria Pia Guermandi (*)

(*)Istituto Beni Culturali della Regione Emilia Romagna, Bologna (Italy)
E-mail: MPGuermandi@ibc.regione.emilia-romagna.it

ABSTRACT
Istituto Beni Culturali (Institute for Cultural Heritage) is the cultural institution of Regione Emilia Romagna. IBC web site [6] has been completely revised in cooperation with Hypermedia Open Center [5], the laboratory of the Electronics and Information Department at Politecnico di Milano. We adopted W2000 model for requirements analysis and for the analysis and organization of the application content.

KEYWORDS: cultural heritage, W2000 model, multi-purpose web-site

INTRODUCTION

The Istituto Beni Culturali (IBC), founded in 1974, is a cultural institution belonging to the Region Emilia Romagna. IBC acts as a research and advisory body to the Region and to local authorities: among its tasks are the cataloguing and safeguarding of cultural property, restoration works, the funding of and the consulting to libraries and museums (in particular to local authority ones) in Emilia Romagna. Ibc is a reference point at national and regional level in the domain of cultural heritage.

The IBC web site has been on-line since 1996. As far as the information available, the IBC web site is similar both to other museum web sites, and to those of research institutions operating in the same fields. It can therefore be regarded as a multi-site, both for its content, and for the audience it addresses.

The contents include general information on cultural events taking place in the region (exhibitions, conferences, presentation of books, restoration works, etc.), as well as information concerning the activities carried out by IBC, on-line publishing (IBC magazine, proceedings, etc.), and also the access to databases (textual information, images, and GIS), pre-information on upcoming conferences, management of museums and library projects.

In the last months our web site has been completely restyled and reorganised in cooperation with Hypermedia Open Center [5], the laboratory of the Electronics and Information Department at Politecnico di Milano. We are trying to explore and employ new patterns of technology-based communication for Cultural Heritage. Methodology used to structure our site is W2000 model as defined by HOC [1, 2, 3, 4].

REQUIREMENTS ANALYSIS

We have analysed IBC web site users profile adopting W5+ model [1]. As many cultural applications, our web site has a large variety of potential users with different needs and expectations. According to W5+ model, we have tried to identify the general purpose of the web-site for each user category and which specific tasks should users perform with the systems to achieve their goals.

We can resume IBC web site main goals according internal expectations, as:

- to acquire a larger visibility for all IBC activities which until now are known only by a restricted number of specialists.
- to promote IBC events (exhibitions, conferences, seminars) to the largest numbers of web users at regional and national level.
- to promote new technologies - in particular web technology - use as a more effective work tool for IBC internal researchers.

As regards as internal constraints, main characteristics IBC new web site should have are:

- easy upkeeping, because our web staff is very limited in resources and people.
- easy updating, because we have new materials to publish almost every day.

Moving from these internal objectives and constraints we have identified two main typologies of users:

- user with generic cultural interests (one specific subset of this typology is represented by schools as web users)
- highly specialised group of users *internal* (IBC researchers) and *external* (librarians, museums curators, cultural heritage operators in public administrations)

For the first typology we have to provide an informative and educational application to inform users about IBC content and activities.

For the second typology we have to create a web site research-oriented and work-oriented. This analysis (which I've heavily resumed) has helped us to structure and to present the information content and in particular to provide useful paradigms of information access (e.g. free and index navigation for the first user typology, query- based access and intranets services for the second one). In this way, on the same semantic domain we have tried to provide different views of the content structure or different functionality to better address specific users and client goals.

HDM - HYPERMEDIA DESIGN MODEL

Hyperbase

To translate this requirements into appropriate design choices, according to HDM Model [2, 3, 4], we have defined our site conceptual design in two main stages.

We have defined the structure, i.e. the organization of the application content and the dynamics, i.e. the application behavior.

In the first stage we have identified our **hyperbase** which is the set of information objects: every information object is an **entity** of hyperbase.

Entities types identified for modelling of IBC web site are the followings:

- IBC Institution
- EVENTS (exhibitions, conferences, presentations, ecc.)

- PUBLISHING (books, CD-Rom, on-line proceedings, IBC review)
- WEB EVENTS
- INSTITUTIONAL ACTIVITIES
- DATA BASES
- Next step, we have defined semantics associations between entities and navigation paths in the Hyperbase.

Access structure

Project second stage has been the identification of access structures with which is possible organize the same information objects, i.e. the same entities, in different ways according to different users typologies and requirements. In this stage we have defined **collections,** i.e. the sets of information objects creating a cognitive context. According to multi-purpose character of our site, we have choosen to provide a large number of collections narrowly focused.

Collections in which our site is structured, are the followings:

Fig. 1 Home page

- HOME PAGE (Fig. 1)
- NEWS
- DOMAIN / AREA OF INTEREST
    1. Libraries and archives
    2. Museums and cultural heritage
    3. Natural and architectural heritage
- CALENDAR (Fig. 2)

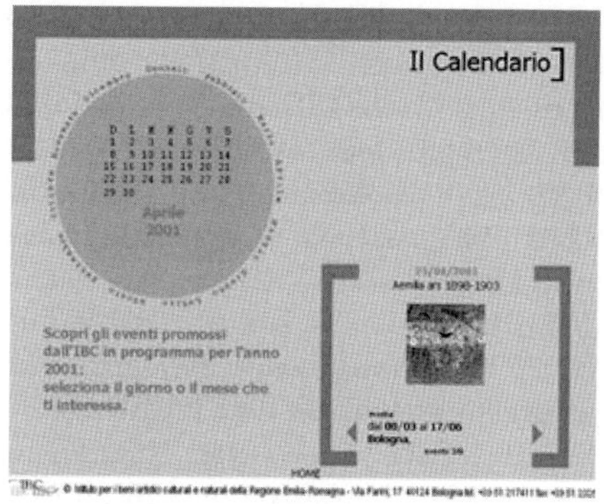

Fig. 2: Events calendar

- EVENTS
    1. Exhibitions
    2. Conferences, courses, seminars
    3. Presentations and other events
- PUBLISHING
    1. Books
    2. CD-Rom /Video
    3. Proceedings and exhibitions catalogues
    4. IBC journal (Figg. 3, 4, 5)
- EDUCATIONAL
- SCHOOL MATERIALS
- WEEK TOPIC
- PROJECTS
- DATA BASES
- LAWS

As you can see, we have defined not only **base collections** (i.e. collections of objects of our Hyperbase), but also **higher collections** (i.e. collections of other collections, as, for example, "DOMAIN", or "EVENTS")

Collections are often split up according time breaks (for example past / current / upcoming exhibitions, recent / past years publications).

We have defined application usability for each of our user typology, so access

structure and navigational path could vary according to user profile (Fig. 6).
According to multipurpose character of our web site, access to hyperbase information is highly structured: in this way the same information unit is accessible from different collections. For example the information objects about an archaeological exhibition could be reached from following collections:
CALENDAR - NEWS - AREA OF INTEREST (*museums and cultural heritage* - events) - EVENTS - PUBLISHING (*exhibition catalogue* and *IBC journal*) - SCHOOL MATERIALS.

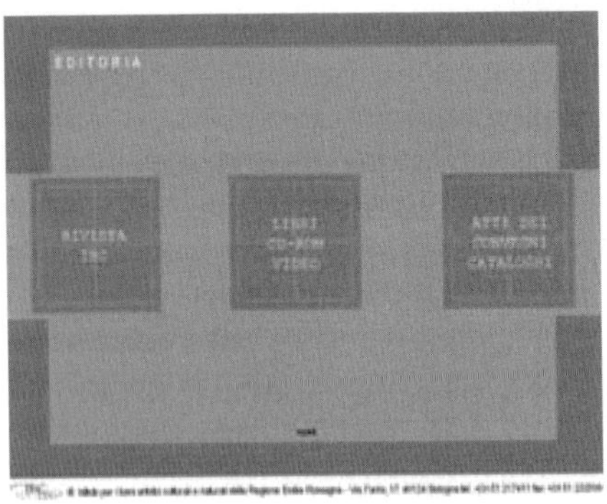

Fig. 3: Publishing Center of Collections

Fig. 4: Publishing. Collection: IBC review

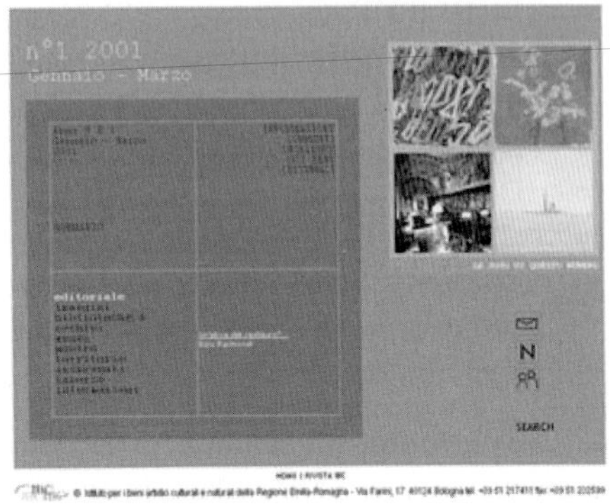

Fig.5:Publishing.Collection member: IBC review issue

Our web site proposes also a number of services like forum, contacts, newsletter: IBC web site user can also send new messages in a choosen forum or register himself in a newsgroup.
To promote internet as a daily work tool we have developed also "areas jobs" restricted to specific user groups (librarians, museum curators, internal researchers) by means of which it will be possible organize a conference or discuss and draw up an administrative deed or simply share projects documentation.

TECHNOLOGICAL TOOLS

To edit our site we have used software tools which characteristics of flexibility and easy maintenance.
To develop Hyperbase from technological point of view we have used Highway, a software tool for the implementation of database. Our databases objects are Hyperbase entities we can search and reorganize in different ways. Highway uses an Information Retrieval System architecture.

As software of editing we use a standard as Macromedia Dreamweaver with which we have developed access structures specialized that make specific queries on our Hyperbase databases.

ACKNOWLEDGMENTS

Current release of IBC web site has been created by the author, Maria Elena Barbieri and Maria Elena Tosi in cooperation with HOC laboratory.

CONCLUSIONS

First stage of new IBC web site planning has been the analysis of old site through reverse-engineering process. This analysis has stressed out a lack in information organization and an insufficient use of access structure. As in many complex interactive applications our web site quality is related to how well it matches the requirements of its potential users: at present we think that our weak point could be usability for specific users group (for example librarians), so in the future we have to improve tools as area jobs.

IBC new web site is on-line from July 2001 in a demo release. In following months we are testing our site with a defined group of users (internal and external) to verify if our main objectives have been - totally or partially - reached. In cooperation with Hypermedia Open Center of Politecnico this group should evaluate IBC web site usability.

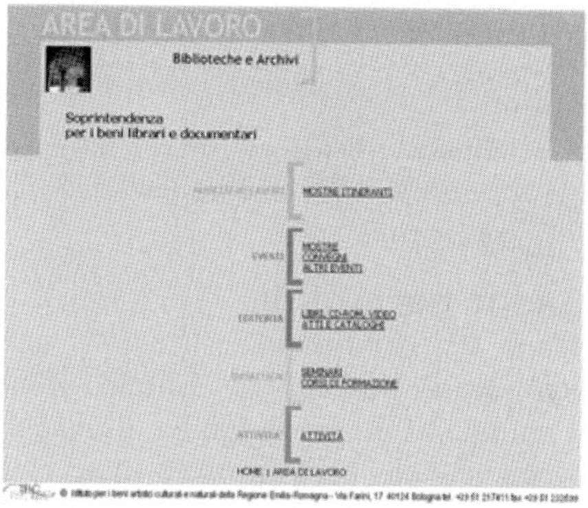

Fig. 6: Area of interest: Libraries and archives

REFERENCES

1. Di Silvestro A., Garzotto F., Paolini P. User Requirements Acquisition for Museum Hypermedia. In Bearman D. and Trant J. Eds. *Cultural Heritage Informatics 1999, Selected Papers from ichim 1999*

2. Garzotto F., Mainetti L., Paolini P. Hypermedia Design, Analysis and Evaluation Issues. In *Communications of the ACM*, 38(8), August 1995, 74-86

3. Garzotto F., Mainetti L., Paolini P. Navigation in hypermedia applications: modeling and semantics. In *Journal of Organizational Computing and Electronic Commerce,* 6 (3), 1996, 211-237

4. Garzotto F., Paolini P., Schwabe D. HDM - A model Based Approach to Hypermedia Application Design. In *ACM Transactions on Information Systems*, 11 (1), January 1993, 1-26

5. Hypermedia Open Center, *http://hoc115.elet.polimi.it/hoc/home.asp*

6. Istituto Beni Culturali: *http://www.ibc.regione.emilia-romagna.it*

ABOUT THE AUTHOR

Maria Pia Guermandi: since 1987 she has been working at Istituto Beni Culturali of Regione Emilia Romagna where she has been involved in several research projects about applications of new technologies to cultural heritage as project leader (main areas of interest: cataloguing data bases, G.I.S., web communications). With this respect she is also involved in European Research Projects. Since 1996 she is chief editor of the IBC (Istituto Beni Culturali) web site. Author of several publications and articles, she is part of Editorial Board of scientific reviews as "Archeologia e Calcolatori" and "IBC. Informazioni, Commenti, Inchieste sui Beni Culturali".
E-mail:
MPGuermandi@ibc.regione.emilia-romagna.it

Demos

Discovernet©: an Education Gateway to Museums and Galleries

Lee Adendorff

Australian Museums and Galleries OnLine, Australia

Australian Museums and Galleries OnLine (AMOL) recently launched Discovernet©, a learning gateway to Australia's cultural institutions. Discovernet© comprises four sections: an interactive game that uses real museum objects to introduce children to the curatorial process, a national database of museum education resources incorporating specialist education metadata, a museum guide and an Australian Tales section that presents contextualisation of museum objects tied to curriculum objectives. Discovernet© was launched together with a web cast from a megafauna fossil dig in the Australian desert.

e-mail: leea@amol.org.au

Teaching and Learning through Exploring and Manipulating 3D Models

Eugenio Morreale (*) *Franco Gambino* (#)

(*) *Univ. di Pisa - Dip. di Informatica, Italy*
(#) *Technology Reply S.r.l., Pisa, Italy*

The Demo will show the methodology and the approach applied to some examples in the areas of historical architectural buildings, naturalistic specimens (Ammonite) in order to support teaching / learning processes through explorations and manipulations of 3D models realized for the chosen examples.

e-mail: morreale@di-unipi.it, f.gambino@reply.it

I-irasshai: an Immersive Cultural Learning Experience

Brian D. Jones

Georgia Institute of Technology - Interactive Media Technology Center, Atlanta (GA), USA

I-irasshai is a unique Japanese cultural learning experience for English-speaking students delivered through a combination of CD-ROM and Internet technology. The goal of the program is for students to gain insight into Japanese culture by allowing them to explore Japan within an authentic context. The immersive environment features twenty 360-degree realistic locations, over 60 cultural activities, and many more web pages containing photographs, videos, and textual information.

e-mail: brian.jones@imtc.gatech.edu

.New Forms of Interaction

A Collaborative Exploration of Leonardo's 3D Virtual City

Thimoty Barbieri (*), *Chiara Losco* (*) *and Giuliano Gaia* (#)
(*)*Politecnico di Milano – Laboratorio HOC, Italy*
(#)*Museo della Scienza e della Tecnica, Milano, Italy*

Leonardo's codexes are full of drawings of palaces and sketches of urban layouts. The historians at the Italian National Science Museum have collected those drawings and directed computer designers to the creation of an exciting virtual reconstruction of an "ideal" city, as Leonardo might have designed it. This digital reconstruction can be explored together with other users, at the same time, and it is possible to collaborate, for example following a guide inside the virtual city.

e-mail: barbieri@elet.polimi.it, losco@hoc.elet.polimi.it, gaia@museoscienza.org

Virtualis: an Interactive Opera

Alain Bonardi
Sorbonne University, Paris, France

Virtualis is a CD-ROM opera specially designed and composed to be "performed" on any personal computer. It uses original interaction models based on physical forces.

e-mail: alain.bonardi@wanadoo.fr

MUSE: an Integrated System for Mobile Fruition and Site Management

Maurizio Malavasi, Enrico Romagnoli
Boconsult Ingegneria dei Sistemi S.p.A., Bologna, Italy

MUSE, a user-centered multimedia fruition system for archaeogical sites and museums, is based on hand-held tablets wireless connected to servers, cameras and large, high performance graphics displays. The demo shows how multiple functions may be integrated to provide a comprehensive tool to augment the fascinating experience of the visit. In this early prototype, functions demonstrated on the tablet include navigation in physical and virtual space, interaction with distributed high performance screens, picture capture for personalized memories and instant access to news provided by the site management. The project is currently under development at Boconsult, a Bologna based SME in cooperation with the University of Bologna, Cineca, and Sinet. It is funded by Murst, the Italian Ministery for University Science and Technology within the framework of the National Research Program "Parnaso".

e-mail: m.malavasi@boconsult.com, eromagnoli@thesis.deis.unibo.it

.Virtual Cities and Architecture

Kotaview: Simulating Traditional Korean Architecture Interactively and Intelligently on the Web

Jie-Eun Hwnag

Korea

KotaView is a web-based tutoring system of traditional Korean architecture. In order to simulate traditional buildings in advanced and intelligent ways, we apply a computer technique, called data modeling. We develop a robust data model to effectively represent the traditional buildings. Therefore the simulation is not only similar to the real world but also effective to experience the structure of traditional Korean architecture with 3D objects interactively.

e-mail: curiozen@yonsei.ac.kr

At the Dawn of European Civilization - Interactive Video, Multimedia and Workshops

Jordan Detev

Bulgarian Multimedia Art Society, Bulgaria

This demo will give new life to the oldest Temple of Fertility (6200 years B.C), which was situated near the modern Bulgarian village Muldava (from the thracian word "mulda" which meant "fortified sanctuary"). 1000 pictures, more than 100 video clips, 2D and 3D animations, photos, original music, etc. contribute to build an immersive experience of the past. Three sources of information are used to reconstruct the Temple and its cult rituals. Archaeological data are taken from the great Bulgarian archaeologist prof. P. Detev's work. The author himself composed the syncretic show (music, dance, video, FX, etc.) upon the myth of the stag-mediator (included in the video). This fascinating work makes it possible to "get in touch" with the Thracian people, their art and spiritual life. www.sokerov.com/detev.

e-mail: detev@sokerov.com

The Virtual Museum of Architecture

Eleonora Cardellini

Università di Roma La Sapienza – Laboratorio Multimediale di Architettura, Italy

This demonstration will guide you through a three-dimensional space for the web which is capable of 'containing' virtual buildings or representations of real buildings (through computer rendering, video or photographs). In addition, the process that led to this museum will be explained: in fact, the virtual museum of architecture is conceived as an open organism, a "vestibule" for a logical sequence of paths and openings.

e-mail: leo.n@libero.it

3D for Communicating Cultural Heritage on the Web: the Palazzo Grassi Web Site

Fabio Pittarello

Università Ca' Foscari di Venezia – Dip. di Informatica, Italy

This demo consists of a presentation of the Palazzo Grassi web site, and focuses on the implementation of guided tours for Internet users, related to the exhibitions promoted by Palazzo Grassi in Venice. In particular the demo will include the presentation of the "Einstein Tower" world, the "Sun Temple", "Baroque Compositions", the "Newton's Cenotaph" and "Discovering the Etruscans".

e-mail: pitt@unive.it

Bologna at a Glance

Maurizio Malavasi[(*)] and *Silvia Summa*[(#)]

[(*)]*Boconsult Ingegneria dei Sistemi S.p.A., Bologna, Italy*
[(#)]*Università di Bologna - Centro di ricerca sui sistemi elettronici (CERSE), Italy*

"Bologna at a glance" is a user centric interactive guide to the city's Cultural Heritage. Every building, museum, author and item can be located on the city map and immediately accessed through a direct link. Pertinent text and image based info is popped from the system database. In this way the users are driven in their exciting discovery of the city and helped in their straightforward planning of a personalised visit. The project is conceived and funded by the Assessorato alla Cultura of Bologna City Council.

e-mail: m.malavasi@boconsult.com, ssumma@deis.unibo.it

Exploring Landscape, Documenting Culture, Constructing Memory. The Loire Valley Internet Workshop

Nicolò Ceccarelli

Politecnico di Milano - III Facoltà di Architettura/Design, Italy

The demo will present a selection of examples of the Val de Loire cultural portal, developed in July 2000 by our team during the 4th edition of the International Workshop "Cities Design & the Internet". The presentation will highlight the main goals of the web site, such as the representation of the sense of Place and the portrail of heritage via *interactive multimedia on the internet.*

e-mail: nicolo.ceccarelli@polimi.it

The Ke EMu Collections Management System

John Doolan

KE Software, Australia

KE EMu is the world's leading collections management system for museums, art galleries, historical societies and herbaria. It integrates an extensible multi-discipline catalogue with interpretative information and multimedia resources while providing sophisticated tools to enable museums to publish their collections on the Internet. KE EMu clients include many of the world's most prestigious institutions, such as the Smithsonian Institution's National Museum of Natural History, the Canadian Museum of Civilization and the National Gallery of Australia.

e-mail: John.Doolan@kesoftware.com

Arts and Humanities Data Service – Digital Collections

Christopher Pressler

King's College - Arts and Humanities Data Service, London, UK

The Arts and Humanities Data Service (AHDS) is a UK national service, based at King's College London. The AHDS holds over two million digital objects, including paintings, spatial data, historical census reports, multi-media music collections, digital texts of major works of literature and many other collections and data types. This demonstration seeks to raise awareness of our collections and services and also to develop partners across Europe in order to enhance digital holdings in arts and humanities.

e-mail: chris.pressler@ahds.ac.uk

Heritage on the Web: Providing Access to Historic Properties and their Contents

Michael Hiley

De Montfort University - Centre for Technology and the Arts, Leicester, UK

In the United Kingdom many extensive collections of paintings etc. remain largely intact located in more than a thousand historic houses which are privately owned. Heritage on the Web maintains a web gateway to historic properties and operates as an association between private owners who have an interest in attracting tourist visitors through the building of web sites and those who wish to visit and study their collections on-line.

e-mail: mrh@dmu.ac.uk

The NZDL Niupepa Collection: Historical Indigenous Language Newspapers on the Internet

Mark Apperley, Te Taka Keegan
University of Waikato, New Zealand

This demonstration will show how the Maori language Niupepa collection can be accessed via the Internet, or from CDROM, using the full-text search and indexing capabilities of the NZDL Greenstone software. The demonstration will also cover examples of some of the many other collections in the NZDL system, including collections of Chinese, French and Arabic documents.

e-mail: m.apperley@cs.waikato.ac.nz

Images of the Early 20th Century. The Photographic Fund of the Istituto di Arti Grafiche of Bergamo for "Emporium" Magazine

Federico Brunetti, Ilaria Alberici
Politecnico di Milano – DITec, Dip. di Disegno Industriale e Tecnologia dell'Architettura, Italy

The I.A.G.B. (I.A.G.B. Istituto di Arti Grafiche) in Bergamo was the first Italian publisher to equip itself, at the beginning of the century, with typographic apparatus for the printing of photographs on the same page as graphics and texts. In particular, the magazine "Emporium" realized an advanced photographic archive. Digital technology allows us to explore these photographic memories with acceptable quality, with the aim of understanding the reasons that led to these interesting reportages.

e-mail: federico.brunetti@polimi.it

The New Norwegian Rock Art Database

Wenche Helliksen, Bjørg Pettersen, May-Liss Sollund, Randi Sønderland
Foundation for Nature Research and Cultural Heritage Research, NINA-NIKU, Norway

The application links text to photographs, tracings and digital maps. The database stores the complete documented history of each site, presenting the user with a list of all previous scientific evaluations and records. The software used is Oracle, Visual Basic and Map Objects. The application runs on a Citrix application server. Only mouse clicks and screen updates are transmitted over the network, allowing the use of inexpensive computers and limited bandwidth.

e-mail: bjorg.pettersen@admosl.ninaniku.no, randi.sonderland@admtrd.ninaniku.no

LeMO: an Internet Multimedia Information System of 20th Century German History

Lutz Nentwig (*), *Arnulf Scriba* (#), *Jürgen Reiche* (§)
(*) *Fraunhofer ISST, Berlin, Germany*
(#) *German Historical Museum, Berlin, Germany*
(§) *Haus der Geschichte of the Federal Republic of Germany, Bonn, Germany*

The aim of the LeMO project was to create a multimedia information system on 20th century German history in the Internet. The LeMO system provides various options for accessing its information. With the need to make cultural content attractive to young people in mind, 3D environments have been developed for each period of 20th century history. Access to specific content is also provided via different search capabilities, including a metadata-based search engine.

e-mail: lutz.nentwig@isst.fhg.de, scriba@dhm.de

Art2LIFE: the Canadian Century

Franca Pernatozzi, Michel Blondeau
Ecentricarts inc., Canada

Initiated by the McMichael Canadian Art Collection and the York Region District School Board and built by Ecentricarts.inc, Art2LIFE brings 100 years of Canadian art, people and events to life. The site combines 100 primary works of art with 2300 primary source assets including: archival images, text, audio and video. It is a journey through the decades and the achievements, stories and personalities that defined 20th century Canada.

e-mail: franca@ecentricarts.com

Zoom Media Lab & the Virtual Collection of Children's Everyday Culture

Sasha Pirker, Stefan Wagner
Uma Information Technology AG, Austria

In partnership between the new Zoom Children's Museum and Uma Information Technology AG, a very special digital collection will be established: the Virtual Collection of Children's Everyday Culture. Using specially designed scanning stations children will archive objects of their everyday life. Installing such scanning stations in museums worldwide the collection will allow comparison of children's everyday culture, internationally. The collection will be accessible through the Internet by Spring 2002.

e-mail: sp@uma.at, sw@uma.at

Facing Challenges of New Media on the Crossroads of East-West and Past-Future (Lithuanian Case Study)

Nerute Kligiene, Laimutis Telksnys

Institute of Mathematics and Informatics, Vilnius, Lithuania

Three aspects of applying ICT in the field of cultural heritage (preservation, dissemination, and training) in case of countries in transition are considered. ICT provides equal opportunities to big or small nations as well as to ethnic minorities, illuminating their cultural diversities and allowing the preservation of their cultural identity. Problems of cultural heritage accessibility and of training in informatics for the humanities are also considered.

e-mail: nerute@ktl.mii.lt, telksnys@ktl.mii.it

The Survival of Traditional Brick Factories in the Baixada Campista

Simonne Teixeira, J. Paulo Gama de Oliveira

Universidade Estadual do Norte Fluminense – UENF, Brazil

The aim of this demo is to show the permanence of brick factories, still in activity, side by side with great ceramic complexes in the Baixada Campista.

e-mail: simonne@uenf.br, Jpgo1@bol.com.br

Oriental Brightness into an European City – Bucharest of '30 years

Laura Ciocoiu

Research Institute for Informatics, Romania

This demo is concerned with the cultural heritage for the reconstruction of the interethnical influences in culture. So, the demo will show the instruments used to reveal to people the oriental influnces into Romanian culture (mentality, architecture, etc.) and to reconstruct, by simulating them, the cities demolished during 19th and 20th century, like the old Bucharest, "the little Paris", as it was known in that period, characterized by its renaissance buildings and monuments, the most part of which were demolished in 30' years and during Communism.

e-mail: ciocoiu@ici.ro

iPaq Virtual Museum Guide

Giorgio Bramati
Compaq Computer, Italia

The application is a simple example on how to use a PDA in a museum environment to access multimedia information related to the showed objects using a wireless lan technology.

e-mail: giorgio.bramati@compaq.com

ICT Based Monument Management

Alfredo Ronchi, Larissa Zoni
Politecnico di Milano - Dept. ISET HMG Lab, Italy

The application aims to support monument maintenance, management, risk evaluation and re-use of compatible functions. Specific modules are devoted to scheduled maintenance and conservation according to a timetable and granularity proposed either by experts or by the information system itself. The distributed information system will collect and redistribute relevant information and case studies. Locally it will offer flexible archiving tools and later on it will operate as historical information source, useful in order to evaluate the consistence of the historical buildings property, keep track of changes and provide support to managers.

e-mail: alfredo.ronchi@polimi.it, larissa.zoni@polimi.it

3D Reconstruction Applied to Virtual Heritage and Cultural Preservation

J. Goncalves, V. Sequeira
European Commission - Joint Research Centre, Ispra (Va), Italy

The demo addresses the main steps in building a complete 3D texture-mapped model and presents the results of some projects involving the 3D modelling of Heritage Buildings, in particular state-of-the-art models of the "Cour d'Honneur" and "Salon Delacroix" at the French National Assembly (Palais Bourbon), Paris, as well as the "Sala dello Scrutinio" at the Doges' Palace, Venice.

e-mail: vitor.sequeira@jrc.it, joao.goncalves@jrc.it

Digital Preservation of Cultural Heritage through Constructive Modeling

Carl Vilbrandt (*), *Galina Pasko* (#), *Janet R. Goodwin* (§), *James M. Goodwin* (+)

(*) University of Aizu, Computer Arts Lab, Aizu-Wakamatsu, Japan
(#) *IT Institute of Kanazawa Institute of Technology, Tokyo, Japan*
(§) *Aizu History Project, Los Angeles (CA), USA*
(+) *University of California Los Angeles - Dept. of Computer Science, Los Angeles (CA), USA*

In this implementation, we use Constructive Solid Geometry (CSG) and Function Representation (FRep) as practical applications of mathematical representations which fit the purposes of long term digital preservation of cultural heritage sites and objects. We demonstrate our methods with CSG models of two Buddhist temples from the Aizu region of Japan and FRep models of traditional Japanese lacquer ware.

e-mail: vilb@u-aizu.ac.jp

Minerva: an Artificial Intelligent System for Composition of Museums

Francesco Amigoni, Viola Schiaffonati

Politecnico di Milano –Dip. Elettronica e Informazione, Artificial Intelligence and Robotics Project, Italy

Minerva is an artificial intelligent system composed of several cooperating computational agents. The global functionality of the system is the automatic composition of a museum. According to criteria given by museum experts, the system starts from a set of works of art (currently from Milan Archeological Museum) and finds an allocation for a sequence of them in an environment previously selected.

e-mail: amigoni@elet.polimi.it, schiaffo@fusberta.elet.polimi.it

Migrator 2000: Jpeg 2000 file format

Andrea de Polo

Fratelli Alinari Spa, Italy

MIGRATOR 2000 will develop, demonstrate and disseminate tools for the JPEG 2000 digital imaging standard. The project focuses on image archiving, image delivery and medical applications, implementing content description and IPR protection including watermarking tools. Additionally the project will produce a range of tools for

JPEG/JPEG 2000 comparison, for examining the differences in bit-rate performance, nature of artefacts, etc. http://www.migrator2000.org.

e-mail: andrea@alinari.it

[Volume 2]

.Index of Authors

.A

.B

.R

.S

.T

.V

.W

.Z